RESEARCH IN
CHEMICAL KINETICS

Volume 3

RESEARCH IN CHEMICAL KINETICS

Volume 3

Edited by

R.G. COMPTON and G. HANCOCK

*Physical Chemistry Laboratory, University of Oxford,
Oxford OX1 3QZ, U.K.*

1995

ELSEVIER

Amsterdam – London – New York – Oxford – Shannon – Tokyo

ELSEVIER SCIENCE BV
Sara Burgerhartstraat 25
P.O. Box 211, 1000 AE Amsterdam, The Netherlands

ISBN: 0-444-82036-1

This book is printed on acid-free paper.

Printed in The Netherlands

PREFACE

This series of volumes aims to produce authoritative review articles on a wide range of newly developing topics in the kinetics of both gaseous and condensed phases. Each article will describe a particular area of the subject pertinent to the research interests and expertise of the contributors, emphasising their recent contributions and putting this work in context of others' progress in the same field. The reviews are aimed at a wide general readership in the kinetics community and are intended to be short, topical accounts of a specific area from the viewpoint of an expert in the field. The Editors and Elsevier are committed to rapid publication of these volumes so as to ensure the highest possible benefit to the kinetics community.

Richard Compton
Gus Hancock

LIST OF CONTRIBUTORS

R. Becerra,
Instituto Quimica Fisica "Rocasolano", C.S.I.C., C/Serrano, 119, 28006, Madrid, Spain.

Dario T. Beruto,
Interdepartmental Centre for Materials Engineering and Institute of Chemistry, Faculty of Engineering, University of Genoa, Piazzale J.F. Kennedy, Fiera del Mare Pad. D, Genova, Italy.

Joseph L. Durant Jr.,
Combustion Research Facility, Sandia National Laboratory, Livermore, Ca 94551, USA.

Marino Giordani,
Interdepartmental Centre for Materials Engineering and Institute of Chemistry, Faculty of Engineering, University of Genoa, Piazzale J.F. Kennedy, Fiera del Mare Pad. D, Genova, Italy.

Nicholas J.B. Green,
Department of Chemistry, King's College London, Strand, London WC2R 2LS, United Kingdom.

Simon M. Pimblott,
Radiation Laboratory, University of Notre Dame, Indiana 46556, USA.

J. Tamm,
Institute of Physical Chemistry, University of Tartu, Tartu, EE2400, Estonia.

L. Tamm,
Institute of Physical Chemistry, University of Tartu, Tartu, EE2400, Estonia.

R.W. Walker,
School of Chemistry, University of Hull, Hull, North Humberside, HU6 7RX, United Kingdom.

R. Walsh,
Department of Chemistry, University of Reading, Whiteknights, P.O. Box 224, Reading RG6 2AD, United Kingdom.

TABLE OF CONTENTS

Research in Chemical Kinetics, Volume 3
R.G. Compton and G. Hancock (editors)
© 1995 Elsevier Science B.V. All rights reserved.

Some Burning Problems in Combustion Chemistry

R. W. Walker

School of Chemistry, University of Hull, Hull, North Humberside, HU6 7RX, United Kingdom

1. INTRODUCTION

The well known [1] extreme complexity of hydrocarbon combustion processes has rendered it extraordinarily difficult to identify quantitative mechanisms and to obtain rate constant parameters even for key elementary reactions in order that important practical systems can be modelled [2]. Very little progress has been made (over nearly 100 years) in understanding the fundamentals of combustion chemistry by direct investigations of the oxidation of hydrocarbons and related compounds. In the period up to about 1970, virtually no reliable kinetic data on the elementary processes involved were obtained in this way despite its prevalence in hydrocarbon oxidation studies. Even the advent of gas chromatography and other sensitive analytical techniques, which enabled quantitative measurement of products in the very early stages of oxidation, did not lead to a significantly increased bank of fundamental kinetic and mechanistic data.[2]

In general, major advances in the understanding of combustion mechanisms have resulted only when single elementary steps have been isolated under carefully selected and controlled conditions, where the reaction products (often formed in multi-channels) can be identified and monitored.[3] Obviously, a reliable and clean source of the radical is essential under the necessary conditions, in terms of (i) the determination of a rate constant of the reaction, and (ii) the use of the rate constant. The extrapolation of the value of a rate constant determined near room temperature to the region of high temperature combustion is often a dangerous operation and frequently carried out with foolish optimism. The realisation that many reactions have non-Arrhenius temperature coefficients, and perhaps a change of mechanism with temperature, has meant that some key reactions have been studied over very wide temperature [3] and pressure conditions [4]. The effect of temperature on the mutual reaction of HO_2 radicals (Fig.1) merits attention and gives an early warning in connection with a process of great importance in combustion. In outline, Foner and Hudson [5] obtained k = 1.8 x 10^9 dm^3 mol^{-1} s^{-1} (1962) at 300 K supported by Paukert and Johnson [6] (2.2 x 10^9, 1972) which when combined with Troe's [7] shock-tube value of 2 x 10^9 at ca 1200 K produced unanimous agreement that A = $10^{9.3\pm0.3}$ dm^3 mol^{-1} s^{-1}, E = 0 over the temperature range 300-2000 K. These parameters were still recommended in 1984, despite some evidence to the contrary [8]. However, a number of workers [3] demonstrated a marked negative temperature dependence between 250 and 550 K, and moreover emphasised that the rate constant was strongly dependent on the bath gas used and the total pressure. With Troe's

2

data point regarded with some suspicion, the uncertainty in the value of the rate constant exceeded an order of magnitude at combustion temperatures. Fortunately, Lightfoot et al [9] demonstrated a minimum was reached at about 700 K, and Troe [10] has obtained new data up to 1100 K and validated his earlier work. The cause of the complex variation is undoubtedly a mechanism change. At low temperatures, re-combination occurs with the H_2O_4 complex becoming unstable at higher temperatures, and an H abstraction process then dominates. The danger of extrapolation in such a situation is clearly apparent.

$$HO_2 \; + \; HO_2 \; + \; M \; \rightarrow \; H_2O_4 \; + \; M$$
$$\downarrow$$
$$H_2O_2 \; + \; O_2$$

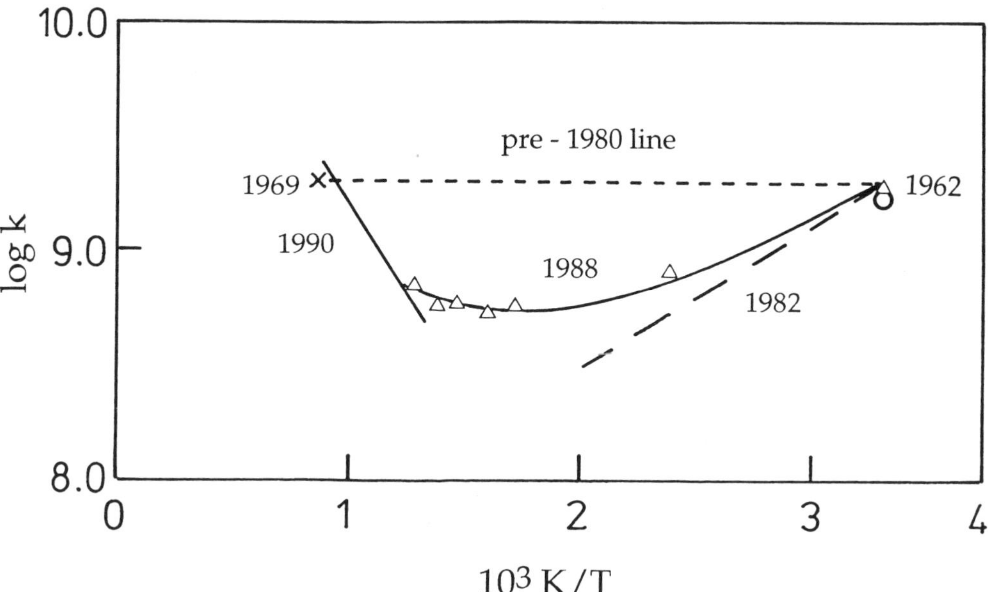

Figure 1   30 years of kinetic data for $2HO_2 \rightarrow H_2O_2 + O_2$
(1 atmosphere pressure, mostly N_2).

A second common pre-requisite is the availability of selective and very sensitive detection of atoms and free-radicals in order to monitor the progress of a reaction. Reactions of small species important in combustion (eg. H, O, OH, HO_2, CH, NH, CH_3 etc) have been studied extensively because of the large variety of ways to generate them (eg. microwave discharge, laser photolysis etc[3]) and the sensitive, although essentially simple, method of monitoring their concentration (eg. resonance fluorescence, laser magnetic resonance[3]). Even within the small species, there is often a marked distinction to be made. Whilst the kinetics of H abstraction from a vast range of RH compounds by the highly reactive OH radical are in good order[3, 11], even to the extent of accurate expressions for attack at specific C-H sites, there is a dearth of information on the **slower** H abstraction reactions of H, HO_2 and CH_3. For example, despite their key role in determining ignition time delays [12], there are **no** data points for HO_2 + C_4H_{10} and only **one** for HO_2 + CH_4 [13] compared with the several hundred data points for OH + CH_4 between 240 and 2000 K with k known to ca 20% [3]. The HO_2 reactions are simply too slow for the normal techniques. However, the reactions largely determine the ignition time delay through the sequence,

$$HO_2 \quad + \quad RH \quad \rightarrow \quad H_2O_2 \quad + \quad R$$

$$H_2O_2 \quad + \quad M \quad \rightarrow \quad 2OH \quad + \quad M$$

and literally the system is 'kept waiting' for the branching agent H_2O_2 to build up in concentration. Even for H + RH, there are few data, despite its importance in oxidation and thermal cracking processes. Cohen [14] comments that "the only reliable rate constants for H + C_nH_{2n+2} where n>3 is that of Baldwin and Walker obtained from their classical H_2 + O_2 addition studies at about 750 K" (see later).

Reactions of O atoms (and other radicals) are of particular importance in high temperature combustion processes such as oxy-acetylene flames which operate at about 3000 K. The reactions occur in collisions with very high initial translational energies ($\geq$ 50 kJ mol^{-1}). Although reactions of ground state O (^{3}P) atoms with translational energies $\leq$ 40 kJ mol^{-1} have been studied extensively [16], a source of more energetic atoms has only just become available [17].

Far less quantitative kinetic information is available on the reactions of the larger alkyl (and related) radicals. Above 1000 K, this is not particularly a handicap because homolysis will normally occur rapidly and most alkyl radicals will be reduced effectively to the prototypical CH_3, C_2H_5 i-C_3H_7 and t-C_4H_9 species which will then undergo oxidation. However, below 1000 K, when the larger radicals come more thermally stable, the lack of data provides a very serious barrier to detailed modelling. The underlying causes arise from the significantly greater difficulty involved in generating 'clean' sources of the larger species under conditions suitable for studying their reactions quantitatively and from the dearth of methods for the quantitative monitoring of the radical concentrations. Beyond C_5 alkyls, few rate constants are available for reactions listed in Table 1. For example, Baldwin, Walker and Walker [18] summarise the relatively few experimentally determined kinetic data for the

4

homolysis of large, branched alkyl radicals. Gutman and Slagle [19, 20] have used the elegant technique of laser photolysis coupled with radical detection by photo-ionisation mass spectrometry to study reactions of unsaturated radicals in an O_2 environment.

2. FOUR CRUCIAL FACTORS IN MODELLING

Apart from the value of a rate constant, four other factors are of crucial importance to modellers concerned with the quantitative interpretation of combustion systems.
(a) Calculation of the rate of attack at specific sites in RH is vital using expressions of the type

$$k_{overall} = n_p \, A_p e^{-E_p/RT} + n_s \, A_s e^{-E_s/RT} + n_t \, A_t e^{-E_t/RT} \qquad\qquad (i)$$

where n_p, n_s and n_t refer to the number of primary, secondary and tertiary C-H bonds, respectively, in RH. The OH + RH data are considered so reliable that realistic attempts have been made to allow for the effect of near-neighbour groups [15]. For example, for H abstraction from a CH_2 group in an alkane, it is possible to assign Arrhenius parameters for the following, *inter alia*, three environments, based on $k = ATe^{-E/RT}$.

	A/dm^3 mol^{-1} s^{-1} K^{-1}	E/kJ mol^{-1}
OH + $CH_3\underline{CH}_2CH_3$	2.90×10^6	1080
OH + $CH_3\underline{CH}_2CH_2$-	2.90×10^6	500
OH + -$CH_2\underline{CH}_2CH_2$-	2.90×10^6	-90

Access to this type of data is crucial when modelling combustion. For example, in propanol combustion, the proportion of CH_2CH_2OH, CH_3CHOH and CH_3CH_2O radicals can be calculated. Their reactions are significantly different, CH_3CHOH reacts rapidly with O_2 to give CH_3CHO and the inert HO_2, and the other two radicals undergo mainly unimolecular decomposition. Unfortunately while the data for OH + alkanes have reached an advanced level, those for attack at other organics are much less developed except at low temperatures because of the importance of the OH radical in atmospheric chemistry [11]. Although Cohen has stressed the paucity of data for H+RH, they have been efficiently used to extrapolate the available rate constants between 250 and 2000 K by use of transition state theory [14]. He suggests that the entropy of activation per attackable H atom increases slightly from CH_4 to C_5H_{12} by about 10 J K^{-1} mol^{-1} 298 K and then stays virtually constant at 100-103 J K^{-1} mol^{-1}. Under these conditions the bond additively concept can be used and Cohen gives a general expression for the reactions of H with large (C>4) alkanes which allows for non-Arrhenius behaviour, equation (ii).

$$k = 5.4 \times 10^3 n_p T^{2.0} \exp(-3540/T) + 4.7 \times 10^3 n_s T^{2.2} \exp(-2640/T)$$
$$+ 3.7 \times 10^3 n_t T^{2.0} \exp(-970/T) \ dm^3 \ mol^{-1} \ s^{-1} \tag{ii}$$

Walker has given a simple database expression based on results obtained from studies of HO_2 radicals with CH_4, $HCHO$, and C_3H_6 [21].

$$k = 2.8 \times 10^9 [n_p \exp(-10355/T) + n_s \exp(-8900/T) + n_t \exp(-8060/T)] \ dm^3 \ mol^{-1} \ s^{-1} \tag{iii}$$

The development of expressions for other atoms and radicals, O + RH apart, which allow for near-neighbour effects is still a distant prospect.

(b) A knowledge of the products of a reaction and of the branching ratios in multi-channelled processes is essential when modelling combustion. This is particularly pertinent when one channel will lead to termination and another to branching, as for example in the important high temperature reaction between OH and CH_3 radicals.

$$
\begin{array}{lclclcl}
OH & + & CH_3 & \rightarrow & H & + & CH_2OH \\
 & & & \rightarrow & H & + & CH_3O \\
 & & & \rightarrow & H_2O & + & {}^1CH_2 \\
 & & & \rightarrow & CH_3OH & &
\end{array}
$$

Although the recombination reaction has been studied in detail [3,22], the first three channels cannot be distinguished and have a combined rate constant of $3.6 \times 10^{10} \ dm^3 \ mol^{-1} \ s^{-1}$ over the range 300-2000 K [3]. Similar problems are found with the reactions $C_2H + O_2$ and H + CHCO, both key reactions in acetylene oxidation. Determination of reaction products is, however, often difficult and time-consuming.

(c) Many of the kinetic data on recombination and dissociation processes, eg. $CH_3 + CH_3$, have been obtained at atmospheric pressure or below. Extrapolation to the higher pressures found in practical combustion units, such as the petrol engine, is extremely difficult despite the availability of several admirable theories of unimolecular reactions. Many modellers have found the theories difficult to apply and are grateful to Troe [23] and others who have developed reliable 'user-friendly' methods, based on fundamental theory [3].

(d) The sheer complexity of many comprehensive chemical mechanisms renders them effectively unusable for the modelling of many practical combustion problems, despite the huge and rapid computing facilities available. The auto-ignition of butane has been modelled by use of a mechanism containing about 500 elementary reactions and 60 different species [24]. Sensitively analysis to isolate key processes in the mechanism **under the conditions used** is vital. In principle, these reactions then become the focus of attention for kineticists. In the butane example, precise data are only needed for about 10 of the 500 reactions involved. Warnatz [25] has been particularly active in this area and has developed computer-aided methods of mechanism reduction. Pilling et al also provide a particularly good example of this approach [26].

3. GENERAL MECHANISM OF HYDROCARBON OXIDATION

It is impossible to give a relatively simple general mechanism of hydrocarbon oxidation under all conditions. At high temperatures (> 2000 K), radical-radical processes are of particular importance because of the very high radical concentrations. However, it is in the auto-ignition region (ca 600-1200 K) that chemistry plays a major role in the combustion phenomena observed, and this review will focus sharply on this temperature range [27]. At the lower temperatures, the chemistry is intimately concerned with cool-flames, 'knock' in engines and combustion-generated pollution [28], and above 1000 K with flames and shock-tube measurements. It is from studies in the middle region (650-820 K) that the author has obtained a vast amount of kinetic and mechanism data relevant to combustion chemistry. The review will be developed around some of the more recent and important of these studies.

Table 1
Basic mechanism for hydrocarbon oxidation between 500 and 1000 K

RH	$+$	O_2	$\rightarrow$	R	$+$	HO_2	(1)
RH			$\rightarrow$	R'	$+$	R''	(1A)
X	$+$	RH	$\rightarrow$	XH	$+$	R	(2)
R	$(+$	$M)$	$\rightarrow$	R'	$+$	$AB'\ (+ M)$	(3)
R	$(+$	$M)$	$\rightarrow$	H	$+$	$AB\ (+ M)$	(4)
R	$+$	O_2	$\rightarrow$	AB	$+$	HO_2	(5A)
R	$+$	O_2	$\rightarrow$	ABO	$+$	OH	(5B)
R	$+$	O_2	$\rightarrow$	OR	$+$	OH	(5C)
R	$+$	$O_2\ (+ M)$	$\rightarrow$	$RO_2\ (+ M)$			(6)
RO_2			$\rightarrow$	AB, ABO, OR	$+$	X	(7A,B,C)
RO_2	$+$	RH	$\rightarrow$	$ROOH$	$+$	R	(8)
RO_2			$\rightarrow$	$QOOH$			(9)
$QOOH$			$\rightarrow$	AB, ABO, OR	$+$	X	(10A,B,C)
$QOOH$	$+$	O_2	$\rightarrow$	O_2QOOH			(10D)
RO_2	$+$	RO_2	$\rightarrow$	chain termination			(11)
$ROOH$			$\rightarrow$	RO	$+$	OH	(12)
H_2O_2	$+$	M	$\rightarrow$	$2OH$	$+$	M	(13)
HO_2	$+$	HO_2	$\rightarrow$	H_2O_2	$+$	O_2	(14)
H	$+$	O_2	$\rightarrow$	OH	$+$	O	(15)
HO_2	$+$	RH	$\rightarrow$	H_2O_2	$+$	R	(16)

AB, alkene; ABO, oxygenated compound; OR, oxygen-ring compound. X is a radical usually OH or HO_2; R is the radical formed from RH. R′, R″ and AB′ contain a smaller number of carbon atoms than RH.

The general mechanism within the auto-ignition region has been accepted since the late 1960s and is shown in Table 1.

Under many conditions, reaction 6 is equilibrated and may be considered as the 'heart of combustion chemistry'.

$$R \; + \; O_2 \; \rightleftharpoons \; RO_2$$

Table 2 shows values of the equibrium constant for a range of alkyl radicals, together with the 'ceiling temperature' when $[R] = [RO_2]$ for $O_2 = 0.1$ atmosphere. At low temperatures, combustion chemistry is centred on reactions of RO_2 radicals, so that reactions (8)-(12), are particularly dominant and lead to large yields of oxygenated products such as O-heterocyclic compounds and organic peroxides [29]. As the temperature is raised the equilibrum position moves towards R radicals and reactions (3)-(5) lead to major amounts of conjugate alkenes and homolysis products [30]. The 'switch' in the chemistry is also closely related to the well-known occurrence of cool flames and negative-temperature coefficients in rates of reaction [31].

Table 2
Data for the equibrium $R + O_2 \; \rightleftharpoons \; RO_2$

R	ΔH^{o}_{298} kJ mol^{-1}	log K/atm^{-1}			Ceiling temp/K $[R]=[RO_2]$ at $O_2=0.1$ atm
		500 K	800 K	1200 K	
H[3]	208	16.8	8.6	4.1	1920
CH_3[3]	- 135	7.4	2.1	- 0.85	930
C_2H_5[*]	- 147	7.6	1.9	- 1.30	900
i-C_3H_7[**]	- 155	7.7	1.6	- 1.75	860
t-C_4H_9[*]	- 153	7.1	1.1	- 2.3	820
CH_2CHCH_2[**]	- 75	1.6	-1.35	- 3.0	550

[*]   I.R. Slagle, E. Ratajczak and D. Gutman, J. Phys. Chem., 90 (1986) 402
[**]   I.R. Slagle, E. Ratajczak, M.C. Heaven, D. Gutman and A.F. Wagner, J.Am. Chem. Soc., 107 (1985) 1838.

Table 2 reveals two other features.

8

(i) The 'ceiling' temperature for HO_2 radicals is considerably higher than that for RO_2 radicals. In consequence, HO_2 chemistry is still important at temperatures in excess of 1000 K.

(ii) The 'ceiling temperatures' for electron-delocalised radicals such as allyl and benzyl are considerably lower (ca 300 K) than for alkyl radicals. As a result, and arising from other aspects of their reactivity, both allyl and benzyl radicals undergo radical-radical reactions under conditions where, for example, ethyl radicals would react almost completely with O_2. With a mixture containing 4, 30 and 26 Torr of C_3H_6, O_2 and N_2, respectively at 750 K, the initial yield of hexa-1, 5-diene is about 20%, indicating that at least 40% of allyl radicals have recombined [32]. As reaction (18) is also important in removing allyl radicals, and nearly half of the C_3H_6 is removed through addition reactions (19) and (20), virtually **all** the allyl radicals are removed in radical-radical processes.

$$X \quad + \quad C_3H_6 \quad \rightarrow \quad XH \quad + \quad CH_2CHCH_2 \tag{2p}$$

$$2CH_2CHCH_2 \quad \rightarrow \quad CH_2=CHCH_2CH_2CH=CH_2 \tag{17}$$

$$CH_2CHCH_2 \quad + \quad HO_2 \quad \rightarrow \quad products \tag{18}$$

$$OH \quad + \quad C_3H_6 \quad \xrightarrow{\;O_2\;} \quad CH_3CHO \quad + \quad HCHO \quad + \quad OH \tag{19}$$

$$HO_2 \quad + \quad C_3H_6 \quad \rightarrow \quad C_3H_6O \quad + \quad OH \tag{20}$$

4. OUTLINE OF SOME 'HULL' APPROACHES

References to the work of the Hull group will be made frequently and it is pertinent to outline the experimental approaches used.

4.1. Addition to $H_2 + O_2$ mixtures at about 750 K

This approach has been used extensively over the last 30 years. Use of trace amounts of additive (RH) and an aged boric-acid-coated vessel permits investigations of the oxidation of alkanes and related compounds in the total absence of surface affects in a constant and controllable radical environment determined by the $H_2 + O_2$ mixture. This is in marked contrast to the direct oxidation method where the radical environment is **controlled** by the **oxidant** and changes constantly as the reaction intermediates are formed and then oxidised. There are two aspects of the approach.

(i) From measurements of the relative rate of consumption of H_2 and RH, kinetic data are obtained for OH + RH and H + RH, (and in some cases HO_2 + RH) [33,34]. As indicated earlier, Cohen [14] has commented on the value of the H + RH data.

(ii) Detailed product analyses over a wide range of mixture composition give mechanistic and kinetic information on the reaction of R radicals in an oxidising environment [33-35].

Limitation to ranges of 730 - 770 K and 250 - 600 Torr restricts the approach, particularly for the study of pressure effects.

4.2. Decomposition of tetramethylbutane (TMB) in the presence of O_2

The central C-C bond in TMB is sterically strained (ca 80 kJ mol^{-1}) so that homolysis occurs at a convenient rate at about 750 K. 99% of the t-C_4H_9 radicals formed react with O_2 to give i-butene and HO_2. The sequence

$$TMB \rightarrow 2t\text{-}C_4H_9$$

$$t\text{-}C_4H_9 + O_2 \rightarrow i\text{-}C_4H_8 + HO_2$$

thus provides a clean and reliable source of HO_2 radicals. The system has been used to obtain the **only** reliable kinetic data for HO_2 + alkane [36]. In the case of C_2H_6 addition, the experimental measurement involves only the relative rate of formation of C_2H_4 and i-butene.

$$HO_2 + C_2H_6 \rightarrow H_2O_2 + C_2H_5 \tag{2e}$$

Over the temperature range 670-770 K, values of $A_{2e} = 10^{10.13}$ dm^3 mol^{-1} s^{-1} and E = 86 kJ mol^{-1} were obtained [37].

This system has also given the **only** kinetic data for the reaction involving addition of HO_2 radicals to alkenes to give oxiranes [38]. Related studies of the decomposition of trimethylbutane and 2,3-dimethylbutane have given accurate values for the heats of formation of i-C_3H_7 and t-C_4H_9 radicals [39-40].

$$HO_2 + C_nH_{2n} \rightarrow C_nH_{2n}O + OH$$

4.3. Decomposition of 4, 4-dimethylpentene-1 in the present of O_2

This system has given the first reliable kinetic data for the oxidation chemistry of allyl radicals in the range 650-800 K.

$$CH_2 = CHCH_2C(CH_3)_3 \rightarrow CH_2CHCH_2 + t\text{-}C_4H_9$$

Arising from the steric strain due to the t-C_4H_9 group and the latent 'stabilisation' energy in the allyl radical, the homolysis is considerably faster than normal. As t-butyl radicals react uniquely with O_2 (99%) to give i-butene + HO_2 (see above), then the chemistry of CH_2CHCH_2 radicals in the presence of O_2 and a clean source of HO_2 can be studied with ease. The rate constants of the multi-channelled reaction between HO_2 + CH_2CHCH_2 have been obtained and it has been shown that reaction (5A) for allyl radicals is very slow [41,42].

$$CH_2CHCH_2 + O_2 \rightarrow CH_2 = C = CH_2 + HO_2 \tag{5Aa}$$

10

4.4. Direct studies of hydrocarbon oxidation

In general, direct studies of hydrocarbon oxidation have produced very little reliable kinetic and mechanistic information. However, in specific cases, under very carefully selected conditions, particular reactions can be isolated, mechanisms elucidated, and kinetic data obtained for key elementary reactions. Of relevance here is the oxidation of C_3H_6 over the temperature range 650-800 K and at pressures no greater than 60 Torr.

At these temperatures, the decomposition of H_2O_2, which corresponds to **secondary initiation** is limited particularly by the use of low pressures. The $CH_2CHCH_2 + O_2 \rightleftharpoons CH_2 = CHCH_2O_2$ is well to the left (Table 2) and **radical branching** via for example the sequence

$$H_2O_2 \quad + \quad M \quad \rightarrow \quad 2OH \quad + \quad M \tag{13}$$

$$CH_2 = CHCH_2O_2 \quad + \quad RH \quad \rightarrow \quad CH_2 = CH_2CH_2OOH \quad + \quad R$$
$$\downarrow$$
$$CH_2 = CH_2CH_2O \quad + \quad OH$$

is negligible. As discussed later, under the carefully chosen conditions, the direct oxidation of propene can be used for the determination of the Arrhenius parameters of the primary initiation reaction (lpe)[32].

$$C_3H_6 \quad + \quad O_2 \quad \rightarrow \quad CH_2CHCH_2 \quad + \quad HO_2 \tag{lpe}$$

The direct oxidation of HCHO [43], i-butene [44] and of other compounds will also be discussed later. The co-oxidation of alkenes (in the presence or absence of tetramethylbutane as a source of HO_2 radicals, section 4.2) has given excellent and unique Arrhenius parameters for the addition of HO_2 radicals to alkenes to form oxiranes [38].

$$HO_2 \quad + \quad C_nH_{2n} \quad \rightarrow \quad C_nH_{2n}O \quad + \quad OH$$

In all these studies, it is possible to deal parochially with only part of the total mechanism independently of the remainder.

5. MAJOR FEATURES OF THE REVIEW

Attention will be focused on three areas of current interest.
(i) The kinetics of the primary initiation reaction. No major review has been made of the homogeneous initiation reactions (1) and (1A).

$$RH \quad + \quad O_2 \quad \rightarrow \quad R \quad + \quad HO_2 \tag{1}$$

$$RH \quad \rightarrow \quad R' \quad + \quad R'' \tag{1A}$$

Modellers have demonstrated the kinetic importance of both reactions, and it is timely to examine the available data critically, particularly as there is sufficient information for k_1 to warrant the establishment of a database.

(ii) The reaction of C_2H_5 radicals with O_2 has exercised the minds of combustion chemists for the last 30 years, and since 1984 has been the subject of a considerable number of experimental and theoretical kinetic studies. Although there is something of a consensus view that the mechanism has been established, a large number of experimental observations cannot be satisfactorily explained. It is important to realise that $C_2H_5 + O_2$ is prototypical of the family of reactions of alkyl radicals with O_2, and indeed is clearly related to the reaction of O_2 with species such as RO, alkenyl, and hydroxyalkyl. In general, attention has unfortunately been focused almost entirely on $C_2H_5 + O_2$, and here relevant experimental results on the related systems will be presented.

(iii) Cyclisation reactions of aliphatic species are of key importance in the formation of soot and polyaromatic hydrocarbons (PAH) at temperatures above about 1200 K. Although soot formation is relatively unimportant below this temperature, the production of benzene and related aromatics is of great concern in terms of pollution and health hazards. However, little attention has been paid to low-temperature cyclisation of aliphatic radicals in the past.

6. PRIMARY INITIATION REACTIONS

Reactions (1) and (1A) are considered the normal primary initiation processes in hydrocarbon oxidation [45] and should be distinguished from secondary initiation process such as (13) where radicals are formed from 'stable' species produced as intermediates. (1) and (1A) are key steps in determining the onset of second-stage ignition [46] and recently Emdee, Brezinsky and Glassman [47], from a modelling study of the oxidation of toluene at about 1200 K, have claimed that the rate is most sensitive to the value of (1t).

$$RH \quad + \quad O_2 \quad \rightarrow \quad R \quad + \quad HO_2 \tag{1}$$

$$RH \quad \rightarrow \quad R' \quad + \quad R'' \tag{1A}$$

$$H_2O_2 \quad + \quad M \quad \rightarrow \quad 2OH \quad + \quad M \tag{13}$$

$$C_6H_5CH_3 \quad + \quad O_2 \quad \rightarrow \quad C_6H_5CH_2 \quad + \quad HO_2 \tag{1t}$$

However, in practice, the picture is considerably more complex.

(i) Below ca 600 K, (1) and (1A) may occur at the surface with considerably reduced activation energies. This has been responsible for dramatic changes in oxidation rate when different surfaces are used [1], the frequently observed lack of reproducibility, and for reported values of k_1 and k_{1A} which are a factor of 10^6 or more higher than the homogeneous value of k_1 [45,48].

Low-temperature studies by Dixon, Skirrow and Tipper [49] gave values of k_{1a} increasing from 7.4×10^{-6} to 2.9×10^{-5} at 345 K and from 6.1×10^{-5} to 5.1×10^{-4} dm^3 mol^{-1} s^{-1} at 393

12

K when the surface to volume ratio of the reaction vessel was increased from 0.6 to 6.1 cm^{-1}. The values are ca 10^9 times too high and correspond to activation energies of about 55 kJ mol^{-1} (see later).

$$CH_3CHO \quad + \quad O_2 \quad \rightarrow \quad CH_3CO \quad + \quad HO_2 \quad\quad (1a)$$

(ii) Reactions (1) (Table 3) and (1A) are very endothermic so that they can be totally masked by the presence of minute amounts of impurities, by photo-initiation and secondary initiation involving the products.

Table 3
Endothermicity of RH + O$_2$ → R + HO$_2$

RH	ΔH^o_{298} / kJ mol^{-1}
C_6H_6	252
CH_4	235
$CH_3\underline{CH_2}CH_3$	199
$(CH_3)_3\underline{CH}$	188
$\underline{CH_3}CH = CH_2$	165
HCHO	170
$CH_2 = CH\underline{CH_2}CH = CH_2$	149

6.1. Difficulties in determining k$_1$ and k$_{1A}$

In complex radical systems such as combustion, experience has shown that there is an inverse relationship between the speed of the elementary reaction and the difficulty of determining its rate constant. This is certainly true for k_1, and several factors apart from surface effects and endothermicity contribute to the difficulty.

(i) The consumption of RH and the formation of products occur predominantly through propagation processes, unless the kinetic chain length is very small.

(ii) Reactions (1) and (1A) may be completely dominated by secondary initiation extremely early in reaction. As has been shown earlier [30], for a mixture containing [CH$_4$] = [O$_2$] = 100 Torr over the range 750-1250 K, the rates of secondary (R$_s$) and primary initiation (R$_p$) will become equal when 10^{-4}% of CH$_4$ has been converted into H$_2$O$_2$ and 10^{-2}% into HCHO.

$$HCHO \quad + \quad O_2 \quad \rightarrow \quad HCO \quad + \quad HO_2 \quad\quad (1f)$$

$$H_2O_2 \quad + \quad M \quad \rightarrow \quad 2OH \quad + \quad M \quad\quad (13)$$

Although less spectacular with other alkanes, even with a tertiary C-H bond, R_p and R_s become equal when $10^{-2}\%$ of i-butane is converted into H_2O_2. Direct determination of k_1 under these conditions is not a practical proposition. As seen later, the more labile C-H bonds in C_3H_6 and aldehydes offer more promise.

(iii) Radical branching through reactions such as

$$H \quad + \quad O_2 \quad \rightarrow \quad OH \quad + \quad O$$

and

$$QOOHO_2 \quad \rightarrow \quad \text{two radicals}$$

must be taken into account. In practice, this is extremely difficult.

6.2. Experimental determination of k_1

6.2.1. Aldehyde + O_2

Although values for k_{1A} have been available for many years [50], attempts to determine k_1 have proved intractable, and (1) is one of the few remaining reactions of importance in combustion for which there are virtually no kinetic data. Early values [51,52] of 3.75×10^{-3} and 1.0×10^{-3} dm^3 mol^{-1} s^{-1} obtained at 400 K for aldehyde + O_2 were clearly too high by factors of at least 10^6, due to surface effects and secondary initiation.

Baldwin, Walker and Langford [48] modelled the oxidation of propionaldehyde at 713 K and obtained $k_{1p} = 0.076$ dm^3 mol^{-1} s^{-1}. The weakness of the aldehydic bond promotes a sufficiently high value of k_{1p} that a measurable initial rate of oxidation is observed (Fig.2). Due to dissociation of the H_2O_2 formed in the reaction, the ΔP-time curves are autocatalytic, and sensitivity analysis shows that the maximum rate is determined largely by k_{13} and $k_{2p}/k_{14}^{1/2}$ and the initial rate by k_{1p} and $k_{2p}/k_{14}^{1/2}$. As k_{13} is known, the maximum rate is used to obtain $k_{2p}/k_{14}^{1/2}$ and hence k_{1p} can be determined accurately from the initial part of the ΔP-time curve. Fig.2. shows the sensitivity of the curves to the value of k_{1p}.

$$C_2H_5CHO \quad + \quad O_2 \quad \rightarrow \quad C_2H_5CO \quad + \quad HO_2 \tag{1p}$$

$$HO_2 \quad + \quad C_2H_5CHO \quad \rightarrow \quad H_2O_2 \quad + \quad C_2H_5CO \tag{2p}$$

$$HO_2 \quad + \quad HO_2 \quad \rightarrow \quad H_2O_2 \quad + \quad O_2 \tag{14}$$

Assuming a value of $A_{1p} = 1.0 \times 10^{10}$ dm^3 mol^{-1} s^{-1} (shown later to be realistic), $E_{1p} = 153$ kJ mol^{-1}, which allowing for experimental error is close to $\Delta H_{1p} = 160\pm4$ kJ mol^{-1}, as would be expected since $E_{-1p} \approx 0$. Although k_{1p} will not be seriously in error, the value may include a contribution from secondary initiation due to the following sequence.

$$C_2H_5 + O_2 \quad \rightarrow \quad C_2H_5O_2 \quad \xrightarrow{\;AH\;} \quad C_2H_5OOH \quad \rightarrow \quad C_2H_5O + OH$$

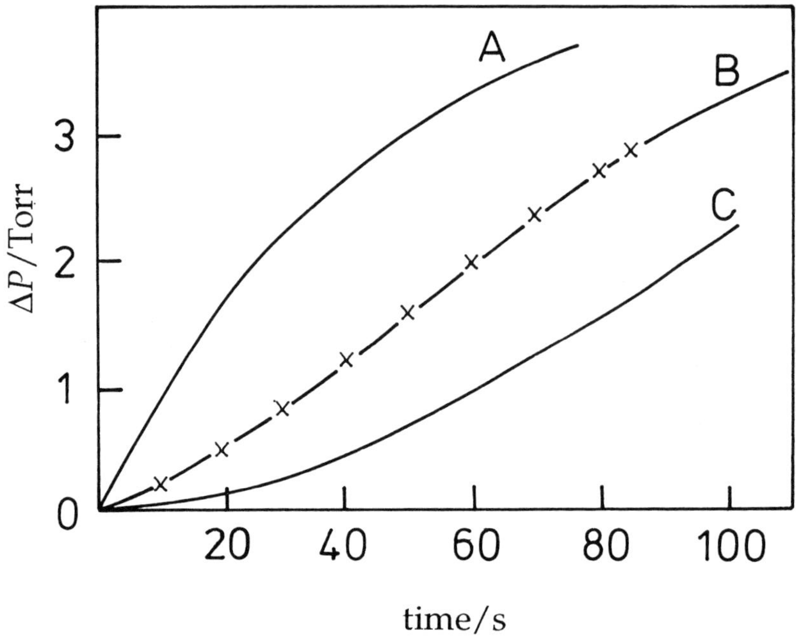

Figure 2 Computed ΔP-time plots at 713K.
$C_2H_5CHO = 4$, $O_2 = 30$, $N_2 = 26$ Torr
$k_{2p} = 0.25$ $(\text{Torr s}^{-1})^{1/2}$, x, expt points.
Curve A, $k_{1p} = 1.0 \times 10^{-4}$ Torr^{-1} s^{-1}, Curve B,
$k_{1p} = 2.0 \times 10^{-6}$ $\text{Torr}^{-1}\text{s}^{-1}$; Curve C,
$k_{1p} = 1.0 \times 10^{-8}$ $\text{Torr}^{-1}\text{s}^{-1}$.

Baldwin, Matchan and Walker [53] used the oxidation of CH_3CHO as a source of CH_3 radicals. A surprising feature of the results was the high yield of CH_4 (60%) and C_2H_6 (6%) in the initial products for a mixture containing 2, 30 and 28 Torr of CH_3CHO, O_2 and N_2 at 813 K. C_2H_6 is formed uniquely in reaction (21), which from an analysis of the results contributes at least 90% of the radical termination. Based on a fundamental principle of chain reactions, namely that on reaching the steady state of radical concentration, the rates of generation and removal of radicals may be equated. This is expressed in equation (iv) where R_p, R_s, R_b and R_t are the rates of primary initiation, secondary initiation, radical branching and radical termination, respectively

$$R_p \; + \; R_s \; + \; R_b \; = \; R_t \tag{iv}$$

Their results show that $[H_2O_2]$ is low so that $R_s = 0$ if initial rates are used (computer interpretation shows a maximum error of 5%). If at this stage $R_b = 0$ is assumed, then with total termination due to reaction (21), equation (iv) may be expressed as (v).

$$k_{1a}\,[CH_3CHO][O_2] \;=\; R_e \tag{v}$$

where R_e is the initial rate of formation of ethane.

$$CH_3 \;+\; CH_3 \;+\; M \;\;\rightarrow\;\; C_2H_6 \;+\; M \tag{21}$$

Although not reported by the authors, their results may be used to obtain a value for k_{1a}. Table 4 shows the values of R_e for a range of mixture composition at 813 K, and use of equation (v) gives k_{1a} for each mixture. The values agree to within an order of magnitude and tend to rise at high pressures of CH_3CHO (AH) and O_2, probably due to secondary initiation in the sequence

$$CH_3 + O_2 \;\rightleftharpoons\; CH_3O_2 \;\;\xrightarrow{\;AH\;}\;\; CH_3OOH \;\;\rightarrow\;\; CH_3O + OH$$
$$(6m) \qquad\qquad (22)$$

At 813 K, CH_3OOH will uniquely undergo homolysis, so that $R_b = k_{22}\,[CH_3O_2]\,[AH]$ and with $R_e = k_{21}\,[CH_3]^2$, then equation (v) becomes (vi) which can be rearranged to give (vii).

$$k_{1a}[CH_3CHO][O_2] \;+\; \frac{K_{6m}k_{22}}{k_{21}^{1/2}}\,R_e^{1/2}[CH_3CHO][O_2] \;=\; R_e \tag{vi}$$

$$k_{1a} \;+\; \frac{K_{6m}k_{22}}{k_r^{1/2}}\,R_e^{1/2} \;=\; \frac{R_e}{[CH_3CHO][O_2]} \tag{vii}$$

$$CH_3O_2 \;+\; CH_3CHO \;\rightarrow\; CH_3OOH \;+\; CH_3CO \tag{22}$$

The values of $R_e/[CH_3CHO][O_2]$ give a good straight line when plotted against $R_e^{1/2}$ in Fig.3 and the intercept gives $k_{1a} = 2.4$ dm^3 mol^{-1} s^{-1} at 813K.

The gradient is of considerable interest. A sensible value would validate the use of (vii) and permit for the first time the determination of a reliable rate constant for H abstraction by CH_3O_2 radicals. At 60 Torr (mostly $O_2 + N_2$), reaction (21) is in its fall-off region at 813 K

Table 4
Values of k_{1a} from CH_3CHO oxidation at 813 K

Mixture composition/Torr			R_e/ Torr s^{-1}	$(k_{1a})_{obs}$/ dm^3 mol^{-1} s^{-1}
CH_3CHO	O_2	N_2		
0.5	30	29.5	0.00121	4.05
2.0	30	28	0.0085	7.10
8.0	30	22	0.060	12.7
2.0	3.0	55	0.00050	4.2
0.5	7.5	52	0.00023	3.1

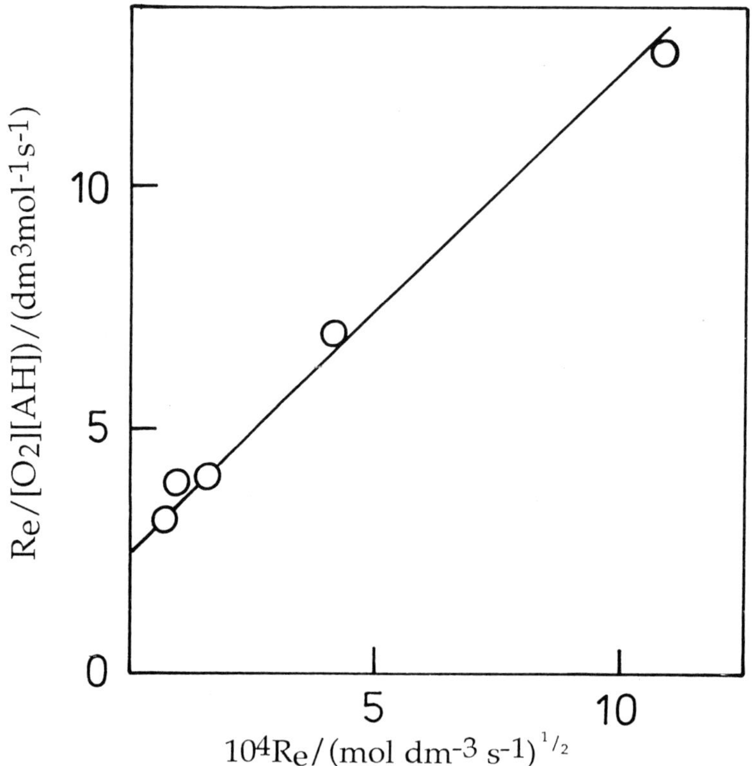

Figure 3 Plot to obtain k_{1a} and k_{22} (see text).

and $k_{21} = 1.40 \times 10^{10}$ dm^3 mol^{-1} s^{-1} from parameters given by Baulch et al [3]. From the same source, $K_{6m} = 6.5 \times 10^3$ dm^3 mol^{-1}, and as the gradient $= K_{6m} k_{22}/k_{21}^{1/2} = 1.05 \times 10^4$ dm$^{9/2}$ mol$^{-3/2}$ s$^{-1/2}$, then $k_{22} = 1.85 \times 10^5$ dm^3 mol^{-1} s^{-1}. No other experimental determination of k_{22} has been made. The nearest analogue is $k_{23} = 1.08 \times 10^6$ dm^3 mol^{-1} s^{-1} obtained [48] from $k_{23}/k_{14}^{1/2} = 43.4$ (dm^3 mol^{-1} s^{-1})$^{1/2}$ at 713 K with $k_{14} = 1.85 \times 10^9$ exp $(-775/T)$ dm^3 mol^{-1} s^{-1}, from studies of propionaldehyde oxidation [3]. Taking $A_{23} = A_{24}$ (per C-H bond) $= 2.2 \times 10^9$ dm^3 mol^{-1} s^{-1} [3], then $E_{23} = 45$ kJ mol^{-1}. As $E_{23} = E_{22}$ is very likely, then from $k_{22} = 1.09 \times 10^5$ (above), $A_{22} = 1.5 \times 10^8$ dm^3 mol^{-1} s^{-1}. A value $A_{23}/A_{22} \approx 15$ is wholly acceptable and the value of k_{22} is validated together with the value of k_{1a} above.

$$HO_2 \quad + \quad C_2H_5CHO \quad \rightarrow \quad H_2O_2 \quad + \quad C_2H_5CO \qquad\qquad (23)$$

$$HO_2 \quad + \quad C_2H_6 \quad \rightarrow \quad H_2O_2 \quad + \quad C_2H_5 \qquad\qquad (24)$$

As $\Delta H_{1p} = \Delta H_{1a} = E_{1a} = E_{1p} = 155 \pm 5$ kJ mol^{-1}, then from $k_{1a} = 2.4$ dm^3 mol^{-1} s^{-1} at 813K, the value at 713 K is 0.096 compared with $k_{1p} = 0.076$ dm^3 mol^{-1} s^{-1}. The agreement is strikingly good.

The most successful and reliable study of (1) for aldehydes has been made with HCHO. In the temperature range 670-815 K, the rate of oxidation of HCHO in Pyrex vessels freshly coated with KCl is **extremely** reproducible [43]. Surface termination of HO_2 and H_2O_2 occurs efficiently in these vessels so that destruction is diffusion-controlled and independent of surface efficiency [54,55]. Using the mixture 2, 30 and 26 Torr of HCHO, O_2 and N_2, respectively, the chain length is only about 2 and can be reduced effectively to zero when pressures down to 0.1 Torr of HCHO are used.

In the initial stages of reaction, the basic mechanism is extremely simple.

$$HCHO \quad + \quad O_2 \quad \rightarrow \quad HCO \quad + \quad HO_2 \qquad\qquad (1f)$$

$$HCO \quad + \quad O_2 \quad \rightarrow \quad HO_2 \quad + \quad CO \qquad\qquad (25)$$

$$HO_2 \quad + \quad HCHO \quad \rightarrow \quad H_2O_2 \quad + \quad HCO \qquad\qquad (26)$$

$$HO_2 \quad \rightarrow \quad surface \qquad\qquad (27)$$

$$H_2O_2 \quad \rightarrow \quad surface \qquad\qquad (28)$$

The rate of formation of CO (measured as an initial rate R_{CO}) is given by equation (viii).

$$R_{CO} = 2k_{1f} [HCHO][O_2] + 2k_{26}\left(\frac{k_{1f}}{k_{27}}\right)[HCHO]^2[O_2] \qquad\qquad (viii)$$

18

Reaction (25) is extremely fast ($k_{25} = 10^{10}$ dm^3 mol^{-1} s^{-1}) [3], and no radical branching occurs through peroxy species generated from HCO radicals. Decomposition of HCO to give H atoms is also negligibly slow. The stationary concentration of H_2O_2 is extremely small, so that reactions of OH radicals are negligible, particularly as initial rates are used.

Rearrangement of (viii) gives (ix).

$$R_{CO} / [HCHO] [O_2] = 2k_{1f} + \frac{2k_{26}k_{1f}}{k_{27}} [HCHO] \tag{ix}$$

Plots of R_{CO}/[HCHO] [O$_2$] against [HCHO] give excellent straight lines and the intercepts are extremely well defined. Variation in the vessel diameter by a factor of 4 gives indistinguishable intercepts, comfirming the homogeneous nature of (1f). Between 670 and 815 K, A_{1f} = 2.04 x 10^{10} dm^3 mol^{-1} s^{-1} and E_{1f} = 163±6 kJ mol^{-1} [43]. It must be emphasised that as experiments are carried at effectively zero chain length, k_{1f} is determined in a **direct** experiment. As the thermochemistry and the kinetics are consistent, the values represent the first totally reliable kinetic parameters for reaction (1).

6.2.2. Alkene + O_2

Studies [32] of C_3H_6 oxidation under carefully selected and controlled conditions have shown that all the difficulties traditionally associated with the determination of k_1 can be reduced to a minor role, essentially because the allyl radical produced in reaction (1_{pe}) is stabilised by electron delocalisation and is hence very unreactive towards O$_2$. The presence of the weak allylic bond in C_3H_6 also means that reaction (1_{pe}) is relatively fast. Table 2 shows that the CH$_2$CHCH$_2$ + O$_2$ equibrium is placed well to the left above 600 K. Allylperoxy reactions are consequently unimportant and moreover ΔH_{29} is about +40 kJ mol^{-1}, with k_{29} = 2.5 x 10^2 dm^3 mol^{-1} s^{-1} [41,56] at 750 K, a factor of 10^6 **slower** than the analogous reaction of alkyl radicals eg (5e).

$$C_2H_5 \quad + \quad O_2 \quad \rightarrow \quad C_2H_4 \quad + \quad HO_2 \tag{5e}$$

[HO$_2$] and [H$_2$O$_2$] are both relatively low, and although an OH chain is important (about one-third of the propene is removed via OH addition), the rates of initiation and termination are not affected. The key experimental observation is that for, say, a mixture composition of 12, 30, and 18 Torr of C_3H_6, O$_2$ and N$_2$, respectively, at 753 K, hexa-1,5-diene (HDE) is a **major** initial product (Fig.4). In fact, up to 50% of the allyl radicals formed from propene recombine to give HDE. The basic chemistry is shown by reactions (2p), (17)-(20) given in Section 3. Secondary initiation is negligible providing initial rates of product formation are considered.

$$C_3H_6 \quad + \quad O_2 \quad \rightarrow \quad C_3H_5 \quad + \quad HO_2 \tag{1pe}$$

$$C_3H_5 \quad + \quad O_2 \quad \rightarrow \quad CH_2 = C = CH_2 \quad + \quad HO_2 \tag{29}$$

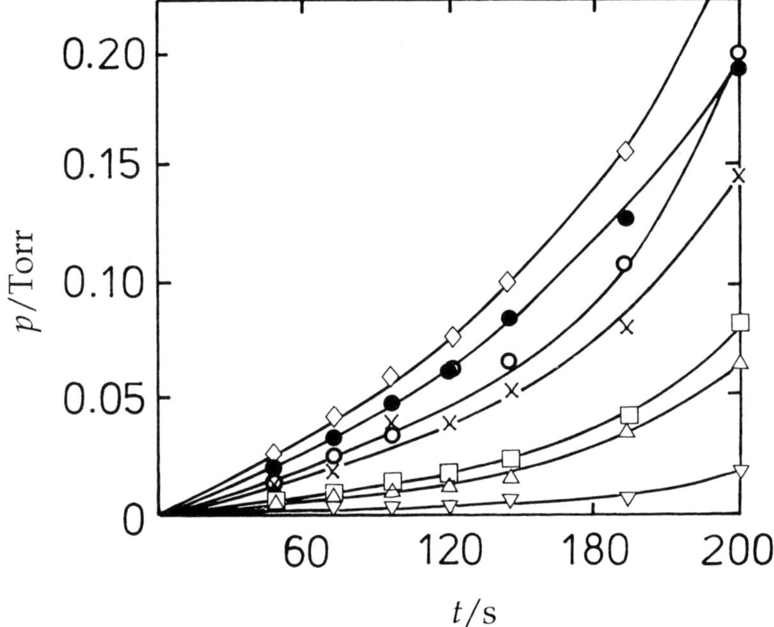

Figure 4 Pressure vs. time profiles for products from
$C_3H_6 + O_2$ at 743K.
$C_3H_6 = 12$, $O_2 = 30$, $N_2 = 18$ Torr. o, CO; ×,
HCHO; □, C_2H_4; △, CH_3CHO; ▽,
$CH_2 = CHCHO$; ●, hexadiene (x 5); ◇,
propene oxide (x 5).

Stothard and Walker [32] used equation (iv) to obtain values for E_{1pe} and A_{1pe}. With
$Rs = 0$, assuming $R_b = 0$ and that termination occurs solely through the recombination of allyl

radicals then k_{1pe} can be obtained directly from equation (x), where R_{HDE} is the initial rate of formation of HDE.

$$k_{1pe} \, [C_3H_6] \, [O_2] = R_{HDE} \qquad\qquad (x)$$

Table 5 shows that the values of k_{1pe} obtained vary by only a factor of 2 over a wide range of mixture composition at 753 K. Taking a mean value and combined with similar results at other temperatures, preliminary values of $(A_{1pe})_{obs} = 10^{9.2} \, dm^3 \, mol^{-1} \, s^{-1}$ and $(E_{1pe})_{obs} = 162 \, kJ \, mol^{-1}$ were reported. It is worth examining these values closely because the study provides an almost 'ideal' example of how to obtain good kinetic data in complex systems such as those found in combustion. Pertinent features may be emphasised.
(i) The system as set up is extremely simple (at least in a relative sense).
(ii) Direct measurement is made of the key experimental kinetic factor (namely here the major chain termination process).
(iii) The system and conditions are **selected** to exclude major complicating factors (in this case radical branching and secondary initiation).
(iv) The relevant part of the chemistry can be isolated as an independent measurement (here the OH chain, although important in product formation, can be ignored).
(v) The mechanism is tested over a wide range of mixture composition. Very frequently in computer simulations, which apparently produce acceptable kinetic data on elementary reactions, this is not the case.

Table 5
Determination of k_{1pe} from $C_3H_6 + O_2$ reaction at 753 K

Mixture composition/Torr			$10^5 R_{HDE}/$	$10^5 R_{po}/$	$10^2 (k_{1pe})_{obs}/$
C_3H_6	O_2	N_2	Torr s^{-1}	Torr s^{-1}	dm^3 mol^{-1} s^{-1}
2	30	28	1.02	4.2	0.80
4	5	51	0.60	1.14	1.40
4	10	46	1.20	1.65	1.41
4	30	26	4.5	2.7	1.76
12	5	43	1.41	1.14	1.10
12	10	38	3.0	2.15	1.17
12	30	18	9.1	1.08	1.19
16	30	14	10.9	1.59	1.08

With regard to the observed Arrhenius parameters for (1pe), the only comparable kinetic data are the values of $A_{1f} = 2.04 \times 10^{10} \, dm^3 \, mol^{-1} \, s^{-1}$ and $E_{1f} = 163 \pm 6 \, kJ \, mol^{-1}$, given earlier. Within experimental error $(E_{1pe})_{obs} = E_{1f} = \Delta H_{1pe} = \Delta H_{1f} = 165 \pm 5$ kJ mol^{-1}, as expected since the activation energies of the reverse reactions will be effectively zero. $A_{1f}/A_{1pe} \approx 10$ is entirely consistent with the extra loss of entropy of activation in reaction (1pe) due to electron delocalisation in the emerging allyl radical. The high degree of consistency in all the data above provides strong support for the approach used. Stothard and Walker [32] refined the treatment by allowing for extra termination due to reactions (13) and (30), and radical branching due to (31).

$$HO_2 \quad + \quad HO_2 \quad \rightarrow \quad H_2O_2 \quad + \quad O_2 \tag{13}$$

$$CH_2CHCH_2 \quad + \quad HO_2 \quad \rightarrow \quad C_3H_6 \quad + \quad O_2 \tag{30}$$

$$CH_2CHCH_2 \quad + \quad O_2 \quad \rightarrow \quad \text{two radicals} \quad + \quad \text{products} \tag{31}$$

Again a simple clean approach was used. By measurement of the initial rate of formation of propene oxide which is formed (here) uniquely in reaction (32), k_{32} being known very

$$HO_2 \quad + \quad C_3H_6 \quad \rightarrow \quad C_3H_6O \quad + \quad OH \tag{32}$$

accurately [57], $[HO_2]$ can be calculated precisely. Similarly, $[CH_2CHCH_2]$ can be calculated accurately from the initial rate of formation of HDE (R_{HDE}). It is important to stress that the correction amounts to only about **20%**, leading to the final values, $A_{1pe} = 10^{9.29 \pm 0.41}$ dm^3 mol^{-1} s^{-1} and $E_{1pe} = 163.5 \pm 6$ kJ mol^{-1}.

Logically, good kinetic data should be forthcoming from a similar type of study of the oxidation of isobutene. The methylallyl (MA) radical produced by H abstraction from one of the two CH_3 groups also has considerable difficulty in reacting, apart from the <u>minor</u> homolysis to give allene and CH_3.

In confirmation, 2,5-dimethylhexa-1,5-diene is found in yields of up to 20% from the mixture containing 4, 30, 26 Torr of isobutene, O_2 and N_2, respectively [44], indicating that MA radicals are removed mainly through radical-radical processes. Given a similar treatment as shown above for propene, with corrections a little larger because of somewhat higher HO_2 concentrations, the basic use of equation (xi) gives $A_{1ib} = 10^{9.68 \pm 0.44}$ dm^3 mol^{-1} s^{-1} and $E_{1ib} = 161.2 \pm 6.4$ kJ mol^{-1} over the range 673-773 K, where R_{DMHDE} is the initial rate of formation of 2,5-dimethylhexa-1,5-diene.

$$k_{1ib}[i\text{-}C_4H_8]\,[O_2] \quad = \quad R_{DMHDE} \tag{xi}$$

$$CH_2C(CH_3)CH_2 \quad \rightarrow \quad CH_2{=}C{=}CH_2 \quad + \quad CH_3 \tag{33}$$

$$(CH_3)_2C{=}CH_2 + O_2 \quad \rightarrow \quad CH_2C(CH_3)CH_2 \quad + \quad HO_2 \tag{1ib}$$

As shown later on in Table 7, the agreement of the Arrhenius parameters with those for $HCHO + O_2$ and $C_3H_6 + O_2$ is remarkable, given the path degeneracy difference of 2 in reaction (1pe) and (1ib).

6.2.3. Difficult alkene systems

It is interesting now to consider the possible determination of k_1 for butene-2 and butene-1. Two experimental features emphasise the large differences in the chemistry involved. Fig.5 shows pressure change - time plots for the 4 alkenes under discussion. Clearly (and direct product analysis confirms the effect), the initial rates of oxidation of butene-1 and butene-2 are considerably faster than those for i-butene and propene (50-100 times). Secondly, the major (50%) initial product with both butene-1 and butene-2 is buta-1,3-diene **and** no evidence is found for any recombination product (and therefore for radical-radical processes) for butenyl radicals. The basic mechanism (focussing on butenyl radicals) for both alkenes shows the importance of HO_2/H_2O_2 chemistry as reaction (5Ab) is the dominant propagation process.

$$C_4H_8 \quad + \quad O_2 \quad \rightarrow \quad C_4H_7 \quad + \quad HO_2 \tag{1b}$$

$$C_4H_7 \quad + \quad O_2 \quad \rightarrow \quad CH_2{=}CH{-}CH{=}CH_2 \quad + \quad HO_2 \tag{5Ab}$$

$$HO_2 \quad + \quad C_4H_8 \quad \rightarrow \quad C_4H_7 \quad + \quad H_2O_2 \tag{34}$$

$$HO_2 \quad + \quad HO_2 \quad \rightarrow \quad H_2O_2 \quad + \quad O_2 \tag{13}$$

With a fuller mechanism to account for the products arising from OH and HO_2 addition to the butenes, the product distribution can be modelled very successfully over a wide range of mixture composition [58]. However, in the absence of a radical branching process, a value of k_{1b} approximately 100 times higher than k_{1pe} is required at about 750 K. Such a factor is totally unrealistic, and indeed experiments (discussed later) involving the addition of butene-1 and butene-2 to $C_3H_6 + O_2$ mixtures at 713 K, show that $k_{1b}/k_{1pe} \approx 5$. Radical branching is undoubtedly of dominating importance. A possible outline sequence involves the addition of HO_2 radicals.

$$HO_2 + CH_3CH{=}CHCH_3 \longrightarrow \underset{\displaystyle OOH}{CH_3CH{-}CHCH_3} \xrightarrow{\ O_2\ } \underset{\displaystyle OOH\ \ O_2}{CH_3CH{-}CHCH_3}$$

$$\downarrow$$

$$\text{branching}$$

This possibility is currently being tested [58]. A similar sequence would be of little importance in the $C_3H_6 + O_2$ mixtures used because (in the early stages) the HO_2 concentration is 50-100 lower, and HO_2 adds more slowly to C_3H_6 [57]. The butene-1 and butene-2 cases provide a clear indication of the difficulty of determining k_1. In the absence of the value for k_{1pe}, the 'high' value for k_{1b} would be acceptable. There is very little doubt that a high value for k_1 can be obtained without any difficulty whatsoever!

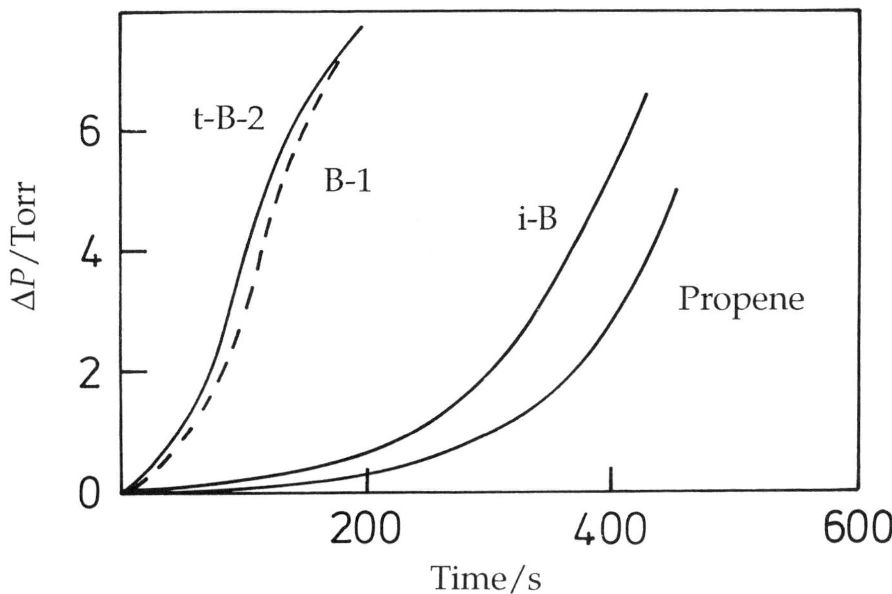

Figure 5   Pressure change - time plots for the oxidation of trans-butene-2, butene-1, i - butene and propene at 753K Alkene = 4, O_2 = 30, N_2 = 26 Torr.

6.2.4. Alkenyl oxidation chemistry

It is pertinent here to comment on four types of alkenyl radical in respect of reaction with O_2.

(i)   The electron-delocalised radical species CH_2CHCH_2 and $CH_2C(CH_3)CH_2$ which react extremely slowly with O_2 as shown (for allyl) by the parameters in Table 6. In consequence the radicals are removed mainly through radical-radical processes.

(ii) Electron-delocalised species which react relatively quickly with O_2 to give a diene. At 750 K, $k \approx 1 \times 10^6$, which is 50-100 times lower than for the analogous reaction of alkyl radicals [59] but about 10^4 higher than for allyl and methylallyl.

$$CH_3CHCHCH_2 + O_2 \longrightarrow CH_2=CH\text{-}CH=CH_2 + HO_2$$

$$\text{(cyclohexenyl radical)} + O_2 \longrightarrow \text{(benzene ring)} + HO_2$$

$$CH_3CHCH=CHCH_3 + O_2 \longrightarrow CH_2=CH\text{-}CH=CH\text{-}CH_3 + HO_2$$

(iii) Alkenyl radicals where the free valency is not at the allyl position, for example.

$$CH_2=CHCH_2CH_2CHCH_3 + O_2 \rightarrow CH_2=CHCH_2CH_2CH=CH_2 + HO_2$$

Here the reaction should proceed with $k \approx 10^8$ dm^3 mol^{-1} s^{-1}, as for an alkyl radical [45]
(iv) Alkenyl radicals where the free valency occurs at a carbon atom in the double bond, for example

$$CH_2=CH + O_2 \rightarrow HCHO + HCO$$

$$CH_3C=CH_2 + O_2 \rightarrow CH_3CO + HCHO$$

and the chemistry is different in character.

Table 6
Arrhenius parameters for $CH_2CHCH_2 + O_2$ pathways [42]

pathway	log (A/dm^3 mol^{-1} s^{-1})	E/kJ mol^{-1}
$CH_2CHCH_2 + O_2 \rightarrow$ 2 radicals + products	8.2 ± 0.6	72.5 ± 8
$CH_2CHCH_2 + O_2 \rightarrow 2HCHO + CO + OH$	9.7 ± 0.4	78.5 ± 4.5
$CH_2CHCH_2 + O_2 \rightarrow CH_2=C=CH_2 + HO_2$	9.0^*	95^*

*   Values based on $k = 2.5 \times 10^2$ dm^3 mol^{-1} s^{-1} at 753K and assumed $A = 10^{9.0}$ dm^3 mol^{-1} s^{-1}.

6.2.5. Alkane + O_2 systems

Even for a tertiary C-H bond in alkanes, the value of ΔH_1 is considerably higher (ca 35 kJ mol^{-1}) than for the allylic C-H in alkenes. At 750 K, the rate constant is thus lowered by about 270 on this account, although compensation for the formation of an electron-delocalised radical reduces the difference to a factor of 30. Nevertheless, the reduction is very significant and any secondary initiation (eg reaction (14)) and radical branching may be greatly increased in relative importance.

Only one reliable estimate for k_1 (alkane) is available in the literature. Baldwin et al [40] studied the decomposition of 2, 3-dimethylbutane in the presence of O_2 (for related work, see section 4.2) in which the major initiation process is (1Ad). Although at low $[O_2]$ a constant value of k_{1Ad} is obtained, the observed value shows an increase at high $[O_2]$, consistent with a contribution from reaction (1d) which effectively involves only abstraction at a tertiary C-H bond.

$$(CH_3)_2CHCH(CH_3)_2 \qquad \rightarrow \qquad 2\,(CH_3)_2CH \qquad\qquad (1Ad)$$

$$(CH_3)_2CHCH(CH_3)_2 \ + \ O_2 \qquad \rightarrow \qquad (CH_3)_2CHC(CH_3)_2 \ + \ HO_2 \qquad\qquad (1d)$$

The treatment is a little complex [40], and values of k_{1d} = 0.135$\pm$0.011 and 0.0180$\pm$0.0027 dm^3 mol^{-1} s^{-1} are obtained at 813 and 773 K, respectively, after allowing for the small amount of attack at the primary C-H bonds (assuming a path degeneracy difference of 6). With ΔH_{1d} = 185 kJ mol^{-1} assumed equal to E_{1d}, then k_{1d} at 773 K gives A_{1d} = $10^{10.7}$ dm^3 mol^{-1} s^{-1} which is at most a factor of 2 too high.

6.2.6. Aromatic + O_2 systems

Emdee, Brezinsky and Glassman [47] modelled the oxidation of toluene at about 1200 K and, based on a computer simulation approach with a comprehensive mechanism, claim that the rate is most sensitive to the value of k_{1t}. From their analysis with E_{1t} = 173 kJ mol^{-1} (their value of ΔH_{1t}), a value of A_{1t} = 3.0 x 10^{11} dm^3 mol^{-1} s^{-1} is obtained. This value is a factor of 100 higher than expected, based on the studies of C_3H_6 + O_2 and i-C_4H_8 + O_2, discussed earlier, as a similar electron delocalisation effect in all three should lead to an A factor of ca 2 x 10^9 dm^3 mol^{-1} s^{-1}. At 773 K, k_{1t}/k_{1pe} = 36 from the respective Arrhenius parameters. This figure is extremely high and later it will be shown that it does not exceed ca 2 at this temperature.

$$C_6H_5CH_3 \qquad + \qquad O_2 \qquad \rightarrow \qquad C_6H_5CH_2 \qquad + \qquad HO_2 \qquad\qquad (1t)$$

6.3. Commentary on the values of k_1

The values obtained for k_1 are summarised in Table 7. With the exception of that for toluene, there is a high degree of consistency. For CH_3CHO, C_2H_5CHO, HCHO and TMB all data are close to A_1 = 1.0 x 10^{10} dm^3 mol^{-1} s^{-1} (per C-H bond) and E_1 = ΔH_1. For C_3H_6 and i-C_4H_8, A_1 is lower by an order of magnitude as expected from the extra loss of entropy of activation in the reactions. Although possibly fortuitous, A_{1ib}/A_{1pe} ~ 2 as expected from path degeneracy differences. The data are limited, but with some confidence it is concluded that they are reliable, particularly those determined over a temperature range. Use of these Arrhenius parameters should give k_1 accurate to a factor of 2 between 600 and 1000 K. A database for k_1 will be given later.

Table 7

Kinetic Parameters for $RH + O_2 \rightarrow R + HO_2$

RH	$k/dm^3\ mol^{-1}\ s^{-1}$	T/K	$\log(A/dm^3\ mol^{-1}\ s^{-1})$	$E/kJ\ mol^{-1}$	Ref
C_2H_5CHO	0.076	713	10.0	153*	[48]
CH_3CHO	2.4	813	10.0	150*	[53]
HCHO	-	673-815	10.31	163±6	[43]
C_3H_6	-	673-793	9.29	163.5±5	[32]
i-C_4H_8	-	673-793	9.68	161.2+6.4	[44]
$(CH_3)_2CHCH(CH_3)_2$	0.135±0.011	813	-	-	[40]
	0.0180±0.0027	773	10.7**	185	[40]
$C_6H_5CH_3$	-	1200	10.5	173†	[47]

*   A (per C-H bond) assumed $= 10^{10.0}\ dm^3\ mol^{-1}\ s^{-1}$
**   For $RH = (CH_3)_2\ CHCH(CH_3)_2$, $\Delta H = E$ assumed
†   Emdee et al [47] assume $\Delta H = E$.

6.4. Use of the $C_3H_6 + O_2$ system to determine k_1.

The success in determining k_1 as described above and the verification of a consistent set of data has led [44] to the use of the 'ideal' $C_3H_6 + O_2$ system to obtain further values of k_1.

When **small** amounts of compound RH containing C-H bonds, which may be attacked by O_2 with at least equal facility as the allyl C-H bonds in C_3H_6, are added to $C_3H_6 + O_2$ mixtures, then additional measureable primary initiation occurs through reaction (1). Providing R has a fast reaction, then [allyl] $>>$ [R], so that the recombination of allyl radicals to give hexadiene (HDE) remains the major termination process

$$RH \quad + \quad O_2 \quad \rightarrow \quad R \quad + \quad HO_2 \tag{1}$$

$$2CH_2CHCH_2 \quad \rightarrow \quad CH_2 = CHCH_2CH_2CH = CH2 \tag{17}$$

Taking a simple case where (1) and (1pe) are the only primary initiation reactions, $R_s = R_b = 0$ and (17) is the only termination, then the ratio k_1/k_{1pe} can be obtained very simply from equation (xii), wher R_{HDE} and $(R_{HDE})_0$ are the initial rates of formation of HDE in the presence and absence of RH. Typically experiments are carried out with mixtures containing 4 to 12 Torr of propene in 30 Torr of O_2 with up to 1 Torr of additive in the range 670 - 770 K. Fig.6 shows some typical results for RH = 2-methylbutene-2 at 693K.

$$R_{HDE}/(R_{HDE})_o = 1 + k_1[RH]/k_{1pe}[C_3H_6]$$ (xii)

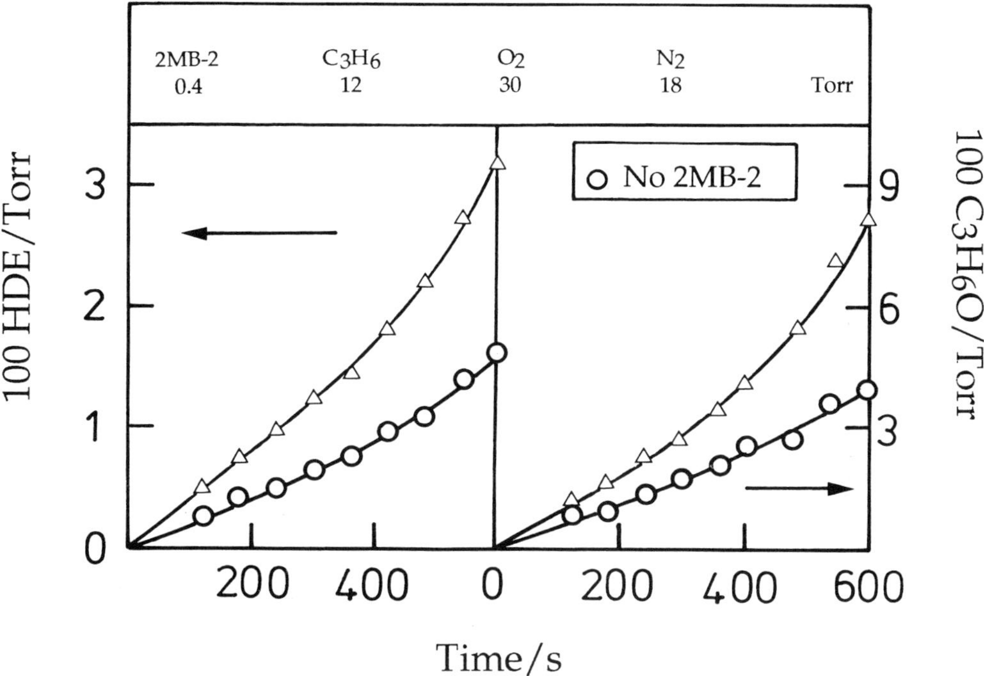

Figure 6   Plots of HDE and C3H6O formation at 693 K, with and without 2MB-2.

By measuring the yields of propene oxide as well, an allowance can be made for radical branching through reactions (31)[32] and for the additional radical termination via (14) and (30).

$$HO_2 \quad + \quad HO_2 \quad \rightarrow \quad H_2O_2 \quad + \quad O_2 \tag{14}$$

$$CH_2CHCH_2 \quad + \quad HO_2 \quad \rightarrow \quad C_3H_6 \quad + \quad O_2 \tag{30}$$

28

$$CH_2CHCH_2 \quad + \quad O_2 \quad \rightarrow \quad \text{two radicals} \quad + \quad \text{products} \quad\quad (31)$$

$$HO_2 \quad + \quad C_3H_6 \quad \rightarrow \quad C_3H_6O \quad + \quad OH \quad\quad (32)$$

The allyl and HO_2 concentrations were calculated from the rates of formation of HDE and propene oxide, respectively, both k_{17} and k_{32} being known accurately [32]. Table 8 summarises the values obtained simply from equation (xii), together with those where the corrections have been made. Typically the corrections do not exceed about 20%. Ingham, Stothard and Walker [44] have shown that the data form a consistent pattern with values of $E_1 = AH_1$.

Three points are emphasised here

(i)   Where the values of k_1/k_{1pe} are less than 5, the increase in HDE is quite small and the accuracy is limited (ca ± 2, i.e. 40%)

(ii)   Within the confines indicated above the value of $k_{1ib}/k_{1pe} = 5.5 \pm 2$ at 693 K obtained from the addition study is in very good agreement with the value of 3.7 from the separate studies of the oxidation of propene and i-butene discussed earlier (Table 7). Validation of the method is thus obtained.

(iii)   It was only possible to obtain a maximum value for k_{1t}

$$C_6H_5CH_3 \quad + \quad O_2 \quad \rightarrow \quad C_6H_5CH_2 \quad + \quad HO_2 \quad\quad (1t)$$

When 1 Torr of toluene was added (the limit to preserve the mechanism) to the mixture specified earlier at 713 K an increase in R_{HDE} was not perceptible leading to $k_{1t}/k_{1pe} < 2$, whereas the use of the Emdee et al [47] parameters would give about 40. However, due to the weaker C-H bonds involved, reasonably accurate values were obtained for ethylbenzene and isopropylbenzene as shown in Table 8. Within experimental error, the weakest C-H bond dissociation energies in toluene, ethylbenzene and i-propylbenzene are 370, 355 and 345 kJ mol^{-1}, and with $E = \Delta H$ and allowing for path degeneracy, the relative values of $k_{1t} : k_{1eb} : k_{1pb}$ are calculated as 1 : 8 : 22 at 750 K, assuming equal A factors per C-H bond. The values of k_{1eb} and k_{1pb} are reasonably consistent with this prediction and imply $k_{1t} \leq k_{1pe}$ in marked contrast to a calculated value of $k_{1t}/k_{1pe} \approx 40$ obtained from Arrhenius parameters suggested by Emdee et al [47]. It is pertinent to point out also that the radicals $C_6H_5CHCH_3$ and $C_6H_5C(CH_3)_2$ do have facile reactions with O_2 so that the condition [allyl] $\gg$ [R] is preserved in the addition studies. In confirmation no evidence was found for radical recombination products such as $C_6H_5CH(CH_3)CH_2CH = CH_2$ or $C_6H_5C(CH_3)_2CH_2 CH = CH_2$.

$$C_6H_5CH_2CH_3 \quad + \quad O_2 \quad \rightarrow \quad C_6H_5CHCH_3 \quad + \quad HO_2 \quad\quad (1eb)$$

$$C_6H_5CHCH_3 \quad + \quad O_2 \quad \rightarrow \quad C_6H_5CH = CH_2 \quad + \quad HO_2$$

$$C_6H_5CH(CH_3)_2 \quad + \quad O_2 \quad \rightarrow \quad C_6H_5C(CH_3)_2 \quad + \quad HO_2 \quad\quad (1pb)$$

Table 8
Kinetic data for $RH + O_2 \rightarrow R + HO_2$ from C_3H_6 addition studies

RH	k_1/k_{1pe} (obs)	k_1/k_{1pe} (corr)	$k_1/dm^3mol^{-1}s^{-1}$	T/K	$\log(A/dm^3mol^{-1}s^{-1})$	$E/kJmol^{-1}$
C_3H_6*					9.29 ± 0.41	163.5 ± 5
i-C_4H_8*	4.8	5.5	0.0051	693	9.68 ± 0.44	161.2 ± 6.4
t-C_4H_8-2	4.0	4.9	0.0045	693	9.6	158
l- C_4H_8	11.1	12	0.011	693	9.1	146
l-C_5H_{10}	12.7	15	0.014	693	9.1	145
$(CH_3)_2C{=}CHCH_3$	25.5	30.5	0.0283	693	9.8	151
$(CH_3)_2C{=}C(CH_3)_2$	117	145	0.134	693	9.9	143
$CH_2{=}CHCH_2CH{=}CH_2$	43	52.5	0.327	743	8.1	122
$C_6H_5CH_3$**	<2	<2	<0.004	713	9.3	163
$C_6H_5C_2H_5$	7.1	8.3	0.0171	713	9.1	149
$C_6H_5CH(CH_3)_2$	16.1	18.5	0.038	713	8.8	140

* Arrhenius parameters from direct studies, see text [32,44].

** Arrhenius parameters recommended are those for $C_3H_6 + O_2$.

$$C_6H_5C(CH_3)_2 \quad + \quad O_2 \quad \rightarrow \quad C_6H_5C(CH_3)=CH_2 \quad + \quad HO_2$$

6.5. Comments on the value of k_{1t} obtained by Emdee et al [47]

Emdee et al modelled the separate experimental oxidations of benzene and toluene at 1000 and 1100 K, respectively. No doubt is cast here on the actual experimental results used to verify the model. The data were obtained in the well-known Princeton flow reactor [60] and included results from a fuel-lean and a fuel-rich mixture. Emdee et al [47] do comment that some adjustments were necessary to the original data following recalibration of flow-meters, resulting in a shortening of the reaction times by up to 20%. The model for toluene contained as a subset the reactions used to model benzene oxidation, and in total 130 elementary reactions, written as reversible steps, were used. It is not proposed to reproduce the mechanism here but to comment on the weaknesses of the approach and interpretation.

(i) The test made of the model is inadequate. Only two mixtures were modelled, the composition of toluene being constant (almost) and the O_2 pressure varied by less than a factor of 2. The total pressure was constant at one atmosphere. Key measurements at low extents of reaction were not made. Given the number of variables in the mechanism and some freedom to change their value, successful prediction is inevitable, even though the mechanism may be seriously flawed.

The practice of modelling with very limited changes of mixture composition is widespread, and to some extent arises from restrictions caused by explosion limits and cool-flames.

(ii) Classical chain reaction theory [54] makes clear that the overall rate of oxidation is strongly dependent on the production of new radicals whether through primary initiation, secondary initiation, radical branching or some other means. It is, however, difficult in practice to assign that production to a particular chemical process, as has been discussed earlier. Primary initiation may well become dominated by secondary initiation or radical branching very early in reaction, and this may be very difficult to detect, particularly if testing is not carried out over a very wide range of mixture composition.

Further, in any mechanism, if one initiation process dominates, then the overall rate of oxidation **must** be sensitive to any change in the rate of that process. In modelling, however, it is extremely easy to draw an incorrect conclusion. The observation by Emdee et al that the rate was very sensitive to k_{1t} (the model predicted overall rate α k_{1t}) cannot be used to confirm the importance of (1t) and to determine k_{1t} if the value in the model has been set much too high, **unless** very stringent tests are carried out usually by making major changes in reaction conditions. What is actually being measured is the total rate of initiation of new radicals, a small part of which may be due to reaction (1t).

(iii) From the results presented by Emdee et al, the rate of oxidation is similar (within a factor of 2) for both toluene and benzene despite (using their parameters) a value of $k_{1t}/k_{1be} = 1.2 \times 10^5$ at 1200 K. The 'chain length' is clearly much higher in the benzene oxidation.

$$C_6H_6 \quad + \quad O_2 \quad \rightarrow \quad C_6H_5 \quad + \quad HO_2 \qquad (1be)$$

According to the mechanism, benzene is removed only by radical attack at the ring leading to the formation of phenyl and phenoxy radicals

$$OH + C_6H_6 \rightarrow H_2O + C_6H_5$$

$$H + C_6H_6 \rightarrow H_2 + C_6H_5$$

$$O + C_6H_6 \rightarrow C_6H_5O + H$$

The phenyl radical reacts rapidly with O_2 in a **branching** reaction to give the 'stable' phenoxy radical which is consumed in two competing reactions (34) which leads to termination and (35) which propagates the chain.

$$C_6H_5 + O_2 \rightarrow C_6H_5O + O$$

$$C_6H_5O + H \rightarrow C_6H_5OH \tag{34}$$

$$C_6H_5O \rightarrow C_5H_5 + CO \tag{35}$$

Not surprisingly both (34) and (35) are key reactions in determining the overall rate of oxidation because the subsequent reactions of C_5H_5 generate the propagating radicals and also involve radical branching reactions (eg $H + O_2 \rightarrow OH + O$) and secondary initiation (eg $H_2O_2 \rightarrow 2OH$). In the ensuing 'long chains' the rate of generation of new radicals through these processes will far outweigh that from primary initiation, and the overall rate will be insensitive to the value of k_{1be}.

According to the mechanism proposed by Emdee et al, H atom abstraction occurs **solely** from the side chain and toluene is removed by the following reactions, where $OC_6H_4CH_3$ is a methylphenoxy radical.

$$C_6H_5CH_3 + O_2 \rightarrow C_6H_5CH_2 + HO_2 \tag{1t}$$

$$H + C_6H_5CH_3 \rightarrow C_6H_5CH_2 + H_2$$

$$OH + C_6H_5CH_3 \rightarrow C_6H_5CH_2 + H_2O \tag{36s}$$

$$O + C_6H_5CH_3 \rightarrow OC_6H_4CH_3 + H$$

The benzyl radicals are stabilised by electron delocalisation and are removed mainly by radical-radical processes, many of which lead to chain termination. With their mechanism, a short chain length is to be expected with the overall rate very sensitive to k_{1t}. However, Emdee et al ignore H-atom abstraction from the ring in toluene. Using their rate constants and taking account of path degeneracy then $k_{36r}/k_{36s} = 0.61$ at 1200 K, so that almost 40% of OH attack occurs at the **ring**. In consequence the **branching** reaction (37) may well be of great importance in generating new centres relative to (1t) and although abstraction at the ring may be of minor importance directly in removing toluene, the subsequent reactions of the methylphenyl radicals are probably very important in determining the overall rate. By analogy, it is the chemistry of phenyl in benzene oxidation and of methylphenyl in toluene oxidation that are important in determining overall rate. By ignoring the ring attack in toluene, a high value for k_{1t} is inevitable.

$$OH \quad + \quad C_6H_5CH_3 \quad \rightarrow \quad C_6H_4CH_3 \quad + \quad H_2O \qquad (36r)$$
$$C_6H_4CH_3 \quad + \quad O_2 \quad \rightarrow \quad OC_6H_4CH_3 \quad + \quad O \qquad (37)$$

(iv) Emdee et al ignore H atom abstraction by HO_2 radicals (38) because 'the reaction has a high activation energy', although very surprisingly HO_2 attack on CH_4 is included. In fact [63], E_{38} is only $\approx$ 45kJ mol^{-1} and E_{39} - $E_{38} \approx$ 60 kJ mol^{-1}. More importantly, as the half-life of H_2O_2 does not exceed 10^{-5} s under their conditions, reaction (38) (and (39)) is effectively a branching process via reaction (13).

$$HO_2 \quad + \quad C_6H_5CH_3 \quad \rightarrow \quad H_2O_2 \quad + \quad C_6H_5CH_2 \qquad (38)$$
$$HO_2 \quad + \quad CH_4 \quad \rightarrow \quad H_2O_2 \quad + \quad CH_3 \qquad (39)$$
$$H_2O_2 \quad + \quad M \quad \rightarrow \quad 2OH \quad + \quad M \qquad (13)$$

In consequence, although (38) may only play a minor role in removing toluene, the important factor is its rate relative to the rate of the primary initiation step (1t). With (38) ignored, a higher value of k_{1t} is necessary to predict the experimentally determined rate of oxidation of toluene.

The difficulty in determining k_{1t} in the way used by Emdee et al may be seen by examining the basic theory underlying their approach as laid out in the classical Semenov theory [55]. Equation (xiii) may be written for a branched chain reaction with a radical concentration n,

$$dn/dt = \theta + fn + Fn^2 - gn - \gamma n - \delta n^2 \qquad \text{(xiii)}$$

where θ is the rate of initiation, fn, gn and γn are the rates of linear branching and linear termination in the gas-phase and at the surface, respectively, and Fn^2 and δn^2 are the rates of quadratic branching and mutual termination, respectively. For illustrative purposes, the quadratic terms will be neglected so that (xiii) becomes equation (xiv), where $\varnothing = f-g-\gamma$

$$dn/dt = \theta + \varnothing n \qquad \text{(xiv)}$$

If $\varnothing$ is positive, the kinetic features of the chain reaction lead to an isothermal explosion with the boundary defined by $\varnothing = 0$. If $\varnothing$ is negative, a stationary centre concentration $n_s = \theta/|\varnothing|$ is reached, so that the stationary state rate (w_s) is given by equation (xv), where p is the propagation constant.

$$w_s = p\theta/|\varnothing| \qquad \text{(xv)}$$

In connecting this abstract equation to a real situation, the nature of p is not difficult to characterise. Uncertainties arise however in θ and $|\varnothing|$.

(i)   Where branching is significant, the value of $|\varnothing|$ may be close to zero, so that small changes in g, y or f may cause large proportional changes in $|\varnothing|$. Indeed the well-known sensitivity of branched chain reactions to traces of foreign materials, surface effects and to slight changes in reaction conditions arises in this way. In the situation where a reaction is modelled, therefore, neglect of any termination or branching reaction can have extremely serious kinetic consequences. In practice, since termination reactions are usually well defined, it is neglect of radical branching that causes the problem. As discussed earlier, Emdee et al appear to ignore a major branching route in their toluene oxidation modelling study, leading to the equivalent of a higher value of $|\varnothing|$, so that a higher value of θ is required to satisfy the experimentally-determined rate. Parochially this is extremely difficult to detect.

(ii)   The chemical nature of the initiation rate must be examined very carefully. For propanal oxidation at 713 K, Baldwin, Walker and Langford [48] showed that only two initiation reactions need be considered.

$$C_2H_5CHO \quad + \quad O_2 \qquad \rightarrow \qquad C_2H_5CO \quad + \quad HO_2 \qquad\qquad (1p)$$
$$H_2O_2 \qquad + \quad M \qquad \rightarrow \qquad 2OH \quad + \quad M \qquad\qquad (13)$$

At any time, $\theta = 2k_{1p}[C_2H_5CHO][O_2] + 2k_{13}[H_2O_2][M]$, so that only for the initial rate of oxidation where $[H_2O_2] = 0$ does $\theta = 2k_{1p}[C_2H_5CHO][O_2]$. As discussed earlier the oxidation is autocatalytic, and the maximum rate is absolutely independent of k_{1p} for any reasonable value of k_{1p} [48]. Only, therefore, by modelling the rate of oxidation in the very early stages of reaction can an accurate value of k_{1p} be obtained. However, in their modelling studies Emdee et al had no experimental data for the first 10-15% of toluene oxidation. Beyond this stage, the relative importance of primary initiation may well be markedly diminished.

(iii) Finally, it is necessary to consider the comment made by Emdee et al that an accurate value of k_{1t} can be obtained because the calculated rate is so sensitive to k_{1t}. As shown by equation (xv), changes in the value of θ reflect directly on the overall rate. However the initiation rate may be a mixture of primary and secondary initiation (indeed it is almost certain to be once the initial stages of reaction are over). Consequently, if the rate of primary initiation (R_p) is set so high as to dominate over legitimate secondary initiation, then variation of R_p must have a direct effect on the rate. Errors in the Emdee et al mechanism almost certainly give rise to this situation.

6.6. Database for the primary initiation reaction

Without doubt it is very easy to obtain inaccurate kinetic parameters for reaction (1), and almost invariably the errors produce a high value for k_1. The data presented here, however, form a consistent set with $E_1 = \Delta H_1$ within experimental error. There is strong evidence, including experimental, to suggest that the value of k_1 for toluene + O_2 obtained by Emdee et all [47] from modelling studies is approximately two orders of magnitude too high between 750 and 1200 K, arising from the neglect of important branching reactions in the mechanism used.

The data shown in Tables 7 and 8 are still relatively limited, but as (1) is a key reaction under a wide range of oxidation conditions, there is an urgent need for a database. Data for RH = alkane will be extremely difficult to determine experimentally, and the following parameters are suggested. They are based on the principle of bond additivity and that the activation energy for the reverse reaction is zero.

	A / dm^3 mol^{-1}s^{-1} (per C - H bond)	E / kJ mol^{-1}
I	7.0×10^9	ΔH
II	7.0×10^8	ΔH
III	7.0×10^7	ΔH

I   Localised electron in R (eg alkanes, oxygenates)
II   Delocalised electron involving the loss of one rotation (eg alkenes, alkylbenzenes)
III   Delocalised electron involving the loss of two rotations (eg $CH_2 = CHC\underline{H}_2CH = CH_2$)

The decreasing value of A between I and III is consistent with the extra loss of entropy of activation (ca 18 J K^{-1} mol^{-1} per lost rotation) in the emerging electron-delocalised radicals.

The database is considered reliable up to about 1200 K (factor of 2) unless there is a marked degree of non-Arrhenius behaviour in reaction (1). At higher temperatures, as shown in Table 9, homolysis of RH becomes dominant.

$$RH \quad \rightarrow \quad R' \quad + \quad R'' \tag{1A}$$

6.7. Primary initiation through homolysis

Although reaction (1A) will not be reviewed in detail, it is the principal alternative primary homogeneous initiation process, particularly at high temperatures. Considerable data have been collected for (1A), particularly by Tsang [50]. In general, k_{1A} can be determined much more easily than k_1, because of the absence of secondary initiation in the systems used. Baldwin, Drewery and Walker, [40] have shown that the activation energy for (1A) can be estimated with considerable accuracy in cases where no experimental data exist. Table 9 gives Arrhenius parameters for a selection of hydrocarbons and Table 10 gives comparisons via k_{1A} and $k_1[O_2]$ of the relative rate of the two initiation reactions with $[O_2] = 1$ atmosphere at 600, 1000 and 1500 K. With linear alkanes, (1) dominates at low temperatures and, for example, $k_1[O_2]/k_{1A} \approx 10^5$ for butane at 600 K. At 1000 K, reactions (1) and (1A) contribute almost equally. As the alkane structure becomes more branched, the strain energy in the weakest bond increases and homolysis becomes relatively more important, despite the presence of tertiary C-H bonds (which increases the value of k_1). As an extreme case, at 600 K, $k_{1A}/k_1[O_2]$ for tetramethylbutane is approximately 10, and reaches about 6000 at 1000 K compared with a value of about 1 for C_2H_6. Even for 2,3-dimethylbutane, $k_{1a}/k_1[O_2] \approx 30$ at 1000 K. Similar effects are observed when electron-delocalised radicals are formed. With hexa-1,5-diene, $k_{1A} \approx k_1[O_2]$ at 600 K and $k_{1a}/k_1[O_2] \approx 700$ at 1000 K. Where $[O_2] << 1$ atmosphere, then clearly initiation through (1A) may be even more important.

Table 9
Arrhenius parameters *, † for initiation processes (1) ** and (1A)‡

Reaction	$\log A_{1A}$	E_{1A}	$\log A_1$	E_1
$C_2H_6 \rightarrow 2CH_3$	16.7	372		
$C\underline{H}_3C\underline{H}_3 + O_2$			10.63	212
$C_4H_{10} \rightarrow 2C_2H_5$	16.5	343		
$CH_3C\underline{H}_2C\underline{H}_2CH_3 + O_2$			10.45	198
$(CH_3)_2CHCH(CH_3)_2 \rightarrow 2iPr$	16.4	319		
$(CH_3)_2C\underline{H}C\underline{H}(CH_3)_2 + O_2$			10.15	190
$(CH_3)_3CCH(CH_3)_2 \rightarrow \text{i-Pr} + \text{t-Bu}$	16.5	305		
$(CH_3)_3CC\underline{H}(CH_3)_2 + O_2$			9.85	190
$(CH_3)_3CC(CH_3)_3 \rightarrow 2\text{t-Bu}$	16.8	290		
$(C\underline{H}_3)_3CC(C\underline{H}_3)_3 + O_2$			11.15	212
$CH_2=CHCH_2CH_2CH=CH_2 \rightarrow 2\text{allyl}$	15.1	241		
$CH_2=CHC\underline{H}_2C\underline{H}_2CH=CH_2 + O_2$			9.45	145
$C_6H_5CH_2CH_3 \rightarrow C_6H_5CH_2 + CH_3$	15.6	306		
$C_6H_5C\underline{H}_2CH_3 + O_2$			9.15	152
$C_6H_5CH_3 \rightarrow C_6H_5CH_2 + H$	15.5•	373•		
$C_6H_5C\underline{H}_3 + O_2$			9.33	166

*   High pressure values; units, dm^3 mol^{-1} s^{-1} and kJ mol^{-1}
†   Weakest H atom abstracted, underlined
** Based on data base given in the text
‡   Based on data given in ref [40]
•   Note parameters for C-H homolysis

With ethylbenzene, the A factors and the activation energies for homolysis and for reaction with O_2 are lowered by almost the same factors by electron delocalisation, so that the values of $k_{1A}/k_1[O_2]$ are essentially the same as those for C_2H_6 at all temperatures. Similar behaviour will be found for aliphatic alkenes such as hex-1-ene where E_{1A} is lowered only by electron delocalisation in the allyl radical produced. However, $k_{1A}/k_1[O_2]$ will be much higher for 4,4-dimethylpent-1-ene where E_{1A} is lowered by both electron delocalisation and strain energy. Toluene is an interesting case as C-H homolysis to form the benzyl radical is faster than C-C homolysis to give the phenyl radical, due to the 50 kJ mol^{-1} stabilisation energy in the benzyl radical.

$$CH_2 = CHCH_2CH_2CH_2CH_3 \quad \rightarrow \quad CH_2CHCH_2 + CH_2CH_2CH_3$$
$$CH_3C(CH_3)_2CH_2CH = CH_2 \quad \rightarrow \quad (CH_3)_3C + CH_2CHCH_2 \tag{40}$$

Table 10
Relative rates* of reactions (1) and (1A) at $[O_2] = 1$ atmosphere

RH	600 K		1000 K		1500 K	
	k_{1A}	$k_1[O_2]$	k_{1A}	$k_1[O_2]$	k_{1A}	$k_1[O_2]$
C_2H_6	2.0×10^{-16}	1.1×10^{-10}	1.8×10^{-3}	2.4×10^{-3}	5.5×10^3	9.5
C_4H_{10}	3.2×10^{-14}	1.5×10^{-9}	2.9×10^{-2}	1.0×10^{-2}	2.8×10^4	21
$(CH_3)_2CHCH(CH_3)_2$	4.1×10^{-12}	8×10^{-9}	0.54	2.0×10^{-2}	1.9×10^5	27
$(CH_3)_3CCH(CH_3)_2$	9.6×10^{-11}	4.2×10^{-9}	3.9	1.1×10^{-2}	7.8×10^5	13.5
$(CH_3)_3CC(CH_3)_3$	3.1×10^{-9}	3.4×10^{-10}	41	7.0×10^{-3}	4.6×10^6	29
$CH_2=CHCH_2CH_2CH=CH_2$	1.3×10^{-6}	6.7×10^{-6}	3.3×10^2	0.46	5.1×10^6	100
$C_6H_5CH_2CH_3$	9×10^{-12}	1.6×10^{-6}	0.42	0.19	8.8×10^4	57
$C_6H_5CH_3**$	1.0×10^{-7}	1.6×10^{-7}	1.05×10^{-4}	5.7×10^{-2}	3.3×10^2	29

*   Units s^{-1}, reactions as specified in Table 9
** Note k_{1A} values for C-H homolysis in CH_3 group

7. THE REACTION OF ALKYL RADICALS WITH O_2

7.1. General introduction

It has long been established that conjugate alkenes are the major initial products in the reaction of alkyl radicals with O_2 (where structurally possible) above ca 550 K [30, 45, 63. 64]. As the carbon chain lengthens, increasing amounts of oxygenates, and in particular O-heterocyclics, are observed in the initial products. At about 750 K, the % of alkene decreases from about 99% for C_2H_5, 85% (1-C_3H_7), 65% (1-C_4H_9) down to less than 50% for C_6 and C_7 alkyls [29, 45]. Structural factors also play a part, and 2-C_3H_7 [40] and t-C_4H_9 [36,65] radicals both give 99% of conjugate alkene because alternative reactions are very limited. Baldwin and Walker [30] indicated in 1972 that the formation of products through oxidation of alkyl radicals is potentially extremely complex as each product can be formed in three ways, the direct bimolecular process (6A), through RO_2 decomposition (8) or via reaction (10) of QOOH radicals, which arise through H atom transfer in RO_2 radicals.

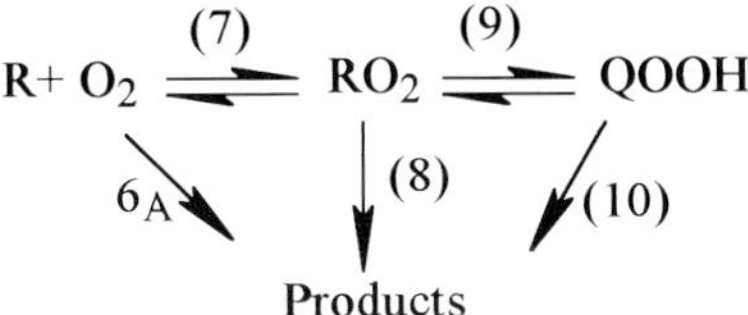

$$R + O_2 \underset{(7)}{\rightleftharpoons} RO_2 \underset{(9)}{\rightleftharpoons} QOOH$$

Products

Very complex expressions may arise for the overall rate constant for product formation if all routes contribute, particularly if (7) and (9) are both equilibrated and RO_2 can form a number of QOOH radicals. For example, 2-pentylperoxy radicals can undergo 5 different H atom transfers, where p, s and t refer to the primary, secondary and tertiary H atom transferred.

$$
\begin{array}{ccccc}
4 & 3 & 4 & 5 & 6 \\
\end{array}
$$

$CH_3CHCH_2CH_2CH_3$ $\quad\xrightarrow{1,4p}\quad$ $CH_2CH(OOH)CH_2CH_2CH_3$

$\quad\xrightarrow{1,3t}\quad$ $CH_3C(OOH)CH_2CH_2CH_3$

$\quad\xrightarrow{1,4s}\quad$ $CH_3CH(OOH)CHCH_2CH_3$

O—O $\quad\xrightarrow{1,5s}\quad$ $CH_3CH(OOH)CH_2CHCH_3$

$\quad\xrightarrow{1,6p}\quad$ $CH_3CH(OOH)CH_2CH_2CH_2$

Walker et al [38,67] have provided strong evidence that formation of conjugate alkene from R + O_2 in the range 600-1000 K does **not** occur through QOOH radicals, but are unable to distinguish between the direct reaction (6A) and reaction (8) because under the conditions used R and RO_2 are fully equilibrated in (7). Benson [66] has argued that the rate of the overall reaction (5A) varies little with alkyl structure and suggested a direct bimolecular abstraction with an activation energy of 15-25 kJ mol^{-1}. Baldwin, Walker and co-workers [59] have obtained a number of values of k_{5A} for C_2-C_5 alkyl radicals at about 750 K. They observed a good correlation between log k_{5A} at a fixed temperature and the enthalpy of reaction, as shown in Table 11 and Fig.7. The correlation extends quite well to the analogous reaction of pentenyl ($\Delta H = -18$) and even to the endothermic reaction ($\Delta H = + 40$ kJ mol^{-1}) of allyl radicals [59, 69].

$$
\begin{array}{lclclcl}
CH_2CHCH_2CH_2CH_3 & + & O_2 & \rightarrow & CH_2{=}CHCH{=}CHCH_3 & + & HO_2 \\
CH_3CHCHCH_2 & + & O_2 & \rightarrow & CH_2{=}CH\text{-}CH{=}CH_2 & + & HO_2 \\
CH_2CHCH_2 & + & O_2 & \rightarrow & CH_2{=}C{=}CH_2 & + & HO_2 \\
\end{array}
$$

The curve shown in Fig.7 is consistent with minimal variation of k at high negative values of ΔH and maximum variation at high positive values [59]. Fish [70] developed an extensive theory for the production of oxygenates, and concluded that O-heterocyclic compounds (often formed in 50% yield with C_7, C_8 alkanes) were formed via QOOH decomposition. Baldwin and Walker [71] have examined this suggestion and obtained quantitative data for the elementary reactions involved in the formation of oxiranes, oxetanes, tetrahydrofuranes and tetrahydropyranes in the sequence

$$R + O_2 \underset{}{\overset{(7)}{\rightleftharpoons}} RO_2 \xrightarrow{(9)} QOOH \xrightarrow{(10)} \text{O-heterocycle} + OH$$

Table 11
Variation of rate constant for $R + O_2 \rightarrow$ alkene $+ HO_2$ with ΔH at 753 K [59,69]

R*	Alkene	k/ (dm³ mol⁻¹ s⁻¹)	k/C-H bond (dm³ mol⁻¹ s⁻¹)	ΔH/ kJ mol⁻¹
$C\underline{H}_3CH_2$	C_2H_4	6.6×10^7	2.2×10^7	-46
$C\underline{H}_3CHC\underline{H}_3$	C_3H_6	1.25×10^8	2.1×10^7	-40
$C\underline{H}_3CHCH_2CH_3$	C_4H_8-1	5.1×10^7	1.7×10^7	-38
$CH_3C\underline{H}_2CH_2$	C_3H_6	7.1×10^7	3.6×10^7	-54
$CH_3CH_2C\underline{H}_2CH_2$	C_4H_8-1	1.6×10^8	8.0×10^7	-56
$CH_3CHC\underline{H}_2CH_3$	trans-C_4H_8-2	7.8×10^7	3.9×10^7	-49
	cis-C_4H_8-2	4.3×10^7	2.2×10^7	-45
$(CH_3)_2C\underline{H}CH_2$	i-C_4H_8	6.8×10^7	6.8×10^7	-62
$(CH_3)_2C\ C\underline{H}(CH_3)_2$	$(CH_3)_2C{=}C(CH_3)$	8.4×10^7	8.4×10^7	-64
$CH_2CHCHC\underline{H}_2CH_3$	$CH_2{=}CHCH{=}CHCH_3$	2.1×10^6	1.05×10^6	-18
$CH_2C\underline{H}CH_2$	$CH_2{=}C{=}CH_2$	2.5×10^2	2.5×10^2	+40

* H atom abstracted is underlined

With R and RO_2 equilibrated, and considerable evidence that $k_{10} \gg k_{-9}$ [67, 71], the rate of formation of O-heterocycle is given by equation (xvi).

$$\frac{d[\text{O-heterocycle}]}{dt} = k_{obs}[R][O_2] = K_7 k_9 [R][O_2] \qquad \text{(xvi)}$$

Following measurements of k_{obs} and use of K_7 (often calculated from Benson's additivity rules [72]), then k_9 can be determined.

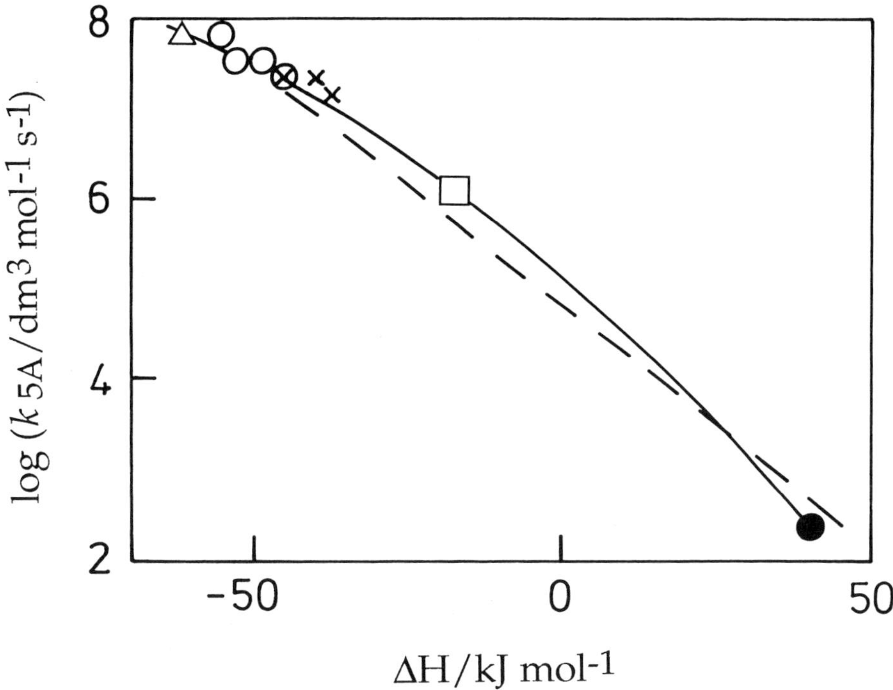

Figure 7   Plot of log k5A (per C-H bond) for the formation of alkenes at 753K.
△, tertiary; ○, secondary; ✕, primary □, pentenyl; ●, allyl C-H.

One example, which has been examined in detail by Baldwin and Walker [71] and more recently by Pilling [73], is the chemistry associated with the formation and removal of neopentylperoxy radicals leading to the formation of i-butene, acetone and 3,3-dimethyloxetane (DMO)

$$(CH_3)_3CCH_2 + O_2 \underset{(e)}{\overset{(a)}{\rightleftharpoons}} (CH_3)_3CCH_2O_2 \xrightarrow{(b)} (CH_3)_2 \overset{\cdot CH_2}{\underset{CH_2OOH}{C}}$$

$$\downarrow (e)$$

$$(CH_3)_2C=CH_2 + CH_3 \qquad\qquad (CH_3)_2CO + 2HCHO + OH$$

with O_2 step (c) and step (d) leading to

$$(CH_3)_2\overset{|}{\underset{CH_2\cdot O}{C}}-\overset{|}{CH_2}$$

$$(DMO)$$

Baldwin and Walker confirmed the following two relative initial rates of product formation from studies of the addition of neopentane to mixtures of $H_2 + O_2$ between 673 and 773K

$$\frac{d([DMO]+[CH_3COCH_3])}{d[i\text{-}C_4H_8]} = \frac{k_b\,K_a\,[O_2]}{k_e} \tag{xvii}$$

$$\text{and } \frac{d[CH_3COCH_3]}{d[DMO]} = \frac{k_c[O_2]}{k_c} \tag{xviii}$$

and were able to determine kinetic data for a number of the elementary reactions involved. In particular, K_a was calculated from Benson's additivity data [72] and $k_b = 1.20 \times 10^{13}$ exp(-14430/T)s^{-1} was obtained for the 1,5p H atom transfer involved. Pilling [73] has repeated the measurements by a **direct** measurement of [OH] and confirmed the accuracy of the k_bK_a/k_e values obtained by Baldwin and Walker. Further, his technique permits the determination of K_a directly and his results show that it is a factor of about 10 greater than the calculated figure, so that Baldwin and Walker's values of k_b are **reduced** by the same factor.

Pilling's direct determination of k_b is of paramount importance because of the set of self-consistent data obtained by Walker and co-workers [71] for a whole series of 1,4, 1,5, 1,6 and 1,7 H atom transfers in alkylperoxy radicals, all based on calculated values of the equilibrium constant for $R + O_2 \rightleftharpoons RO_2$. Pilling's work on the 1,5p H atom transfer in neopentylperoxy radicals effectively calibrates and validates the complete set of data which has been used extensively to model many practical combusion problems.

Of relevance to later discussion are the Arrhenius parameters (particularly the activation energies) obtained for 1,4 H atom transfers to give oxiranes, and these are summarised in Table 12. Further it is important to emphasise the existence of extensive evidence that the subsequent decomposition of QOOH radicals is sufficiently fast that $RO_2 \rightarrow QOOH$ isomerisation is **effectively irreversible**. [67, 71, 73]

7.2. The reaction of C_2H_5 radicals with O_2

Over the last 10 years, particular attention has been focussed on the $C_2H_5 + O_2$ reaction as its chemistry is regarded as simple but prototypical, although this view may be contested. In 1980, Baldwin, Pickering and Walker [64] generated C_2H_5 radicals by the addition of C_2H_6 to mixtures of tetramethylbutane and O_2, and in confirming earlier work showed that in the initial stages of reaction C_2H_4 and oxirane were the only products (at least 99.9%). Over the temperature range 673 - 813 K they obtained $A_{5A}/A_{5C} = 13.6 \pm 2.1$ and $E_{5C} - E_{5A} = 12.5 \pm 1$ kJ mol^{-1}

$$C_2H_5 \quad + \quad O_2 \quad \rightarrow \quad C_2H_4 \quad + \quad HO_2 \qquad (5A)$$

$$C_2H_5 \quad + \quad O_2 \quad \rightarrow \quad C_2H_4O \quad + \quad OH \qquad (5C)$$

These parameters correspond to variation in the **initial** product ratio $[C_2H_4]/[C_2H_4O]$ from 88 at 813 K to 127 at 673 K. Slagle, Feng and Gutman [74] in an elegant investigation produced C_2H_5 radicals photochemically in the presence of O_2 and monitored their decay in real time mass spectrometrically. Between 300 and 900 K, the overall rate constant for $C_2H_5 + O_2$ decreased monotonically and below 500 K the rate constant increased with pressure over the range 1-13 Torr. Above this temperature, C_2H_4 was the only product observed.

Later work by Gutman et al [75] using a similar technique confirmed that the overall rate constant was effectively independent of temperature between 750 and 1000 K. Slagle, Feng, and Gutman [74] argued that C_2H_4 formation by a direct bimolecular abstraction route was inconsistent with a zero or slightly negative energy barrier and proposed a coupled reaction path between 300 and 1000 K which also accounted for the observed pressure effects.

$$C_2H_5 \quad + \quad O_2 \; \rightleftharpoons \; C_2H_5O_2{}^* \quad \rightarrow \quad C_2H_4 \quad + \quad HO_2$$
$$\downarrow M$$
$$C_2H_5O_2$$

At low temperatures and high pressures, $C_2H_5O_2{}^*$ radicals are mainly stabilised by collision with M to form the product (in real time) $C_2H_5O_2$. As the pressure is lowered, an increasing proportion of $C_2H_5O_2{}^*$ radicals reform the reactants $C_2H_5 + O_2$ and the overall rate constant falls, as observed experimentally. The energy of $C_2H_5O_2{}^*$ increases with temperature and with it the concomitant decomposition to give C_2H_4. However Plumb and Ryan [76] in their flow-tube study of $C_2H_5 + O_2$ using mass-spectrometric determination of $[C_2H_5]$ at 295 K showed that the percentage of C_2H_4 formed decreased from 15% to a limiting value of 6% as the pressure (mostly He) increased from 0.5 to 10 Torr. They concluded that $C_2H_5 + O_2$ occurs in two uncoupled paths, one being pressure dependent and the other, a direct bimolecular reaction (5d), being independent of pressure.

$$C_2H_5 \quad + \quad O_2 \quad + \quad M \; \rightleftharpoons \; C_2H_5O_2 \quad + \quad M$$
$$C_2H_5 \quad + \quad O_2 \quad \qquad \rightarrow \quad C_2H_4 \quad + \quad HO_2 \qquad (5d)$$

42

However, Wallington et al [77-79] in a comprehensive series of papers showed that the C_2H_4 yield at 298 K decreased with increasing pressure from 12% of the C_2H_5 consumed by O_2 at 1 Torr (mostly N_2) to 0.02% at 600 Torr, corresponding to a $P^{-0.8}$ dependence. They suggested a **maximum** value of 8 x 10^5 dm^3 $mol^{-1}s^{-1}$ for the direct reaction (6A) at 295 K. In effect C_2H_4 is not formed at high pressures, which validates the view that it is produced in the coupled set of reactions through an excited ethylperoxy intermediate and that there is no evidence for the direct bimolecular reaction (6A). McAdam and Walker [80] produced C_2H_5 radicals by oxidising C_2H_5CHO between 593 and 753 K, and determined the initial product ratio $[C_2H_4]/[C_2H_6]$ over a range of mixture composition. With C_2H_4 and C_2H_6 formed uniquely in the overall reactions (5A) and (41), k_{5A}/k_{41} can be determined from equation (xix).

$$d[C_2H_4]/d[C_2H_6] \qquad = \qquad k_{5A}[O_2]/k_{41}[C_2H_5CHO] \qquad \text{(xix)}$$
$$C_2H_5 \quad + \quad O_2 \quad \rightarrow \quad C_2H_4 \quad + \quad HO_2 \qquad \text{(5A)}$$
$$C_2H_5 \quad + \quad C_2H_5CHO \quad \rightarrow \quad C_2H_6 \quad + \quad C_2H_5CO \qquad \text{(41)}$$

Use of literature values of k_{41}, and correction for H abstraction from the C_2H_5 group in the aldehyde, gave E_{5A} = -6.3±3.5 kJ mol^{-1} and A_{5A} = $10^{7.05±0.18}$ dm^3 mol^{-1} s^{-1}. The values of k_{5A} were also in good agreement with those obtained by Slagle, Feng & Gutman [74]. A later study by Gulati and Walker [81] gave a similar small negative activation energy for the analogous reaction of i-C_3H_7 radicals with O_2

$$i\text{-}C_3H_7 \quad + \quad O_2 \quad \rightarrow \quad C_3H_6 \quad + \quad HO_2$$

Gutman and co-workers (WSSG) [75] extended their earlier investigations on $C_2H_5 + O_2$ through a well designed experimental and theoretical study over the range 296-850 K at pressures between 0.5 and 15 Torr, by use of laser photolysis coupled to a photo-ionisation mass spectrometer for the measurement of C_2H_5 and C_2H_4 product profiles in real time. They were able to measure both the rate constant for the overall loss of C_2H_5 and the branching ratio F, $([C_2H_4])/([C_2H_4] + [C_2H_5O_2])$. The experimental data were concordant with those from the earlier study and with those obtained at room temperature by a number of workers [75-79]. In addition WSSG used RRKM theory to model their experiments and the data of independent workers, based on the mechanism proposed by Slagle, Feng & Gutman [74].

Two points should be emphasised.

(i) Above about 650 K, the step $C_2H_5 + O_2 \rightarrow C_2H_5O_2$ is fully equilibrated, and the overall rate constant for consumption of C_2H_5 is effectively independent of pressure and temperature. (ii) The branching ratio F is effectively unity above 650 K, but below this temperature falls as the pressure is increased because the $C_2H_5O_2^*$ radicals are increasingly stabilised by collision.

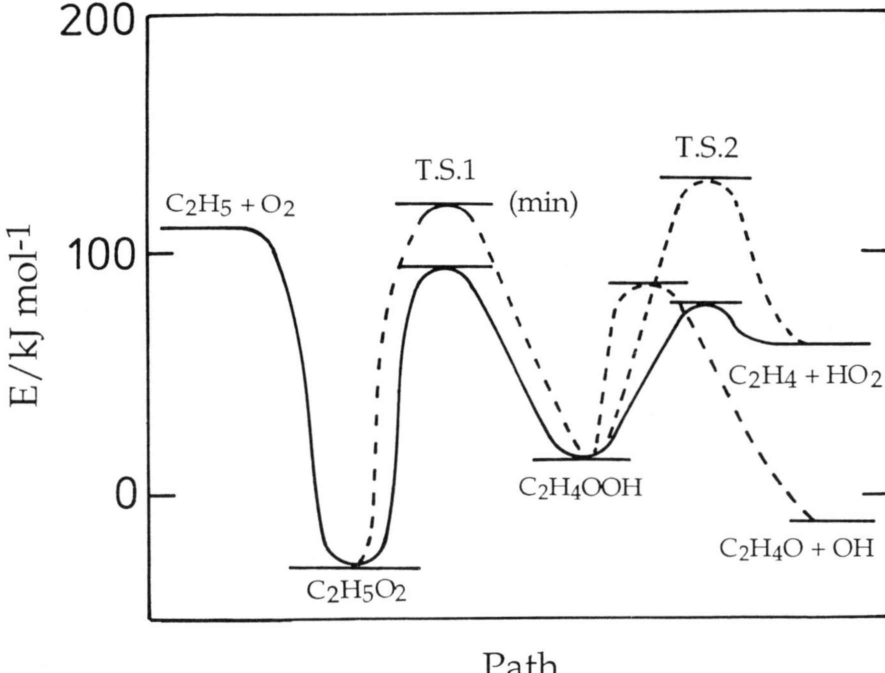

Figure 8 Energy diagram for $C_2H_5 + O_2 \rightarrow C_2H_4 + HO_2$
Full line as given by Wagner *et al* (ref. 75).
Dashed line as amended by the author (see text).

The potential energy surface used in the modelling is shown in Fig.8. Following addition of O_2 to C_2H_5, the $C_2H_5O_2$ radical passes through a 5-centre cyclic transition state (TS1) which leads to the ethylhydroperoxide radical C_2H_4OOH which then loses an HO_2 radical, and the formation of C_2H_4 occurs via transition state (TS2). Two key quantitative thermochemical features in the energy surface must be stressed. The enthalpy changes for the transition $C_2H_5 + O_2 \rightarrow$ TS1 and for $C_2H_5 + O_2 \rightarrow$ TS2 are -16.5 and -25 kJ mol⁻¹, respectively at 298 K. Consequently, in the path from C_2H_5 to the final products $C_2H_4 + HO_2$, no barrier has an energy higher than the energy of the original reactants. With the mechanism proposed, then at low temperatures and very low pressures, a fraction will return to the reactants, very few will be stabilised and significant numbers will decompose to give C_2H_4. As the pressure is increased, the stabilisation efficiency increases and eventually the branching ratio F will fall to zero. As the temperature is raised, $C_2H_5O_2^*$ will have higher energy and, for a particular pressure, the branching ratio will increase. Further the $C_2H_5O_2^*$

radicals will increasingly return to the reactants and eventually (above about 650 K) an established equilibrium $C_2H_5 + O_2 \rightleftharpoons C_2H_5O_2$ will be observed. Under equilibrium conditions, $C_2H_5O_2$ will have sufficient energy to decompose to C_2H_4, so that the branching factor will automatically assume a pressure-independent value of unity. This switch in mechanism from low to high yields of alkenes over a narrow temperature range is well known in the combustion chemistry of alkanes [1, 29, 30, 45].

WSSG pointed out that the overall rate constant is particularly sensitive to the enthalpy change for $C_2H_5O_2 \rightarrow$ TS1, whereas the temperature dependence of the branching factor F depends on the tightness of the cyclic transition state. It is clear that the rate constant is most sensitive to $\Delta H(C_2H_5 + O_2 \rightarrow$ TS2) and the only requirement is that the potential energy of TS2 does not exceed that of the reactants $C_2H_5 + O_2$. With both barriers below $\Delta H(C_2H_5O_2 \rightarrow C_2H_5 + O_2)$, then when F = 1, the overall rate constant (which equals that for C_2H_4 formation) should be virtually independent of temperature.

Bozzelli and Dean [82] added to the wealth of support for the coupled mechanism by using a theoretical treatment which incorporated literature rate constants, transition state theory for isomerisation, thermochemical data to calculate reverse rate constants and bimolecular QRRK theory [83,84] to calculate reaction probabilities for the initially formed chemically activated complexes such as $C_2H_5O_2{}^*$. A full range of possible reactions are considered for each radical intermediate involved. Their calculated rate constants are in good agreement with experimental data [75,80], and they conclude that there is no evidence for a direct bimolecular $C_2H_5O_2{}^*$ reaction below 1000 K.

Bozzelli and Dean comment that "the energised hydroperoxy radical $C_2H_4OOH^*$, **if formed**, dissociates to C_2H_4 and HO_2 almost completely at low pressures due to the high β-scission A factor (compared to that for reverse isomerisation), whilst at high pressures large fractions can be stabilised. This well is very shallow (they estimate ca 65 kJ mol^{-1}), however, and the stabilised radical again dissociates almost completely to $C_2H_4 + HO_2$". This statement requires clarification and appears at odds with experimental fact. In the 'real-time' experiments of WSSG, C_2H_4 formed from stabilised C_2H_4OOH radicals would not be observed as the process would be too slow. However, even in the experiments carried out by Kaiser, Lorkovic and Wallington [79], where photolysis lasted up to 2 minutes, C_2H_4 yields were reduced to zero at high pressures. At temperatures below 500 K, a barrier of 65 kJ mol^{-1} will allow competing reactions unless the pressure is very low and it is likely that removal of C_2H_4OOH by O_2 addition would be dominant even at relatively low O_2 pressures.

Two other aspects of Bozzelli and Dean's paper merit discussion here.
(i) The channels (a) and (b), with much higher energy barriers, become competitive at about 2000 K (each contributing about 10% to the total rate constant)

$$C_2H_5 \quad + \quad O_2 \quad \rightarrow \quad CH_3CHO \quad + \quad OH \qquad \text{(a)}$$
$$C_2H_5 \quad + \quad O_2 \quad \rightarrow \quad C_2H_5O \quad + \quad O \qquad \text{(b)}$$

At this temperature, however, the $C_2H_5 + O_2$ reaction will not be an important step in combustion mechanisms, and (a) and (b) may be ignored.

(ii) Given the coupled mechanism and the proposed potential energy surface, the variation of the predicted rate constant k_{C2H4} for C_2H_4 formation with temperature should show a marked dependence on pressure. As Bozzelli and Dean show, at high pressures (above 1 atmosphere) k_{C2H4} is relatively low at 298 K ($C_2H_5O_2^*$ stabilisation) and rises with temperature until the branching factor F is unity, after which it falls steadily by a factor of 2-3 between 700 and 1800 K. At a lower pressure (ca 1 Torr), high yields of C_2H_4 are observed at 298K (little $C_2H_5O_2^*$ stabilisation), and the predicted value of k_{C2H4} remains virtually constant to 700 K and then falls by a factor of 2-3 as the temperature is increased to 1800 K.

The fall in k_{C2H4} between 700 and 1800 K, which is independent of pressure, can be associated with the thermochemistry of the coupled mechanism shown in Fig. 8.

$$C_2H_5 \;+\; O_2 \;\underset{}{\overset{(a)}{\rightleftharpoons}}\; C_2H_5O_2 \;\overset{(b)}{\longrightarrow}\; C_2H_4OOH$$

As C_2H_5 and $C_2H_5O_2$ are considered to be in equilibrium, F = 1, and as the step forming C_2H_4OOH is effectively irreversible, then $k_{C2H4} = K_a k_b$, and the activation energy for C_2H_4 formation will be $\Delta H_a + E_b = -16.5$ kJ mol^{-1}, a dominant factor in determining the temperature coefficient.

7.3. Comments on $C_2H_5 + O_2$

There is a general consensus view that the coupled mechanism gives a quantitative account of the reaction between $C_2H_5 + O_2$ over a wide pressure and temperature range. Certainly below 1000 K, there are no grounds for including the direct bimolecular route (except Plumb and Ryan's limited pressure data at 298 K). However, as pointed out repeatedly by Walker and co-workers [67, 68, 80, 81, 85] there are problems, not with a coupled mechanism, but with specific aspects of the one proposed, namely, first, the requirement that C_2H_4 formation occurs via C_2H_4OOH radicals and secondly, the thermochemistry attributed to the potential energy surface used by WSSG and by Bozzelli and Dean. These concerns extend to other alkyl $+ O_2$ systems, and it is unfortunate that they have not received the detailed modelling attention given to $C_2H_5 + O_2$. It is appropriate to review some of the features, mostly associated with other alkyl $+ O_2$ systems, which give cause for concern about the validity of the mechanism proposed by WSSG for the $C_2H_5 + O_2$ reaction.

7.4. Possible difficulties associated with the thermochemistry of $C_2H_5 + O_2$

The potential energy 'surface' favoured by Walker and co-workers is shown by the dotted line in Fig. 8. In particular, the two barriers to C_2H_4 formation are considerably higher, and critically, the transition states TS1 and TS2 are markedly higher than the initial energy of $C_2H_5 + O_2$. Clearly, such a surface would have serious implications for the validity of the WSSG mechanism. Additionally a path, not considered by them or Bozzelli and Dean, leads to the formation of oxirane (dashed line).

7.4.1. Barrier for $RO_2 \rightarrow QOOH$

It is generally accepted that O-heterocyclic compounds are formed in the region 600-1000 K from alkyl radicals by the peroxy radical isomerisation and decomposition (PRID) mechanism [67, 70, 71]

$$R + O_2 \xrightarrow{\text{(a)}} RO_2 \xrightarrow{\text{(b)}} QOOH \xrightarrow{\text{(c)}} \text{O-heterocycle} + OH$$

With (a) equilibrated, as observed by Gutman [75], and (b) and (c) effectively irreversible, then the overall rate constant k_{obs} for heterocycle formation is given by $k_{obs} = K_a k_b$. From studies (mainly) of the addition of traces of alkane to slowly reacting mixtures of $H_2 + O_2$, Walker and co-workers [71] have measured the value of k_b for a full range of 1, 4 to 1, 7 H atom transfers in reaction (b). When originally determined via measurement of the yields of oxiranes, oxetanes, tetrahydrofuranes and tetrahydropyranes, respectively, the values of K_a were calculated from Benson's additivity rules [72]. Arrhenius parameters for k_b are shown in Table 12. The values of the activation energies are determined by three factors:
(i) the activation energy for $RO_2 + RH$ (abstraction),
(ii) the type of H atom transferred,
(iii)the strain in the ring transition state.

As discussed by Baldwin et al [71], the variation of the activation energies is totally consistent with these factors. Further, the fall in A factor (by a factor of about 10) with increase in the size of the ring transition state by each added atom is totally consistent with the increased loss of entropy of activation due to an extra loss of a rotation.
The consistency of the data is confirmed by Pilling's [73] work on the H atom transfer in neopentylperoxy radicals. As pointed out in Section (7.1) the value of $K_a k_b$ was in almost exact agreement with that obtained by Walker and co-workers, and further the value of K_a was determined experimentally. Use of this in place of the value obtained from Benson's additivity rules gives a reliable value of E equal to $123 + 8$ kJ mol^{-1} for the 1,5p step $RO_2 \rightarrow$ QOOH for neopentylperoxy. This is significantly greater than the barrier proposed by WSSG for the 1,4p $C_2H_5OO \rightarrow C_2H_4OOH$ step. When allowance is made for the increased strain in the 1,4p ring transition state, then a value of at least 150 kJ mol^{-1} is required for

$$CH_3CH_2O_2 \qquad \rightarrow \qquad CH_2CH_2OOH$$

compared with the value of 128 kJ mol^{-1} used by WSSG. Clearly the formation of C_2H_4 should involve a strong positive activation energy (about 25 kJ mol^{-1}) and moreover the rate constant would be very sensitive to the height of the barrier (Table 12). For the three cases involving 1,4p, 1,4s and 1,4t H atom transfers ($\Delta H(R + O_2 \rightleftharpoons RO_2)$ effectively constant) which involve the same sized ring transition state so that the A factors should only differ through path degeneracy, the relative rates of alkene formation should be 1:7:40 at 750 K, compared to the very small changes shown in Table 11.

Table 12
Arrhenius parameters* for $RO_2 \rightarrow QOOH$ for alkyl radicals

Type of H transfer**	A/s^{-1} (per C-H bond)	$E/kJ\ mol^{-1}$
1,4p	1.4×10^{12}	153
1,5p	1.75×10^{11}	123
1,6p	2.2×10^{10}	105
1,7p	2.75×10^{9}	90
1,3s	1.15×10^{13}	176
1,4s	1.4×10^{12}	133
1,5s	1.75×10^{11}	110
1,6s	2.2×10^{10}	90
1,7s	2.75×10^{9}	75
1,3t	1.15×10^{13}	160
1,4t	1.4×10^{12}	118
1,5t	1.75×10^{11}	93
1,6t	2.2×10^{10}	75
1,7t	2.75×10^{9}	62

*   Based on comprehensive data obtained by Walker et al [71] and Pilling's [73] absolute data for the 1,5p transfer
** See text

7.4.2. Barrier for QOOH homolysis to form an alkene

No experimental data exist for the decomposition $QOOH \rightarrow$ alkene $+ HO_2$, but Walker and co-workers [85] argue that they have determined extensive information from the reverse addition of HO_2 radicals to alkenes. Use of the decomposition of tetramethylbutane in the presence of O_2 as an HO_2 radical source over the temperature range 650-790 K, together with the measurement of the appropriate oxirane permits extremely reliable values to be obtained for the overall reaction. Additionally, competitive studies have been used under conditions where the oxiranes are formed uniquely through HO_2 addition to alkenes [36, 68].

$$HO_2 \quad + \quad alkene \quad \rightarrow \quad oxirane \quad + \quad OH$$

The overall reaction occurs in two stages, first HO_2 addition and then homolysis of the peroxide bond and cyclisation, as shown for butene-2.

$$HO_2 \quad + \quad CH_3CH = CHCH_2 \quad \rightarrow \quad CH_3CH\!-\!CHCH_3 \quad \rightarrow \quad CH_3CH\!-\!CHCH_3 \quad + \quad OH$$
$$\qquad\qquad\qquad\qquad\qquad\qquad\qquad\qquad | \qquad\qquad\qquad\qquad\quad \backslash \ / \qquad\qquad$$
$$\qquad\qquad\qquad\qquad\qquad\qquad\qquad\quad OOH \qquad\qquad\qquad\qquad\quad O \qquad\qquad$$

Table 13 shows the Arrhenius parameters for a wide range of alkenes. The activation energies correlate extremely well with the ionisation energy of the alkenes, which provides strong evidence that the measured parameters refer to the HO_2 addition step. Further evidence is discussed later. A value of 72 kJ mol^{-1} for $HO_2 + C_2H_4 \rightarrow C_2H_4OOH$ results in the dotted line in Fig 8, and if correct rules out formation of C_2H_4 via C_2H_4OOH, as proposed by WSSG. Comparable data (Table 13) for the addition of CH_3O_2 and i-$C_3H_7O_2$ radicals to alkenes lends further support to the high energy barrier for HO_2 addition.

7.5. Experiments with butane and butene-2

Scheme I summarises the possible reactions involved in the formation of trans-butene-2 and 2,3-dimethyloxirane when n-C_4H_{10} and cis-butene-2 are added separately in trace amounts to slowly reacting mixtures of $H_2 + O_2$ at 753 K [67]. As shown in Section 4.1, this system is an excellent source of H, OH and HO_2 radicals.

Scheme I

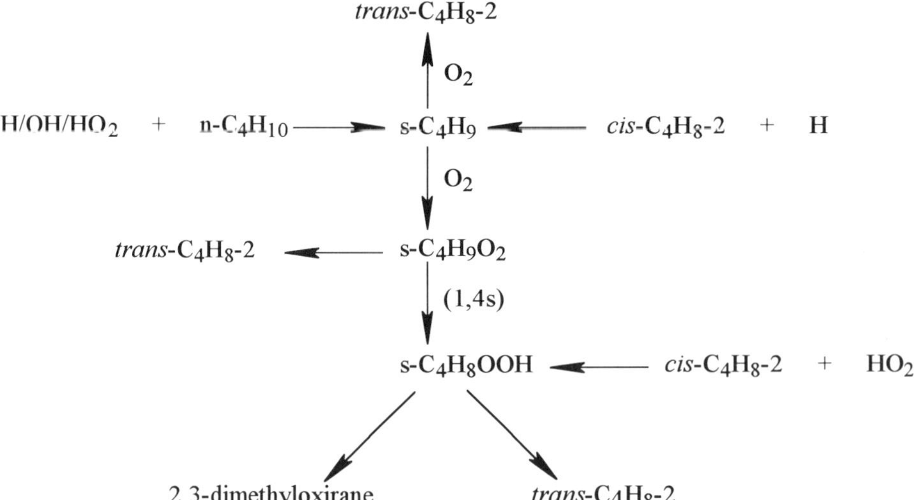

With cis-butene-2 as additive to a mixture containing 70, 140 and 290 Torr of O_2, H_2 and N_2, respectively, the initial product ratio [2,3-dimethyloxirane]/[trans-butene-2] = 1.85 compared with a value of 0.15 when butane is the additive. If trans-butene-2 is only formed from the s-C_4H_8OOH radical, then the product ratio should be the **same** for both additives. The much higher value for cis-butene-2 can only be explained if trans-butene-2 is formed predominantly in the direct bimolecular process (rejected above) or in a reaction of s-$C_4H_9O_2$ radicals which does **not** involve the QOOH species s-C_4H_8OOH. The large difference in the two values also confirms that the 1,4s step to s-C_4H_8OOH must be irreversible.

Even more emphatic evidence that the decomposition of QOOH radicals does not yield mainly alkene was obtained by Stothard and Walker [38] from further studies of the addition of the separate addition of cis- and trans-butene-2 to slowly reacting mixtures of $H_2 + O_2$ at 713 K and 753 K and to tetramethylbutane + O2 mixtures. The relevant chemistry is shown below for trans-C_4H_8-2.

$$HO_2 \quad + \quad t\text{-}C_4H_8\text{-}2 \quad \rightleftharpoons \quad CH_3CHCHCH_3 \quad \rightarrow \quad cis\text{-}C_4H_8\text{-}2 \quad + \quad HO_2$$

$$\underset{\displaystyle \downarrow}{OOH}$$

$$CH_3CH\text{-}CHCH_3 \quad + \quad OH$$
$$\underset{\displaystyle O}{\diagdown \ \diagup}$$

If the $CH_3CH(OOH)CHCH_3$ radicals predominantly give alkene, then a high proportion should reform t-C_4H_8-2, but as the trans-form is only marginally favoured [67], then the initial product ratio [2,3 dimethyloxirane]/[cis-C_4H_8-2] should be very low. Conversely Walker and Stothard report average values of ca 4 at 753 K and 5 at 713 K over a wide range of mixture composition. The results of more recent experiments [59] at 713 K with cis-C_4H_8 as additive are summarised in Table 14. Again a consistent value favouring the oxirane is obtained and is in good agreement with that from the earlier study reported above. In reality, the ratios for both cis- and trans- additives are likely to be considerably higher than indicated because OH addition is important, and is well known to be reversible under the conditions used. If accepted, then it must be concluded that $CH_3CH(OOH)CHCH_3$ radicals decompose almost completely to the oxirane.

Further, Walker & Stothard argue that as the thermochemistry of the participating reactions is essentially independent of the alkene system, this conclusion can be extended to C_2H_4OOH radicals, to the $C_2H_5 + O_2$ system and, to the $R + O_2$ system in general. Not only does this confound the WSSG coupled mechanism, but the results also confirm that $HO_2 + C_nH_{2n} \rightarrow C_nH_{2n}OOH$ is the rate-determining step in the formation of oxiranes and that therefore (Table 13) the barrier for $HO_2 + C_2H_4$ is considerably higher than 25 kJ mol^{-1} as proposed by WSSG.

50

Table 13
Arrhenius parameters for addition of HO_2, CH_3O_2 and i-$C_3H_7O_2$ to alkenes

Alkene	E_i/kJ mol^{-1}	$\log(k/dm^3mol^{-1}s^{-1})$	E/kJmol^{-1}	T/K
$\underline{HO_2[38]}$				
C_2H_4	1013	9.35	71.9	670-770
C_3H_6	939	9.01	62.6	670-770
pent-1-ene	917	8.88	59.7	670-790
hex-1-ene	912	8.91	58.6	670-770
i-butene	891	8.80	53.1	670-790
cis-hex-2-ene	884	8.41	53.4	670-770
trans-hex-2-ene	884	8.41	53.4	670-770
trans-but-2-ene	881	8.61	50.0	670-790
2,3-dimethylbut-2-ene	801	8.58	35.4	650-770
$\underline{CH_3O_2}$ *				
2-methylbut-1-ene	880	8.6	53	373-403
2-methylbut-2-ene	837	8.2	42	373-403
2,3-dimethylbut-2-ene	801	8.1	36	373-403
$\underline{i\text{-}C_3H_7O_2}$ **				
C_3H_6	939	8.9	68	373-408
i-C_4H_8	891	8.6	63	373-408
2-methylbut-1-ene	880	8.3	55	363-408
2-methylbut-2-ene	837	8.0	48	303-363
2,3-dimethylbut-2-ene	801	8.0	41	303-363

* Data from D. A. Osborne and D. J. Waddington, J.Chem.Soc., Perkin Trans 2, (1980) 925.

** Data from M. I. Sway and D. J. Waddington, J.Chem.Soc., Perkin Trans 2, (1988) 139.

Table 14
Initial values of [2,3-dimethyloxirane]/[cis/trans-C_4H_8-2] from C_4H_8-2 oxidation studies

Mixture/Torr

trans-C_4H_8-2	O_2	H_2	[2,3DMO]/[cis-C_4H_8-2]
		T = 753 K	
2.5	70	140	3.8
2.5	210	140	3.9
2.5	358	140	4.2
10	70	140	4.5
2.5	70	428	4.6
10	30	0	3.9
4	30	0	4.3
		T = 713 K	
2.5	20	140	5.5
2.5	310	140	5.8
6.3	20	140	4.6
10	30	0	5.0
4	30	0	5.4

cis-C_4H_8-2	O_2	H_2	[2,3DMO]/[trans-C_4H_8-2]
		T = 713 K	
2.5	70	140	3.2
10	140	140	3.4
10	30	0	3.7
4	30	0	3.7
4	3	0	4.1

7.6. Experiments with isobutylperoxide radicals

A study of the kinetics and mechanism of QOOH radical chemistry in the **absence of O_2** offers considerable advantages because a number of complicating features are removed. Batt used radical (X) attack on t-butylhydroperoxide in the absence of O_2 to produce isobutylhydroperoxide radicals in the temperature range 370-470 K, and then measured the yields of 2,2-dimethyloxirane and i-butene, considered to be formed in competing reactions of the hydroperoxide radical. Although the results are inconclusive, the work illustrates the type of approach necessary to elucidate QOOH chemistry. Batt proposed the following mechanism.

$$X + \underset{\underset{CH_3}{|}}{\overset{\overset{CH_3}{|}}{CH_3\,C-OOH}} \longrightarrow \underset{\underset{CH_3}{|}}{\overset{\overset{CH_3}{|}}{CH_2\,C-OOH}} + XH$$

$$\Big\downarrow k_a \qquad\qquad \Big\downarrow k_b$$

$$(CH_3)_2C = CH_2 + HO_2 \qquad\qquad (CH_3)_2C \overset{}{\underset{\diagdown O \diagup}{—}} CH_2 + OH$$

He reported a value of $\log(k_a/k_b) = 10^{2.6\pm0.4}\exp(-2620/T)$ which corresponds to [i-butene]/[2,2-dimethyloxirane] about 0.06 at 300 K and about 10 at 750 K. The oxirane formation thus dominates at low temperature, but the evidence points to i-butene as the major product from isobutylhydroperoxide at 750 K, which appears to contradict the conclusion reached in the last section. However, Batt's results suggest that $E_a-E_b = 22$ kJ mol^{-1}, i.e. homolysis to i-butene has the higher energy barrier, and this is not consistent with the WSSG view of QOOH chemistry for C_2H_4OOH. The mechanism for oxirane formation from HO_2 addition to C_2H_4 consists of a two-stage sequence.

$$HO_2 + C_2H_4 \underset{k_a}{\overset{k_{-a}}{\rightleftharpoons}} C_2H_4OOH \overset{k_b}{\longrightarrow} C_2H_4O + OH$$

If, as suggested by Walker and co-workers, $k_b \gg k_a$, then step (-a) is rate determining and $E_{-a}-E_{obs} = 72$ kJ mol^{-1} where E_{obs} is the overall activation energy for oxirane formation (see Table 13). With the WSSG view that C_2H_4OOH decomposes to give mainly $C_2H_4 + HO_2$, then $k_a \gg k_b$, so that the addition step is equilibrated and $E_{obs} = E_{-a} + E_b -E_a$. Assuming $E_a-E_b = 22$ kJ mol^{-1} (the value above for isobutylhydroperoxide radicals), then with $E_{obs} = 72$, then $E_{-a} = 94$ kJ mol^{-1}, compared with a maximum value of ca 25 kJ mol^{-1} required by WSSG (see Fig. 8).

Unfortunately, the system may be more complex than suggested by Batt. Abstraction of the peroxidic H in t-butylperoxide will produce t-butylperoxy radicals which may contribute to isobutene formation.

7.7. Formation of larger O-heterocyclic rings

It is extremely unfortunate that direct experiments on $C_2H_5 + O_2$ have not been repeated on larger alkyl $+ O_2$ systems. Writing a similar coupled mechanism for pentene-1 from 1-pentyl radicals, it is clear that stabilisation of the excited $1\text{-}C_5H_{11}O_2^*$ radicals would occur at very much lower pressures than observed for $C_2H_5O_2^*$.

$$1\text{-}C_5H_{11} + O_2 \rightleftharpoons 1\text{-}C_5H_{11}O_2^* \longrightarrow CH_2 = CHCH_2CH_2CH_3 + HO_2$$

$$\downarrow M$$

$$1\text{-}C_5H_{11}O_2$$

At low temperatures, pentene-1 would be seen only at very low pressures in time-resolved experiments. At high temperatures, the pentyl, pentylperoxy equilibrium would be observed and formation of pentene-1 through 1,4H transfer would be in direct competition with 1,5, 1,6 and 1,7 transfers leading to the analogous larger O-ring compounds.

$$CH_2CH_2CH_2CH_2CH_3 \rightarrow CH_2(OOH)CHCH_2CH_2CH_3 \quad (1,4s)$$
$$|$$
$$O\text{-}O \rightarrow CH_2(OOH)CH_2CHCH_2CH_3 \quad (1,5s)$$

$$\rightarrow CH_2(OOH)CH_2CH_2CHCH_3 \quad (1,6s)$$

$$\rightarrow CH_2(OOH)CH_2CH_2CH_2CH_2 \quad (1,7p)$$

Table 15 shows the relative rate of the 4 isomerisations obtained from data in Table 12 and that the percentage of isomerisation corresponding to 1,4 H atom transfer is only about 4%, whereas experimentally [87], the percentage of pentene-1 formed is at least 10 times higher. Similarly in the oxidation of butanal at 713 K, which is an excellent source of $CH_3CH_2CH_2$ radicals, oxetane and propene are formed uniquely from a reaction between 1-C_3H_7 and O_2. If they are formed in 1,5p and 1,4s transfers, respectively, then the product ratio $[C_3H_6]/[\text{oxetane}]$ should be 1.0 compared with the observed value of 35 [67]. Similar observations may be made about a number of alkyl $+ O_2$ systems, and it appears that if the conjugate alkene-forming process has to pass through the very tight transition state involved in a 1,4 H atom transfer, then it is not possible to account for the very high yields of alkene observed. A possible explanation could be that all $RO_2 \rightarrow QOOH$ isomerisations except that involving a 1,4 transfer are effectively reversible, so that although the other transfers are faster, the only effective path onwards is via the QOOH leading to an alkene. However, this explanation would completely destroy the pattern of rate constants observed for $RO_2 \rightarrow$ QOOH (see Table 12 and ref [71]), and experimental evidence has shown that the isomerisation is effectively reversible [73].

Table 15
Relative rate* of $RO_2 \rightarrow$ QOOH for $CH_3CH_2CH_2CH_2CH_2O_2$ radicals at 750 K

QOOH	transfer	$k/dm^3\ mol^{-1}\ s^{-1}$	% contribution
$CH_2(OOH)CHCH_2CH_2CH_3$	1,4s	7.7×10^2	4
$CH_2(OOH)CH_2CHCH_2CH_3$	1,5s	3.8×10^3	20.5
$CH_2(OOH)CH_2CH_2CHCH_3$	1,6s	1.2×10^4	63.5
$CH_2(OOH)CH_2CH_2CH_2CH_2$	1,7p	2.2×10^3	12

* Data from Table 12

7.8. Concluding remarks

The present position on the detailed mechanism for $R + O_2$ is extremely unsatisfactory.
(i) Agreement exists that for $C_2H_5 + O_2$ the direct bimolecular reaction is probably unimportant below 800 K, and accepting a maximum value of $10^{7.0}$ at 750 K, values of $A = 10^{8.5}\ dm^3\ mol^{-1}\ s^{-1}$ and $E = 22\ kJ\ mol^{-1}$ are suggested for this route.
(ii) A coupled mechanism written as

$$C_2H_5 + O_2 \longrightarrow C_2H_5O_2{}^* \longrightarrow C_2H_4 + HO_2$$
$$\downarrow M$$
$$C_2H_5 + O_2$$

is acceptable, providing the reaction path does not pass through the C_2H_4OOH radical. Presumably, the $C_2H_5O_2{}^*$ would have to undergo a concerted reaction to form C_2H_4 and HO_2.
(iii) Walker and McAdam [80] have suggested that O_2 has particular characteristics as an abstracting species. It is suggested that as O_2 attacks an H atom on the CH_3 group in the C_2H_5 radical, the high density of unpaired electrons on the remote O atom may lead to bonding between this atom and the carbon atom with the localised free electron. As the resulting ring would be relatively unstrained, the intermediate may have a degree of stability so that the competition between reverse decomposition and forward formation of C_2H_4 could give rise to a negative temperature coefficient. The exact relationship between the

intermediate here and that possibly undergoing a concerted reaction in (ii) is, however, obscure.

(iv) Bozzelli and Dean [83] attempted to remove the problem over the height of the barrier between C_2H_4OOH and $C_2H_4 + HO_2$ by suggesting that oxirane is not formed in the reaction between $HO_2 + C_2H_4$. Their suggestion that, due to a high concentration of OH, the oxirane is formed in the sequence

$$2OH \rightarrow H_2O + O$$

$$O + C_2H_4 \rightarrow C_2H_4O$$

is totally unacceptable because of the extremely high reactivity of OH radicals with alkanes and the consequent **very** low concentration of OH.

The comprehensive series of studies of HO_2 addition to alkenes [68] leave absolutely no doubt that the major product is oxirane. It is interesting to note that Waddington and co-workers [88, 89] have studied the addition of CH_3O_2 and $2\text{-}C_3H_7O_2$ radicals to alkenes. Oxirans are the only major product and, the activation energies for the addition are very similar to those for HO_2, as shown in Table 13.

(v) It should be noted that analogous reactions to $C_2H_5 + O_2$ are widespread outside alkyl systems. Table 16 gives a few simple examples where the kinetic data are available. Any general mechanism should explain the data shown. Reactions (42)-(44) all show positive temperature coefficients and all are sufficiently fast to be important in atmospheric chemistry. The rate expression given for reaction (42) in Table 16 reflects a complex variation of k_{42}, which decreases in value by about 15% between 298 and 474 K and then increases by a factor of 2 up to 684 K. Grotheer *et al* [90] interpreted this behaviour as a change in mechanism, with mainly an association reaction $CH_2OH + O_2 \rightarrow O_2CH_2OH$ at low temperatures being replaced by an abstraction process at high temperatures. Reactions (43) and (44) differ in at least two ways.

(i) The available data suggest a monotonic increase in k_{43} and k_{44} from room temperature.

(ii) The values of k_{43} and k_{44} are a factor of at least 10^3 lower than k_{42} at 300 K.

CH_2OH	+	O_2	$\rightarrow$	HCHO	+ HO_2	(42)
CH_3O	+	O_2	$\rightarrow$	HCHO	+ HO_2	(43)
CH_3CH_2O	+	O_2	$\rightarrow$	CH_3CHO	+ HO_2	(44)

A likely explanation is that the association $RO + O_2 \rightarrow RO_3$ is not favoured even at 300 K, so that the only mechanistic route for $RO + O_2$ is through H abstraction. If this view is accepted then the Arrhenius parameters for reactions (43) and (44) may be taken as typical for a direct H-abstraction route. Given that $\Delta H = -46$ kJ mol^{-1} for $C_2H_5 + O_2 \rightarrow C_2H_4 + HO_2$, then Arrhenius parameters of $A = 1.0 \times 10^8$ dm^3 mol^{-1} s^{-1} and $E = 15$ kJ mol^{-1} would appear appropriate for a direct bimolecular abstraction route. At 298 K, the value of the rate constant would be significantly below that detectable by SWWG, and Bozzelli and Dean. At

750 K, however it would contribute about 20-30% to the overall rate constant for C_2H_4 formation.

Table 16
Kinetic Data for Reactions $R + O_2 \rightarrow P + HO_2$*

R	P	A	E	ΔH	k_{300}
CH_2OH	HCHO	$(k=1.5 \times 10^{12}T^{-1}$	$)$		
		$+ 2.4 \times 10^{11}\exp(-2525/T)$		-88	5×10^9
CH_3O	HCHO	4.0×10^7	8.9	-122	1.1×10^6
CH_3CH_2O	CH_3CHO	6.0×10^7	6.9	-152	3.8×10^6

* Units, dm^3 mol^{-1} s^{-1} and kJ mol^{-1}; data obtained from reference [3].

(vi) Parochially the WSSG interpretation of $C_2H_5 + O_2$ is totally consistent with the experimental data. Unfortunately observations on a number of related reactions, HO_2 + alkene and $RO_2 \rightarrow QOOH$, are inconsistent with the mechanism used for $C_2H_5 + O_2$. Clearly theoretical treatments and time-resolved experiments on other alkyl + O_2 reactions are absolutely necessary for further clarification. Direct experiments on HO_2 + alkene are also vital.

However, the whole basis of the interpretation may have been oversimplified. Many of the discussions of $C_2H_5 + O_2$ (and related systems) assume that only one potential energy surface is involved. However, two recent theoretical treatments for $C_2H_5 + O_2$ and $CH_3 + O_2$ indicate the need to consider two potential energy surfaces. Fig. 9 shows a sketch of the two surfaces $^2A'$ and $^2A''$, suggested by Walch [91] for the production of two sets of products $CH_3O + O$ and $HCHO + OH$. He suggests that the $^2A''$ surface correlates with the addition of ground state O_2, and that the O distant from the carbon atom has an in-plane **doubly-**occupied 2p-like orbital. In consequence, H transfer to form CH_2OOH is not favoured and the only product channel available gives $CH_3O + O$. The $^2A'$ surface corresponds to the excited state of O_2 and here the distant O atom has a **singly** occupied in plane 2-p orbital, so that H transfer is favourable and $HCHO + OH$ may be formed. As Fig 9 shows, formation of $HCHO + OH$ involves two potential energy surfaces with crossing from one to the other.

Quelch, Gallo and Schaefer [92] have investigated the potential energy surface for $C_2H_5 + O_2$ and examined in detail the conformations of the ethylperoxy radical. A point very germane to the present discussion is that they point out the existence of two states for the ethylperoxy radical, namely a $^2A''$ ground state and the $^2A'$ excited state, as discussed by Walch [90] for $CH_3 + O_2$. The possibility now exists that the $C_2H_5 + O_2$ side of one potential energy surface is not now linked directly to the addition of HO_2 to C_2H_4, in which case a number of the apparently contradictory pieces of evidence would be removed. Until a deeper theoretical insight into both $C_2H_5 + O_2$ and many other similar systems is available, little progress can be made.

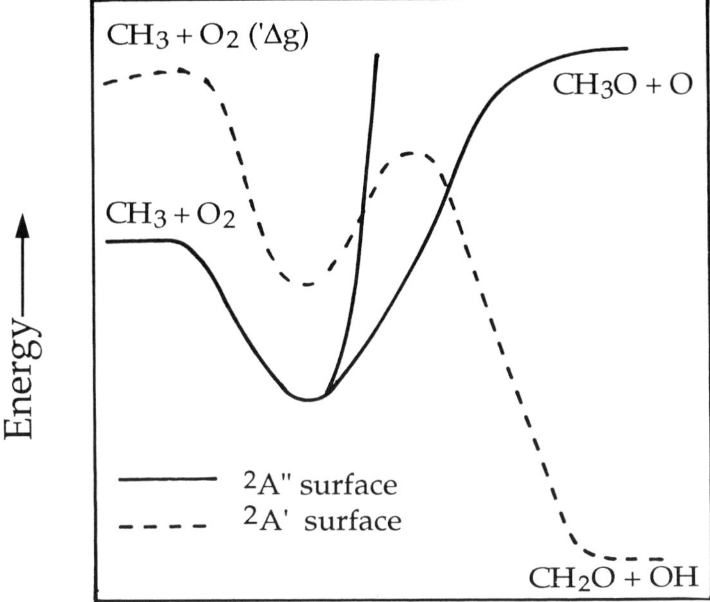

Figure 9   Sketch of energy surfaces for CH3 + O2.

8. CYCLISATION REACTIONS OF NON-AROMATIC SPECIES

8.1. Cyclisation at high temperatures

The formation of benzene, polyaromatic hydrocarbons (PAH), aromatics in general and soot in flames from non-aromatic sources has long been a subject of interest in combustion, but no widespread acceptance of mechanism has emerged. Westmorland et al [93] give an excellent summary of the main routes suggested for the formation of benzene and simple aromatics which presumably are precursors to soot. They range from 'polymerisation' of acetylene, Diels-Alder additions, recombination of allyl radicals, radical addition to highly unsaturated hydrocarbons to, more recently, the recombination of propargyl ($CH_2 = C = CH$) radicals. With advanced developments in both experimental and computer techniques, mechanisms have been subjected to stringent tests of the applicability in particular systems [93]. Two particular mechanisms merit further discussion because of the principles that underly the proposals.

58

8.1.1. Formation of benzene from the recombination of propargyl radicals.

Stein et al [94] put forward a mechanism based on the combination of propargyl radicals ($CH_2 = C = CH$]. They emphasise that particular attention has been given to the addition of vinyl and similar radicals to unsaturated molecular species. Although this route to benzene is relatively simple, at the higher temperatures of flames (> 1500 K) where soot formation is at its most significant, the thermochemistry involved renders many of the steps very reversible, so dramatically reducing the rate of aromatic formation.

Stein et al consider that the propargyl radical is an 'ideal' precursor for benzene formation.

(a)   There is strong evidence for high concentrations in flames.

(b)   The radical is easily formed.

(c)   Arising from its structure and electron delocalisation effects, the propargyl radical is stable and has very few possible reaction channels.

(d)   Addition to unsaturated molecules is highly reversible.

They emphasise that following the recombination of propargyl radicals, there is a cascading series of reactions which finally lead to benzene formation, the sequence being markedly irreversible. The overall reaction is highly exothermic with an enthalpy change of about -710 kJ mol^{-1}. They propose a mechanism as summarised below (two other recombinations of propargyl are, of course, possible), and the potential energy diagram is sketched in Fig 10. The cascading nature of the benzene formation is apparent. Stein et al measured the Arrhenius parameters of the two steps hexa-diyne-1,5 to 1,2 dimethylenecyclobutene and from the latter to benzene and fulvene, and obtained k/s^{-1} = $10^{11.7}$ exp(-17870/T) and k/s^{-1} = $10^{12.9}$(-25200/T), respectively. With these data, Stein et al successfully simulated the formation of benzene in acetylene flames.

$$2CH_2 = C = CH \longrightarrow H \equiv C\text{-}CH_2\text{-}CH_2\text{-}C \equiv H \longrightarrow [CH_2 = C = CHCH = C = CH_2]$$

In their own experiments in the range 520-1000 K, they showed significant differences in product distributions when the isomerisations were studied at atmospheric pressure and at very low pressures (VLPP). In particular, at low pressures, benzene was observed directly from hexadiyne-1,5 at temperatures as low as 600 K. They attribute this to chemical activation whereby at the low collision frequencies of VLPP experiments, 1,2-dimethylenecyclobutene is formed from hexadiyne-1,5 with about 230 kJ mol^{-1} excess energy which allows the cascade (see Fig 10) over each energy barrier right through to benzene. At the higher pressures, thermal stabilisation occurs and the intermediate products can be detected.

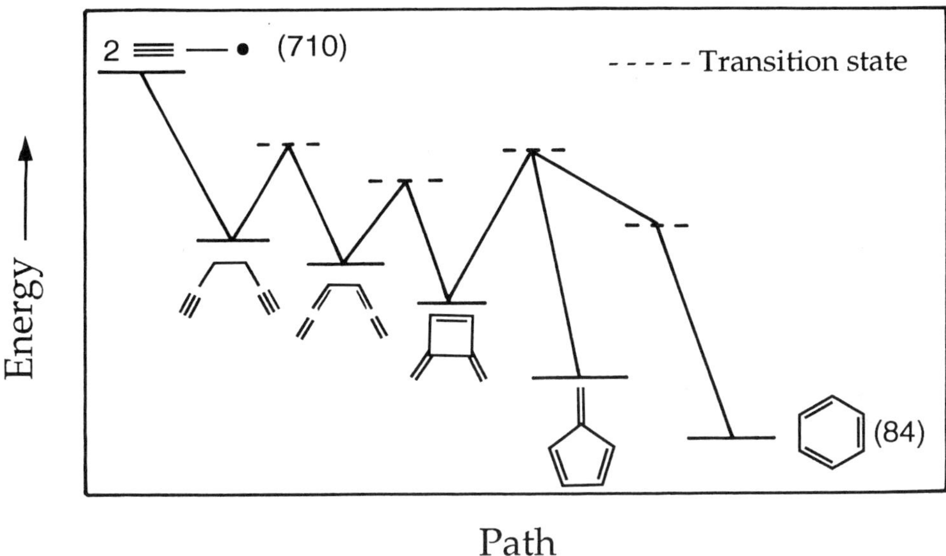

Figure 10   Sketch of energy cascade from propargyl radicals to benzene.

Westmoreland et al [93] in testing several mechanisms for benzene and soot formation consider that propargyl recombination is feasible only for acetylene flames. They also argue that subsequent cyclisation and necessary H atom shifts (frequently a very limiting factor because of the very high energy barriers involved) are sufficient to rule out $C_3H_3 + C_3H_3$ and, for that matter, $C_3H_3 + C_3H_4$ reactions as a source of benzene.

8.1.2. Formation of benzene by chemically-activated isomerisation

As pointed out earlier, many suggestions for benzene formation in flames involve radical addition processes. For example, Frenklach and co-workers [95] proposed the sequence involving addition of $1\text{-}C_4H_3$ to acetylene

60

$$HC \equiv C - CH = \overset{\bullet}{C}H \; + \; HC \equiv CH \rightarrow HC \equiv C - CH = CH - CH = \overset{\bullet}{C}H \rightarrow c \text{-} C_6H_5$$

was the major route to the aromatic ring, and successfully simulated soot formation in pyrolysis experiments of C_2H_2 at 1700-2300 K and 7 atmospheres.

However, Westmoreland and co-workers [93] have argued that benzene is not formed by high pressure limit radical addition processes, but by chemically-activated addition and isomerisation reactions. In a simple form, they can be summarised below, where R is a radical, X is an unsaturated molecule, and A_1 and A_2 are intermediates, A_2 being an isomer of A_1.

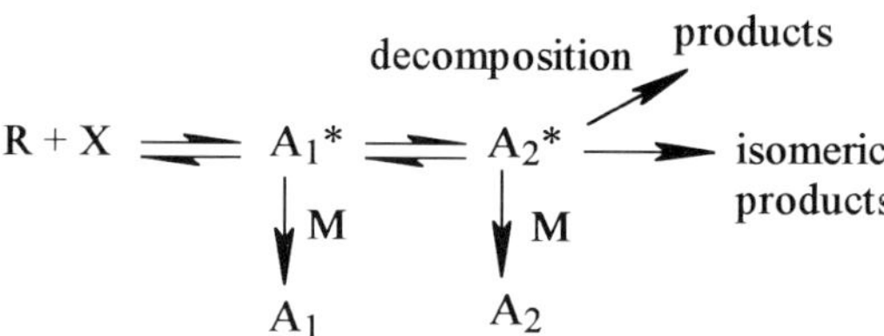

Westmoreland et al point out that the chemically-activated route is fundamentally different from radical addition and isomerisation involving thermalised species in the following ways.
(i) Formation of <u>assumed</u> thermalised intermediates is impeded.
(ii) The oxidation of thermalised intermediates is averted.
(iii) New channels, not normally considered, become available.

As the 'normal' chemistry of A_1 and A_2 does not occur, the formation of A_1^* leads very rapidly to A_2^* and hence to the final product with no direct observation of the stable intermediates A_1 and A_2. This suggestion is not novel, as indeed the formation of C_2H_4 from $C_2H_5 + O_2$ occurs in a similar fashion through $C_2H_5O_2$, as discussed in Section 7. Only if the pressure is sufficiently high to thermally stabilise the vibrationally-excited species will a dramatic reduction in rate be observed.

Westmoreland et al [93] tested their chemical-activation theory on a number of flames where benzene is formed. They concluded that only additions of vinylic $1\text{-}C_4H_5$ and $1\text{-}C_4H_3$ radicals to acetylene were sufficiently fast to account for the highest rates of formation of benzene, by forming benzene and phenyl 'directly' (ie. without passing through stabilised intermediates) by chemically-activated channels.

8.2. Concluding remarks on high-temperature cyclisation

It is most unlikely that any single mechanism is solely responsible for soot formation at temperatures above 1000 K. It is sensible to assume that radical addition involving highly unsaturated species and thermalised adducts will be important up to about 1500 K, particularly where electron-delocalised species are involved. Above this temperature, the recombination of propargyl radicals is a favourable route because of the 'cascading' energy fall to the final products which are formed overall in a very exothermic sequence and thus avoid thermodynamic bottlenecks that reduce growth in normal radical-molecule additions. Miller and Melius [96] have examined the kinetic and thermodynamic aspects of the recombination and support this view. Cyclisation through chemical activation clearly must occur, but is unlikely to be important in high pressure flames even at very high temperatures. Particularly in real fuels such as gasoline, where complex mixtures of hydrocarbons are found it should be emphasised that highly substituted aromatics will be formed from radicals related to propargyl and C_4H_3 etc, for example as shown below.

(a) Recombination

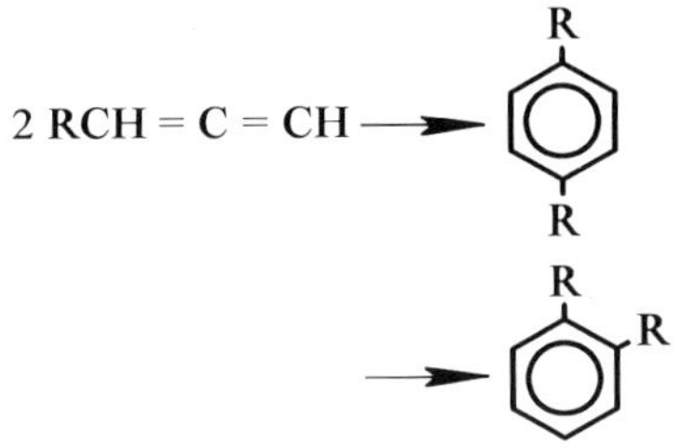

$$2\ RCH = C = CH \longrightarrow$$

(b) Radical addition

$$RCH{=}CH{-}CR'{=}CH \quad + \quad HC \equiv CH \longrightarrow RCH{=}CH{-}CR'{=}CH{-}CH{=}\dot{C}H$$

$$c{-}C_6H_5RR'^*$$

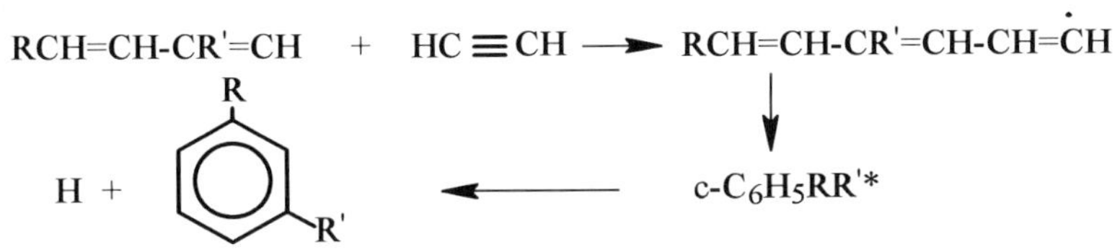

Stein et al [94] pointed out that substituted hexadiynes could undergo 'cascade' sequences to form polyaromatic hydrocarbons.

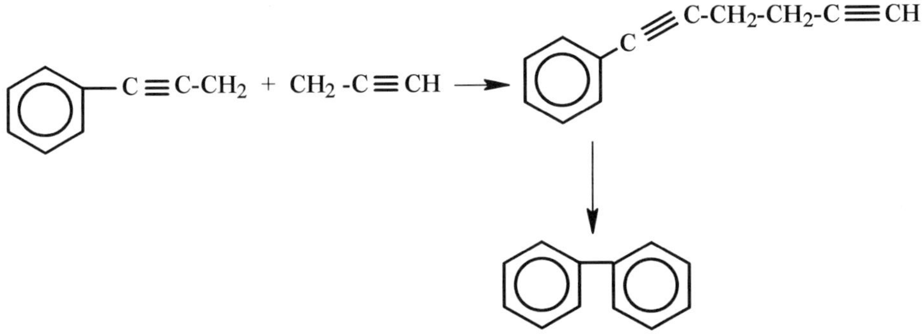

$$-C \equiv C{-}CH_2 \ + \ CH_2\ {-}C \equiv CH \longrightarrow$$

8.3. Cyclisation reactions below 1000 K

Although temperatures below 1000 K are not regarded as important for the formation of soot, production of benzene and aromatics would be very important in terms of pollution from exhaust emissions in gasoline engines. The cyclisation mechanisms discussed above are totally acceptable below 1000 K, but the species involved such as propargyl and highly unsaturated radicals (eg C_4H_3) are not present in high concentration. Moreover, acetylene is not regarded as a low-temperature product, presumably because (45) is the major reaction of vinyl radicals, rather than the high temperature processes such as (46) [3]

$$C_2H_3 \quad + \quad O_2 \quad \rightarrow \quad HCHO \quad + \quad HCO \tag{45}$$

$$C_2H_3 \quad + \quad O \quad \rightarrow \quad HC{\equiv}CH \quad + \quad OH \tag{46}$$

For the overall reaction (47), for which a very reasonable multi-stage mechanism can be written, involving the cyclisation of the hexadienyl radical $CH_2CHCHCH_2CH = CH_2$, the overall exothermicity is about -740 kJ mol^{-1}, which provides a strong drive for the formation of benzene. As pointed out earlier, the electron-delocalised allyl radical is present in high concentration in C_3H_6 oxidation, and hence in the oxidations of many longer chain alkanes.

$$2\,CH_2CHCH_2 \quad + \quad O_2 \quad \rightarrow \quad C_6H_6 \quad + \quad 2H_2O \tag{47}$$

Formation of cyclic compounds has been observed in the oxidation of penta-1,4-diene (PDE) and hexa-1,5-diene (HDE) [58]. In a mixture containing 4, 30 and 26 Torr of PDE, O_2 and N_2, respectively, at 753 K, cyclo-pentadiene is observed in an initial yield of about 25%, which implies that the fully electron-delocalised pentadienyl radical rapidly cyclises.

$$OH \quad + \quad CH_2 = CHCH_2CH = CH_2 \quad \rightarrow \quad CH_2CHCHCHCH_2 \quad + \quad H_2O$$

$$CH_2CHCHCHCH_2 \quad \rightarrow \quad c\text{-}C_5H_7 \xrightarrow{O_2} c\text{-}C_5H_6 \quad + \quad HO_2$$

When PDE is replaced in the mixture by HDE, then significant amounts of cyclohexadiene, benzene and buta-1, 3-diene are observed as initial products, the combined yield of the cyclic products being about 20% with the ratio [cyclic products]/[butadiene] $\approx$ 1. The three products are formed in the sequence as follows [33, 58].

$$CH_2CHCHCH_2CH = CH_2 \xrightarrow{(49)} c\text{-}C_6H_9 \xrightarrow[70\%]{O_2} c\text{-HDE} \xrightarrow[80\%]{OH/O_2} C_6H_6$$

$$\downarrow (48)$$

$$CH_2 = CHCH = CH_2$$

The two competing reactions (48) and (49) determine the relative yields of cyclic compounds and butadiene, and from the initial yields of the products (allowing for the %

conversions above) then $k_{49}/k_{48} \approx 2$. No kinetic data are available for (48) and (49), but k_{48} = 10^{14} exp (-16600/T) s^{-1} is reasonable [39], so that k_{49} = 5.5 x 10^4 s^{-1} at 753 K is estimated. With A_{49} = 10^{11} s^{-1} [72], then E_{49} = 90 kJ mol^{-1}. E_{49} may be predicted almost precisely given the three contributions to its value,
(i) the 'normal' activation energy of a radical addition to an alkene (ca 35 kJ mol^{-1}),
(ii) the strain energy in the transition state (effectively zero given the six-membered ring),
(iii) the loss of virtually all the delocalisation energy in a 'tight' transition state (ca 55 kJ mol^{-1}).
For the pentadienyl radical cyclisation, the activation energy should be about 30 kJ mol^{-1} higher (ring strain energy), and the loss of one less rotor in the transition state should increase the A factor by about 10, giving A = 10^{12} s^{-1} and E = 120 kJ mol^{-1}.

The observation that pentadienyl and hexadienyl radicals readily cyclise is of considerable importance in the oxidation chemistry of the larger radicals derived from alkenes. Gulati and Walker [33] studied the addition of cyclohexane to H_2 + O_2 mixtures at 753 K and reported that the ring was retained in virtually all of the initial products. In particular, they did not observe homolysis products such as C_2H_4, buta-1,3-diene or butenes. Similar observations have been made when cyclopentane was added to H_2 + O_2 mixtures [97]. In the case of cyclohexane, following H abstraction from the ring, any C_2H_4 formed would arise from the homolysis of the CH_2 = $CHCH_2CH_2CH_2CH_2$ radical produced by isomerisation of the c-C_6H_{11} radical.

$$c\text{-}C_6H_{11} \underset{(-50)}{\overset{(50)}{\rightleftharpoons}} C_6H_{11} \overset{(51)}{\longrightarrow} CH_2 = CHCH_2CH_2 + C_2H_4$$

Fig.11 shows the potential energy surface for this sequence, which is based on the known enthalpies of the species together with
(i) an energy barrier of 35 kJ mol^{-1} for the addition of butenyl to C_2H_4 [98],
(ii) the same energy barrier for cyclisation of C_6H_{11} to c-C_6H_{11}, based on zero ring strain energy in the transition state (the barrier is thus effectively higher than that for radical addition to a double bond [97]).

For the cyclisation, k_{-50} = 10^{10} exp(-4210/T)s^{-1} is very reasonable. As the decomposition of C_6H_{11} to butenyl + C_2H_4 (51) will have Arrhenius parameters of A_{51} = $10^{13.5}$ s^{-1} and E_{51} = 125 kJ mol^{-1}, then it is clear that (50) is fully reversible (in the absence of other competing reactions) since at 753 K, k_{-46} » k_{47}. Walker and Handford-Styring [97] considered the alternative reactions of C_6H_{11} radicals in the presence of one atmosphere of O_2 and showed that none are as fast as the cyclisation process (-50). Consequently, although isomerisation of c-C_6H_{11} radicals to the linear form does occur in cyclohexane oxidation [33], the reverse reaction completely dominates all others. Only above about 1300 K will reaction (51), with its higher activation energy, become dominant.

A similar set of kinetic parameters will apply to any radical of the type RCH = $CHCH_2CH_2CH_2CHR$ (the C atoms could be further substituted) which will rapidly cyclise even under oxidising conditions between 600 K and 1300 K. In the presence of O_2, a

64

substituted cyclohexene will be formed which will (in similar fashion to cyclohexene oxidation chemistry [33]) rapidly give a substituted cyclohexadiene and then a substituted benzene. Formation of aromatics in this way from linear and branched alkenes may well be very significant in the temperature region 600-1000 K.

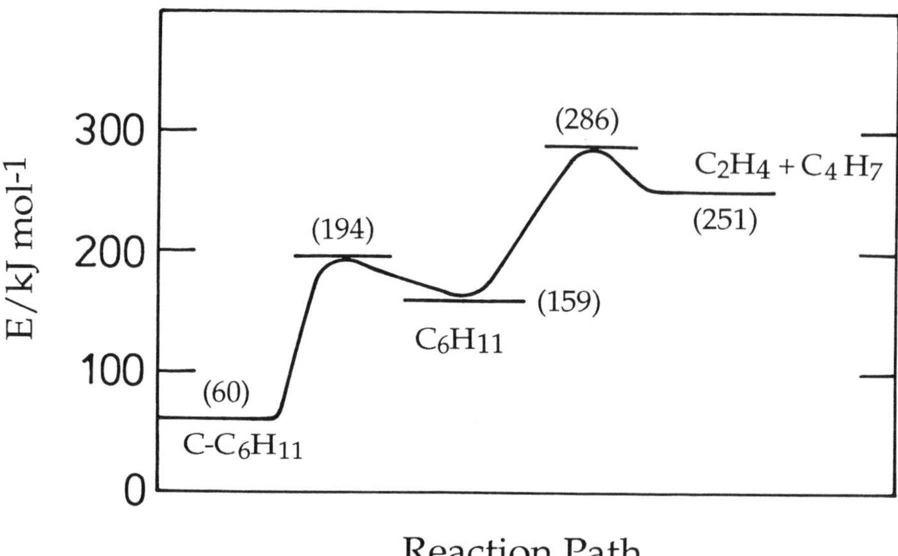

Figure 11   Energy path C2H4 for formation from C-C6H11 radicals.

REFERENCES

1.   G.J. Minkoff and C.F.H. Tipper, Chemistry of Combustion Reactions, Butterworths, London, 1962.
2.   D.J. Hucknall, Chemistry of Hydrocarbon Combustion, Chapman and Hall, London, 1985.
3.   D.L. Baulch, C.J. Cobos, R.A. Cox, C. Esser, Th.Just, J.A. Kerr, M.J. Pilling, J. Troe, R.W. Walker, and J. Warnatz, J.Phys.Chem., Ref. Data, 21 (1992) 411 - see commentary attached to many of the reactions covered.

4.   A.B. Trenwith, J.Chem.Soc., Faraday Trans. 1, 75 (1979) 614 and references therein.

5.   S.N.Foner and R.L. Hudson, Advances in Chemistry Series, 36 (1962), 34.

6.   T.T. Paukert and H.S. Johnston, J.Chem.Phys., 56 (1972) 2824.

7.   J.Troe, Ber. Bunsenges. Phys. Chem., 73 (1969) 946.

8.   J. Warnatz, in Combustion Chemistry (ed. W.C. Gardiner Jr.) Springer Verlag, New York, 1984.

9.   P.D. Lightfoot, B.Veyret and R. Lesclaux, Chem. Phys. Lett., 150 (1988) 120.

10. H. Hippler, J.Troe and J.Willner, J. Chem. Phys., 93 (1990) 1755.

11. D.L. Baulch, M.Bowers, D.G. Malcolm and R.T.Tuckerman, J. Phys. Chem. Ref. Data., 15 (1986) 465.

12. W.J. Pitz and C.K. Westbrook, Combustion and Flame, 63 (1986) 113.

13. R.R. Baldwin, P.N. Jones and R.W. Walker, J. Chem. Soc., Faraday Trans., 2, 84, (1988) 199.

14. N. Cohen, Int. J. Chem. Kin., 23 (1991) 683.

15. R.W Walker, Int. J. Chem. Kin., 17 (1985) 573.

16. R.W.P. White, D.J. Smith and R. Grice, Chem. Phys. Lett., 193 (1992) 269.

17. J.E. Pollard, Rev. Sci. Instrum., 63 (1992) 1771.

18. R.R. Baldwin, R.W. Walker, and R.W. Walker, J.Chem.Soc., Faraday Trans., 1, 77 (1981) 2157.

19. I.R. Slagle, A. Bencsura, S.B. Xing and D. Gutman, Proc. Int. Symp. Combustion, 24 (1992) 653.

20. I.R. Slagle, J.R. Bernhardt and D. Gutman, Proc. Int. Symp. Combustion, 22 (1989) 953.

21. R.W. Walker, Proc. Int. Symp. Combustion, 22 (1989) 883.

22. K.J. Hughes, A.R. Pereira and M.J. Pilling, Ber. Bunsenges. Phys. Chem., 96 (1992) 1352.

23. R.G. Gilbert, K. Luther, and J. Troe, Ber. Bunsenges. Phys. Chem., 87 (1983) 169.

24. W.J. Pitz, C.K. Westbrook, W.M. Proscia and F.L. Dryer, Proc. Int. Symp. Combustion, 20 (1985) 831.

25. U. Maas and J. Warnatz, Combustion and Flame, 74 (1988) 53.

26. A.S. Tomlin, M.J. Pilling, T. Turanyi, J.H. Merkin and J. Brindley, Combustion and Flame, 91 (1992) 107.

27. R.W. Walker, Science Progress, 74 (1990) 163.

28. C.T. Bowman, Fossil Fuel Combustion - A Source Book, (ed W. Bartok & A.F. Sarofim), Wiley, New York, 1991.

29. R.T. Pollard, Comprehensive Chemical Kinetics, vol.17, (ed. C.H. Bamford & C.F.H. Tipper), Elsevier, Amsterdam, p.358.

30. R.R. Baldwin and R.W. Walker, Proc. European Int. Symp. Combustion, 14 (1973) 241.

31. J.F. Griffiths and S.K. Scott, Prog. Ener. Combust. Sci., 13 (1987) 161.

32. N.D. Stothard and R.W. Walker, J. Chem. Soc. Faraday Trans., 87 (1991) 241.

33. S.K. Gulati and R.W. Walker, J.Chem.Soc. Faraday Trans. 2, 85 (1989) 1799.

34. R.R. Baldwin, G.R. Drewery and R.W. Walker, J. Chem. Soc. Faraday Trans. 2, 82 (1986) 251.

35. R.R. Baker, R.R. Baldwin and R.W. Walker, Combustion and Flame, 14 (1970) 31.

36. G.M. Atri, R.R. Baldwin, G.A. Evans and R.W. Walker, J. Chem. Soc., Faraday Trans. 1, 74 (1978) 366.

37. R.R. Baldwin, C.E. Dean, M.R. Honeyman and R.W. Walker, J. Chem. Soc. Faraday Trans. 1, 82 (1986) 89.

38. N.D. Stothard and R.W. Walker, J. Chem. Soc. Faraday Trans., 86 (1990) 2115.

39. R.R. Baldwin, R.W. Walker and R.W. Walker, J. Chem. Soc. Faraday Trans. 1, 76 (1980) 825.

40. R.R. Baldwin, G.R. Drewery and R.W. Walker, J. Chem. Soc. Faraday Trans. 1, 80 (1984) 2827.

41. Z.H.. Lodhi and R.W. Walker, J. Chem. Soc. Faraday Trans., 87 (1991) 2361.

42. N.D. Stothard and R.W. Walker, J. Chem. Soc. Faraday Trans., 88 (1992) 2621.

43. R.R. Baldwin, A.R. Fuller, D. Longthorn and R.W. Walker, J. Chem. Soc. Faraday Trans. 1, 70 (1974) 1257.

44. T. Ingham, R.W. Walker and R.E. Woolford, Proc. Int. Symp. Combustion, 25 (1994), to be published.

45. R.W. Walker, Specialist Periodical Report, Chemical Society, Reaction Kinetics, 1, (1975) 161.

46. J. Warnatz, Paper presented at the 12th International Symposium on Gas Kinetics, Reading, September 1992.

47. J.L. Emdee, K. Brezinsky and I. Glassman, J. Phys. Chem., 96 (1992) 2151.

48. R.R. Baldwin, R.W. Walker and D.H. Langford, Trans. Faraday Soc., 65 (1969) 792.

49. D.J. Dixon, G. Skirrow and C.F.H. Tipper, Combustion Institute European Symposium, Academic Press, London (1973) p.94.

50. W. Tsang, Int. J. Chem. Kin., 1 (1969) 245.

51. J.B. Farmer and C.A. McDowell, Trans. Faraday Soc., 48 (1952) 624.

52. A. Combe, M. Niclause and M. Letort, Rev. Inst. France Petrole, 10 (1955), 786, 929.

53. R.R. Baldwin, M.J. Matchan and R.W. Walker, Combustion and Flame, 15 (1970) 109.

54. N. Semenov, Chemical Kinetics and Chain Reactions, Clarendon Press, Oxford, (1935) p.41.

55. R.R. Baldwin and R.W. Walker, Essays in Chemistry (ed J.N. Bradley, R.D. Gillard and R.F. Hudson), Academic Press, London, 3 (1972) 1.

56. Z.H. Lodhi, N. Stothard and R.W. Walker, Proc. Int. Symp. Combustion, 23 (1991) 123.

57. R.R. Baldwin, C.E. Dean and R.W. Walker, J. Chem. Soc., Faraday Trans. 2, 82 (1986) 251.

58. R.W. Walker, unpublished work.

59. R.R. Baldwin, J.P. Bennett and R.W. Walker, J. Chem. Soc. Faraday Trans. 1, 76 (1980) 2396.

60. K. Brezinsky, T.A. Litzinger and I. Glassman, Int. J. Chem. Kin., 16 (1984) 1053.

61. W. Tsang, J. Phys. Chem., 76 (1972) 143.

62. J.A. Walker and W. Tsang, Int. J. Chem. Kin., 11 (1979) 867.

63. R.R. Baldwin, D.E. Hopkins and R.W. Walker, Trans. Faraday Soc., 66 (1970) 189.

64. R.R. Baldwin, I.A. Pickering and R.W. Walker, J. Chem. Soc. Faraday Trans. 1, 76 (1980) 2374.

65. R.R. Baldwin, M.W.M. Hisham, A. Keen and R.W. Walker, J. Chem. Soc. Faraday Trans. 1, 78 (1982) 1165.
66. S.W. Benson, Adv. Chem. Series, 76 (1968) 143.
67. R.R. Baker, R.R. Baldwin and R.W. Walker, J. Chem. Soc. Faraday Trans., 1, 71 (1975) 756.
68. R.W. Walker, Specialist Periodical Reports, The Chemical Society, 2 (1977) 296.
69. Z.H. Lodhi and R.W. Walker, J. Chem. Soc. Faraday Trans., 87 (1991) 681.
70. A. Fish, Organic Peroxides, (ed. D. Swern), Wiley, New York, 1 (1970) 141.
71. R.R. Baldwin, M.W.M. Hisham and R.W. Walker, J. Chem. Soc. Faraday Trans. 1, 78 (1982) 1615.
72. S.W. Benson, Thermochemical Kinetics, Wiley, New York, 976.
73. K.J. Hughes, P.A. Halford-Maw, M.J. Pilling and T. Turanyi, Proc. Int. Symp. Combustion, 24 (1992) 645.
74. I.R. Slagle, Q. Feng and D. Gutman, J. Phys. Chem., 88 (1984) 3648.
75. A.F. Wagner, I.R. Slagle, D. Sarzynski and D. Gutman, J. Phys. Chem., 94 (1990) 1853.
76. I.C. Plumb and K.R. Ryan, Int. J. Chem. Kin., 13 (1981) 1011.
77. T.J. Wallington, J.M. Andino, E.W. Kaiser and S.M. Japor, Int. J. Chem. Kin., 21 (1989) 1113.
78. E.W. Kaiser, L. Rimai and T.J. Wallington, J. Phys. Chem., 93 (1989) 4094.
79. E.W. Kaiser, I.M. Lorkovic and T.J. Wallington, J. Phys. Chem., 94 (1990) 3352.
80. K.G. McAdam and R.W. Walker, J. Chem. Soc. Faraday Trans. 2, 83 (1987) 1509.
81. S.K. Gulati and R.W. Walker, J. Chem. Soc. Faraday Trans. 2, 84 (1988) 199.
82. J.W. Bozzelli and A.M. Dean, J. Phys. Chem., 94 (1990) 3313.
83. A.M. Dean, J. Phys. Chem., 89 (1985) 4600.
84. L.S. Kassel, J. Phys. Chem., 32 (1928) 225, 1065.
85. R.R. Baldwin, C.E. Dean and R.W. Walker, J. Chem. Soc., Faraday Trans. 2, 82 (1986) 1445.
86. L. Batt, private communication.
87. R.R. Baldwin, J. P Bennett and R.W. Walker, J. Chem. Soc. Faraday Trans. 1, 76 (1980) 825.
88. D.A. Osborne and D.J. Waddington, J. Chem. Soc., Perkin Trans. 2, (1980) 925.
89. M.I. Sway and D.J. Waddington, J. Chem. Soc., Perkin Trns. 2, (1983) 139.
90. H.H. Grotheer, G. Riekart, D. Walter and Th. Just, J. Phys. Chem., 92 (1988) 4028.
91. S.P. Walch, Chem. Phys. Lett., 215 (1993) 81.
92. G.E. Quelch, M.M. Gallo and H.F. Schaefer III, J. Am. Chem. Soc., 114 (1992) 8239.
93. S.E. Stein, J.A. Walker, M.M. Suryan and A. Fahr, Proc. Int. Symp. combustion, 23 (1990) 85.
94. P.R. Westmoreland, A.M. Dean, J.B. Howard and J.P. Longwell, J. Phys. Chem., 93 (1989) 8171.
95. M. Frenklach, D.W. Clary, W.C. Gardiner, Jr., and S.E. Stein, Proc. Int. Symp. Combustion, 20 (1984) 887.
96. J.A. Miller and C.F. Melius, Combustion and Flame, 91 (1992) 21.

97. S. Handford-Styring and R.W. Walker, submitted for publication in J. Chem. Soc. Faraday Trans.

98. J.A. Kerr and S.J. Moss (Ed), Handbook of Bimolecular and Termolecular Gas Reactions, CRC Press, Boca Raton, Florida, Vol.II (1981).

Research in Chemical Kinetics, Volume 3
R.G. Compton and G. Hancock (editors)
© 1995 Elsevier Science B.V. All rights reserved.

Reactions of NHx Species

Joseph L. Durant, Jr.

Combustion Research Facility
Sandia National Laboratory
Livermore, Ca 94551, USA

1 INTRODUCTION

The chemistry of NHx species is both interesting and important to a wide variety of processes. In combustion, NHx species are intermediates in the conversion of alkyl fuel bound nitrogen to NOx and N_2 products. Ammonia, urea and cyanuric acid have been used as flue gas additives to remove NOx in a variety of DeNOx processes.[1-4] Central to this conversion is the reaction of NH_2 with NO, which has a primary product channel forming N_2 and H_2O. The secondary product channel from this reaction appears to be N_2H + OH, which acts as a chain branching step, allowing the DeNOx process to continue. This reaction has also been claimed to be an important NH_2 sink in the troposphere.

In a 1979 paper, Hack, Schacke, Schröter and Wagner noted: "It was a great success when Haber and Bosch first broke the N-N bond to produce NH_3 from N_2 molecules. Nowadays the destruction of N_2 and the resulting production of NOx leads to severe problems since the production of energy by fuel-air combustion has increased so rapidly in the last twenty years."[5]

Reactions of NHx species most often involve formation of adducts, which undergo extensive rearrangement before proceeding on to final product formation. For example, the NH_2 + NO reaction forms an H_2NNO adduct, which undergoes a 1-3 H shift, a cis-trans isomerization about the NO bond, a trans-cis isomerization about the NN bond and a four-center elimination to produce N_2 + H_2O products.

The reactions of NHx species also typically have more than one set of possible products. For example, the reaction of NH with NO can yield H + N_2O products or OH + N_2 products. Under conditions typical in thermal DeNOx (temp = 1250 K, excess O_2 present) H atoms are rapidly converted via the H + $O_2 \rightarrow$ OH + O reaction, making the H + N_2O channel chain branching, and the OH + N_2 channel merely chain propagating. Additionally, N_2O produced via the H + N_2O channel reduces the DeNOx efficiency. It is clear that, in order to successfully model systems containing NHx species, one must know something of the product branching fractions. The presence of multiple products on the potential energy surfaces for these reactions also means that the surfaces can be accessed from a variety of reactants. To continue consideration of the NH + NO reaction, the

HNNO surface can also be accessed by the H + N_2O reaction. This allows additional information to be obtained about the potential energy surface by experimentally accessing the surface in different ways.

An additional complexity encountered in NHx reactions is the existence of multiple energetically accessible potential energy surfaces. This may take the form of a surface crossing connecting reactants and products or the existence of multiple surfaces accessible to reactants, such as the $^{1,3}A'$, $^{1,3}A''$ and $^5\Pi$ surfaces accessible to NH + O reactants. In the former case, calculations of reaction rate constants become recast as calculation of rates of surface crossing. In the latter case careful consideration has to be given to the correlations between reactants and product asymptotes. In the case that reactants and products are connected by multiple electronic states we lack a detailed understanding of the effect of the excited state surface on the reaction dynamics. Typically in cases of this type the reaction is modeled by assuming that the reactive flux is separable into the various channels, and there are no interference effects between the channels.

1.1 Scope of this review

The chemistry of NHx species has been reviewed by various authors, with various emphasis, in the past. Lesclaux reviewed chemistry of the NH_2 radical in 1984,[6] covering sources for and detection of NH_2, as well as its reactions with a variety of species. Miller and Bowman[7] produced a review focusing on combustion chemistry of nitrogen-containing species. They discussed at some length mechanisms for NOx formation and removal in combustion systems, as well as producing a mechanism for nitrogenous species in combustion. Hanson and Salaam[8] also reviewed chemistry of nitrogen-containing species, focusing on a careful examination of experimental results at high temperatures, and giving evaluated rate constant data for reactions of species containing N, H and O.

Recent years have seen a continuation in the efforts to provide compilations and evaluations of reaction rate constants. The compilation of reaction rate constants in the NIST database[9] stands out for being current (new editions are issued roughly yearly), and for its ease of use. It is a computerized database allowing searches on both reactants and products, together with graphical presentation of the cited results. It draws on experimental and theoretical results, and includes recommended reaction rate coefficients from various evaluations.

Baulch, et al[10] and Cohen and Westberg[11] have evaluated kinetic data for combustion modeling, including some reactions of nitrogen containing species. There are also recent evaluations by Tsang and Herron[12,13] focusing on the nitrogen chemistry important for energetic materials (i.e. explosives and propellants).

IUPAC[14] and NASA/JPL[15] evaluations focus on atmospheric chemistry, and provide evaluations of the experimental and theoretical results on a given reaction. Because of their focus on atmospheric conditions the reactions selected (and rates recommended) are appropriate for $200 \leq T \leq 300$ K and pressures at and below 1 atm. Extrapolation to combustion conditions should be done with great care.

These sources, and others I may have omitted, have lessened the need for a review to compile rate constant information, and, in many cases, to evaluate that information. We shall instead focus on a few reactions of NHx species in an attempt to highlight the richness of the chemistry. In particular, we will examine the reactions of NH with O, NH_2 with O, NH with NO, and NH_2 with NO. We shall emphasize direct determinations of reaction rate coefficients and products, and de-emphasize results obtained by extensive modeling of complex reacting systems (most often flames). While modeling complex systems is important from both an engineering perspective and in highlighting regions of ignorance with respect to elementary chemical processes, it has a poor track record for accurate quantification of elementary chemical processes. We will similarly de-emphasize purely theoretical calculations of elementary chemical processes.

We will draw heavily on recent results of electronic structure calculations, which are now capable of predicting energies of reactive intermediates and transition states to within a few kcal/mole. Fitting into this category of calculations are those carried out using the Gaussian-2 (G2) and G2Q methods[16,17], and those carried out using various CASSCF/MRCI or CCSD(T) schemes with large basis sets.[18] These methods, when carefully used, are found to agree quite well with one another; as a further test, results of the G2 method have been compared to experimentally measured heats of atomization for 55 well-established species containing first and second-row elements. The average absolute deviation in heats of atomization was found to be 1.1 kcal/mole.[16] Comparable results were found in an examination of a number of well-characterized transition states using a modification of the G2 method, dubbed G2Q.[17] While it is arguable whether this is truly "chemical accuracy", these results speak to the quality of modern theoretical calculations, and the added insight these calculations can bring into the often complex reaction mechanisms encountered in studying NHx species.

1.2 Thermochemistry

Before we begin in earnest, it is important that we are all "playing by the same rules," which in this case we take to mean the use of well defined, and hopefully accurate, thermochemistry. Table 1.1 presents the experimentally known thermochemistry for the $H_xN_yO_z$ system. In general we follow the recommendations of the most recent JANAF tables.[19] Anderson[20] has recently re-evaluated data relating to the heats of formation of NH and NH_2; we choose to use his recommended value for NH. For NH_2 we have chosen the value recommended in the recent evaluation of RH bond strengths by Berkowitz, Ellison and Gutman,[21] which is in excellent agreement with Anderson's recommended value. We have chosen to revise the JANAF heat of formation for HNO to reflect a refined value of 16450 ± 10 cm^{-1} for the predissociation threshold for the HNO (A)-state.[22] This has the effect of lowering the heat of formation of HNO by 1.6 kcal/mole. The heats of formation and classical well depths, D_e(H-NO), for the excited states of HNO have been derived from that of the ground state, making use of the known spectroscopy of NO,[23] and HNO,[24-26] and calculation of the vibrational frequency for the v_1 mode in HNO (A)^{3}A".[27] The classical well depths for the three states of HNO have been included to simplify comparison with theoretical results. The thermochemistry for the majority of the $H_xN_yO_z$ adduct species has not been experimentally determined, although there is considerable spectroscopic data on H_2NO, and limited spectroscopic data on other of the adducts.

Table 1.1: Experimental Thermochemistry for the H, N, O system

species	ΔH_f (0 K) (kcal/mole)	D_e(H-NO) (kcal/mole)
H	51.634 ± 0.001	
N	112.53 ± 0.02	
O	58.98 ± 0.02	
OH	9.2 ± 0.3	
NO	21.46 ± 0.04	
NH	85.4 ± 0.3	
NH$_2$	45.8 ± 0.3	
N$_2$O	20.4 ± 0.1	
H$_2$O	-57.10 ± 0.01	
HNO (X)^{1}A'	$26.06 + 0.05$	52.59 ± 0.03
HNO (A)^{3}A"	44.0 ± 0.4	34.6 ± 0.4
HNO (A)^{1}A"	63.67 ± 0.05	14.22 ± 0.03

2 THE H, N, O SYSTEM

The reaction of NH with O accesses NHO/HON potential energy surfaces, and a wealth of interesting chemistry emerges. If we look in detail we find that there are three bimolecular reactions linking HNO containing species,

$$NH + O \leftrightarrow H + NO \tag{2.1}$$

$$NH + O \leftrightarrow N + OH \tag{2.2}$$

$$H + NO \leftrightarrow N + OH \tag{2.3}$$

plus four association/dissociation reactions,

$$H + NO\ (+M) \leftrightarrow HNO\ (+M) \tag{2.4}$$

$$H + NO\ (+M) \leftrightarrow HON\ (+M) \tag{2.5}$$

$$NH + O\ (+M) \leftrightarrow HNO\ (+M) \tag{2.6}$$

$$N + OH\ (+M) \leftrightarrow HON\ (+M) \tag{2.7}$$

and one isomerization reaction,

$$\text{HNO (+M)} \leftrightarrow \text{HON (+M)} \tag{2.8}.$$

Additionally, there are singlet, triplet and quintuplet surfaces which can be accessed by these reactions. A rich reactive system, indeed.

2.1. The HNO Potential Energy Surface

The HNO potential energy surface has been of theoretical interest for a number of years. Bruna and Marian[28] and Bruna carried out MRD-CI calculations on the lowest singlet and triplet states of HNO and NOH and the transition states for isomerization. Later workers have refined and extended their calculations, including other states and reaction pathways.[29-36] There have been several surfaces generated including ones by Carter, Mills and Murrell,[37] Dixon and coworkers,[22] and Colton and Schatz.[38] Recently Guadagnini, Schatz and Walch have produced a global potential energy surface for the lowest $^1A'$, $^3A''$ and $^1A''$ states of HNO based on CASSCF/ICCI calculations with large basis sets.[27] The results of the ab initio calculations were scaled to reproduce the experimental thermochemistry; the N + OH asymptote was adjusted by 4.4 kcal/mole, and the O + NH asymptote by 2.6 kcal/mole. With this scaling they calculate D_e(H-NO)'s

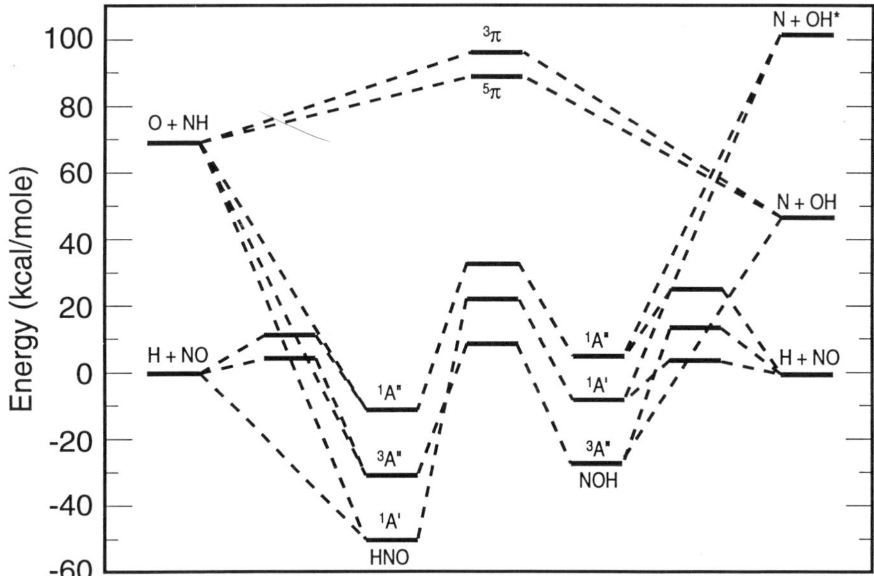

Figure 2.1: Potential energy surfaces for HNO species.

for the three lowest states of HNO of 50.1, 31.7 and 10.9 kcal/mole, respectively. Comparing these D_e's with the well depths tabulated in Table 1.1 suggests that their surface is accurate to no better than perhaps 3.5 kcal/mole.

A schematic of the HNO potential energy surfaces is shown in Figure 2.1, with correlations indicated, as well as energies of transition states linking various reagents, with energies taken from Guadagnini, Schatz and Walch[27] and Walch.[34]

2.2. NH + O → Products

We will consider the two reactions

$$NH + O \rightarrow H + NO \qquad \Delta H_{rxn} = -71.3 \text{ kcal/mole} \qquad (2.1)$$

and

$$NH + O \rightarrow N + OH \qquad \Delta H_{rxn} = -22.7 \text{ kcal/mole} \qquad (2.2)$$

together. The interaction of O ^3P with NH $^3\Sigma$ gives rise to 18 surfaces, 2 singlet, 2 triply degenerate triplets and 2 five-fold degenerate quintets. The two singlet surfaces correlate with the $^1A'$ and $^1A''$ singlet states of HNO. The association reaction proceeds without a barrier,[27] forming highly excited HNO. The lifetime of this excited HNO adduct has been calculated to be of the order of 10^{-14}s;[39] collisional stabilization to the intermediate should not be expected under anything less than very extreme conditions. The spin-allowed, direct H-atom transfer between N and O occurs on triplet and quintet surfaces. The process has been examined theoretically by Walch.[34] He carried out CASSCF/CCI calculations, and located the transition states for the direct abstraction, assuming a collinear geometry on the $^5\Pi$ and $^3\Pi$ surfaces. He found that the classical barrier for the reaction on the triplet surface was 11.7 kcal/mole, while the classical barrier on the quintet surface was 5.6 kcal/mole. These barriers have also been calculated by Durant and Rohlfing[17] using G2Q. They obtained values of 11.9 kcal/mole on the triplet surface, and 5.4 kcal/mole on the quintet surface, in excellent agreement with Walch's results. The linear $^5\Pi$ configuration is lower in energy than the nonlinear configurations, and thus represents the unconstrained transition state on the quintet surface. However, as the linear triplet state is bent, it splits into a $^3A''$ surface, which is stabilized by the deformation and correlates with the $^3A''$ HNO surface, and a $^3A'$ surface, which is destabilized by the deformation.[34] Presumably, then, the reaction occurs mainly on the $^1A'$, $^1A''$ and nonlinear $^3A''$ surfaces, since they connect reactants to products without the presence of barriers higher than the reactant asymptote. This picture of the reaction leads to the expectation of a fast radical-radical reaction with little temperature dependence.

Experimental data on these reactions are sparse. Temps has measured a reaction rate coefficient of $k_{(2.1+2.2)} = 8 \times 10^{-12}$ cm^3/molecule/s at room temperature, using the discharge-flow technique with LMR detection.[40] Adamson and coworkers,[41] using infrared kinetic spectroscopy, have measured an overall rate of $k_{(2.1+2.2)} = (6.5 \pm 1.5) \times 10^{-11}$ cm^3/molecule/s. Hack, Wagner and Zasypkin[42] have used a quasi-static laser flash photolysis technique to measure an upper bound for reaction 2.2 of $k_{2.2} \leq 1.7 \times 10^{-13}$ cm^3/molecule/s. The small size of the branching fraction for OH production is what we would expect based on the reaction of NH with O proceeding through a HNO/HON adduct which then dissociates, without a barrier, to products. The reactions are highly exothermic, and we do not expect the back reactions, $k_{-2.1}$ and $k_{-2.2}$, to be important. Mertens et al.[43] studied the reaction in shock-heated HNCO/N$_2$O/Ar mixtures in the 2730$\leq$ T $\leq$ 3380 K range. NH concentrations was monitored by cw, narrow-linewidth laser absorption. By modeling a full reaction scheme they extracted a temperature independent reaction rate coefficient of $k_{(2.1+2.2)} = (1.5 \pm 0.9) \times 10^{-10}$ cm^3/molecule/s in the temperature range 2730 $\leq$ T $\leq$ 3380 K. They found that NH concentrations in their HNCO/N$_2$O/Ar system were relatively insensitive to the relative importance of reactions 2.1 and 2.2, hence they were unable to extract a product branching fraction for the reaction. The roughly factor of 20 increase in reaction rate between Temps' room temperature rate coefficient and Mertens et al.'s 3000K rate coefficient is at odds with our expectations based on the potential energy surface. Clearly more work is called for on this system.

The nascent product state distribution of NO from the O + NH $\rightarrow$ H + NO reaction has been studied by Huang and Dagdigian.[39] The reaction was studied in a reaction cell at 60 mTorr. O-atoms were produced by microwave discharge of O$_2$ or O$_2$/Ar mixtures, and NH was produced by 193 nm photolysis of NH$_3$. They obtained a nascent NO v=1 rotational temperature of 1130 $\pm$ 50 K; their relative NO vibrational populations are listed in Table 2.1. Huang and Dagdigian calculated that, on average, 8% of the total reaction exothermicity appeared as NO vibrational excitation, and 3% of the total reaction exothermicity appeared as NO rotational excitation. This leaves, on average, 92% of the reaction exothermicity funneled into product translational excitation. For a **H + HL** $\rightarrow$ **HH + L** reaction, the reagent orbital angular momentum is funneled into **HH** product rotational excitation, and the product rotational distribution is determined by the velocity dependent opacity function for the reaction. The low degree of rotational excitation suggests that reaction is favored for small impact parameters and/or low collision velocities. The vibrational distribution is much colder than predicted from a statistical prior distribution, but is well modeled by calculation of the Franck Condon overlap between HNO $v_{NO}=0$ and the free NO molecule.

Table 2.1: Relative NO vibrational populations from the NH + O $\rightarrow$ H + NO reaction.

v	relative population	v	relative population
1	1	5	0.045 ± 0.017
2	0.43 ± 0.04	6	0.012 ± 0.005
3	0.20 ± 0.02	7	0.028 ± 0.019
4	0.17 ± 0.02	8	0.008 ± 0.007

2.3. N + OH $\leftrightarrow$ NO + H

The reaction of N with OH is the third step in the Zeldovich mechanism for conversion of N_2 to NOx in combustion,[44]

O + N_2 $\leftrightarrow$ NO + N
N + O_2 $\leftrightarrow$ NO + O
N + OH $\leftrightarrow$ H + NO

This "thermal mechanism" accounts for a fraction of the NOx produced in combustion of fuels which do not contain nitrogen. Reaction 2.3 is quite exothermic, with

$$N + OH \rightarrow H + NO \qquad \Delta H_{rxn} = \text{-48.7 kcal/mole} \qquad (2.3)$$

The interaction of N ^{4}S with OH $^2\Pi$ will give rise to 16 electronic surfaces. If one examines the potential energy surfaces in Figure 2.1 one sees that the relevant state for reaction 2.3 is the triply-degenerate ^{3}A" state of HON. The ^{5}A' and ^{5}A" states of NOH are expected to be repulsive, and will not correlate with H ^{2}S + NO $^2\Pi$ products. Left out in the accounting (and in calculations of the potential energy surface) is the ^{3}A' state of HON. It is of higher energy than the states which have been calculated, but may offer a direct path from reactants to products.

The reaction rate coefficient of reaction 2.3 has been studied directly by Howard and Smith, for $250 \leq T \leq 515$ K,[45,46] and by Brune, Schwab and Anderson at room temperature.[47] Howard and Smith used flash photolysis coupled with discharge flow to measure the rate coefficient. They photolyzed H_2O in the presence of N atoms created by a microwave discharge. The OH radical concentration profiles were measured with lamp-induced fluorescence, while the N-atom concentrations were derived using chemiluminescent titrations with NO. Analysis of their data yielded a rate constant expression of $k_{2.3} = (2.21 \pm 0.18)$ x 10^{-10} $T^{-0.25 \pm 0.17}$ cm^3/molecule/s for the temperature range $250 \leq T \leq 515$ K. Brune, Schwab and Anderson used LMR, resonance absorption, and resonance fluorescence to monitor species in their flow tube study of this reaction. Their measured rate at room temperature, $k_{2.3} = (4.2 \pm 0.8)$ x 10^{-11} cm^3/molecule/s is in good agreement with Howard and Smith's room temperature rate of $k_{2.3} = (4.7 \pm 0.4)$ x 10^{-11} cm^3/molecule/s.[46] Very recently Rowe, Sims and Smith[48] have extended measurement of this reaction rate coefficient down to 10 K, using laser-photolysis/laser induced fluorescence technique in the discharge from a modified Laval nozzle. They found that the rate continued to rise at lower temperatures.

The negative temperature coefficient for $k_{2.3}$ suggests that the reaction proceeds without a barrier, passing through the HON well. The barrier heights calculated by Guadagnini et al.[27] suggest that isomerization of HON to HNO, followed by dissociation to H + NO, may compete with direct dissociation of HON to H + NO.

At high temperatures the endothermic reverse reaction

$$H + NO \rightarrow N + OH \qquad (-2.3)$$

begins to become important. This reaction serves to convert NO back into atomic N in the context of the Zeldovich mechanism. This reaction was immensely popular in 1975 through 1977, with no fewer than 7 experimental studies of its rate coefficient in the 2000 - 4000 K temperature range[49-55]. Results of these determinations are plotted in Figure 2.2. Since the initial flurry of interest, the rate coefficient has not been remeasured. The various experimental studies are in fair agreement as to the temperature dependence of the reaction, but the rate coefficients vary by roughly a factor of 10. The measured Arrhenius activation energies are all slightly in excess of the reaction endothermicity. As important as this reaction is, it is surprising that it has not been the subject of continued experimental attention.

By use of the equilibrium constant one can predict $k_{-2.3}$ based on measurement of $k_{2.3}$. It is interesting to use this to extrapolate Howard and Smith's data to shock tube temperatures. Despite the length of the extrapolation, the predicted rate falls within the range of the measured high temperature rates, as shown in Figure 2.2.

2.4. H + NO (+M) $\leftrightarrow$ HNO (+M)

From Figure 2.1 one sees that addition of H-atom to NO correlates with the lowest three states of HNO. The other three surfaces arising from combination of

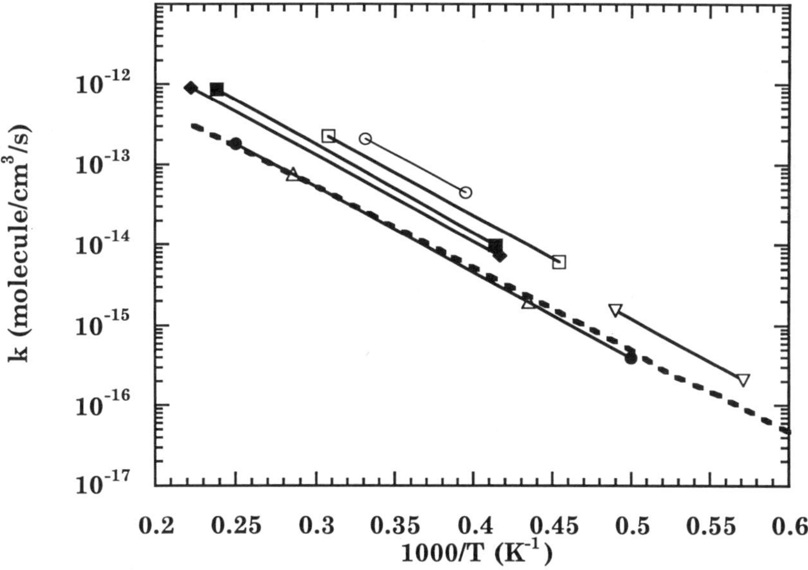

Figure 2.2: Experimentally measured rate coefficients for H + NO $\rightarrow$ Products. O - ref. 49, □ - ref. 50, ◆ - ref. 51, ● - ref. 52, Δ - ref. 53, ■ - ref. 54, ∇ - ref. 55, --- rate calculated from equilibrium constant and low temperature data of Howard and Smith[46] for N + OH $\rightarrow$ Products. .

the ^2S H-atom with the $^2\Pi$ NO correspond to a repulsive triplet state, with the unpaired spins being those in the H-N (non) bond. CASSCF/CCI calculations by Walch and Rohlfing[33] as well as the calculations of Guadagini et al.[27] predict that the reaction on the $^1A'$ ground state surface occurs without an energetic barrier. Walch and Rohfing found classical barriers of 4.1 kcal/mole on the $^3A''$ surface and 10.2 kcal/mole on the $^1A''$ surface. The barrier on the $^3A''$ surface has also been calculated by Durant and Rohlfing, using the G2Q method.[17] They find a classical barrier height of 4.1 kcal/mole, in perfect agreement with the results of Walch and Rohlfing. The surface of Guadagnini et al. has barriers of 4.6 kcal/mole and 11.1 kcal/mole for these transition states.

The association reaction of H-atom with NO has been subject of experimental study for decades. In 1929 Smallwood[56] investigated the NO catalyzed recombination of H-atoms, which occurs when reaction 2.4 is followed by the fast reaction

$$H + HNO \rightarrow H_2 + NO \tag{2.9}$$

In 1933 Harteck[57] condensed product from the reaction of H-atoms with NO, and collected a solid of empirical formula HNO. The explosive nature of the adduct was demonstrated by multiple detonations in the course of the experimental study. In the '50s Cashion and Polanyi[58] and Clement and Ramsay[59] observed chemiluminescence arising from the reaction. Transitions were found to originate from both the electronically excited $^1A''$ state and vibrationally excited levels of the ground $^1A'$ state.

The presence of excited HNO A^1A'' products from the reaction of H with NO is interesting in light of theoretical predictions of a barrier for the adiabatic reaction forming HNO A^1A'' from H and NO. A closer look at the spectroscopy of HNO reveals the fact that the low lying states of HNO arc mixed. Most apparent is the rotational-level dependent predissociation in the HNO A^1A'' state.[26,60] This J-dependent predissociation has been analyzed by Dixon et al.[22] and found to arise from a electron orbital-rotational coupling with continuum levels of the X^1A' state. The variation in the predissociation threshold is linked to the variation in the height of the J-dependent centrifugal barrier. Their detailed analysis of this predissociation allows determination of D_0(H-NO) = 16450 ± 10 cm^{-1}. Dixon and Rosser[61] and Petersen[62] also investigated the analogous predissociation threshold in DNO, and found that it was shifted from the HNO value by the difference in HNO and DNO zero point energies.[61] This is the behavior to be expected for a dissociation whose J = 0 surface is barrierless. Additionally, the variation of the predissociation threshold with J is very sensitive to the local curvature of the surface in the region around the centrifugal barrier; existence of a J = 0 barrier would change the observed J-dependence of the threshold, and is inconsistent with the experimental observations.

Dixon and coworkers[22] also observed the onset of a rotation-independent predissociation mechanism between 860 and 970 cm^{-1} above the dissociation limit. They ascribed it to spin-orbit coupling with continuum levels in the a^3A''

state, and suggested that the barrier to HNO a^{3}A" formation was of this magnitude. The barriers calculated by Walch[33] and Durant and Rohlfing,[17] and Guadagnini et al.[27] are somewhat higher than Dixon et al.'s range of $2.5 \leq E \leq 2.8$ kcal/mole, but the difference is within expected error bounds.

There is also evidence for mixing of the states below the H + NO dissociation energy. There are numerous perturbations in the A^{1}A" state, and the lifetime of the A^{1}A" state is found to be level-dependent, with efficient collisional mixing of various levels.[63] The results of the MRD-CI calculations by Bruna[64] and the potential surfaces constructed by Dixon et al.[22] and Guadagnini et al.[27] show that the a^{3}A" HNO surface crosses the X^{1}A' HNO surface as the molecule is distorted toward linearity or toward HON, with the ^{3}A" state becoming the ground state for those geometries.

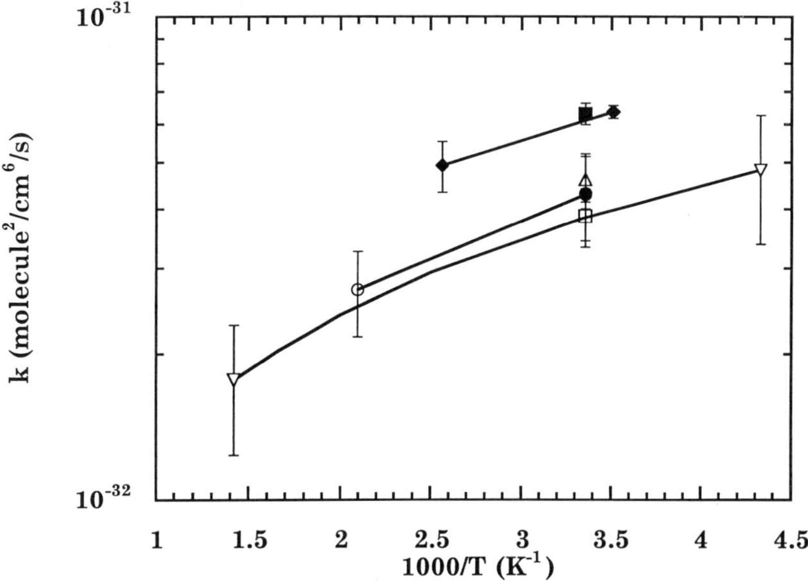

Figure 2.3: Experimentally measured rate coefficients for H + NO + H$_2$ → HNO + H$_2$ reaction. ∇ - ref. 65, □ - ref. 66, ■ - ref. 67, ◆ - ref. 68, ○ - ref. 69, ● - ref. 70, Δ - ref. 71.

The HNO (A→X) chemiluminescence, with its origin at ~766 nm, has been used by researchers to monitor H-atom concentrations in the course of the H + NO + M → HNO + M reaction. In the early '60s Clyne and Thrush carried out the first of these studies, using the discharge-flow technique to measure the low pressure recombination rate coefficient for the reaction with H$_2$ as the third body.[65,72] They measured the reaction rate coefficient from 231 to 704 K, and found that it exhibited the negative temperature dependence characteristic of recombination reactions, with their data being well represented by $k(H_2) = 3.83 \times 10^{-23}$ $(T/298)^{-0.9}$ $cm^6/molecule^2/s$. The only subsequent temperature dependent studies are due to Cvetanovic and coworkers[68,69] and were only for H_2 as the third body. They used a modulated, Hg-sensitized technique to measure the rate coefficients. In this technique H-atoms were formed by exciting Hg-atoms using a modulated mercury lamp, and then reacting the excited Hg-atoms with H_2. The phase and intensity of the HNO chemiluminescence was monitored with a lock-in amplifier. This information was used to derive the kinetic parameters. Cvetanovic and coworkers confirmed the decrease in reaction rate with increasing temperature, although their measured rate coefficients are slightly larger than that measured by Clyne and Thrush. The deuterium kinetic isotope effect has been measured by Hartley and Thrush,[73] who obtained a value of $k_H/k_D = 0.9 \pm 0.2$ with Ar as the third body, and by Ishikawa, Sugawara and Sato[71] who obtained a value of $k_H/k_D = 1.0 \pm 0.3$ with H_2 as the third body.

The relative efficiencies of H_2, He, Ar and Ne as third bodies in the recombination reaction was studied by Clyne and Thrush,[65] and later by Ahumada, Michael and Osborne[67] who added Kr to the list of third bodies. Hartley and Thrush[73] extended the study of third body efficiencies to a number of polyatomics. Unfortunately, all these workers restricted their work to room temperature, and did not examine the differing temperature dependencies for the relative third body efficiencies. A single study of the temperature dependence of N_2 was undertaken by Campell and Handy.[74] They examined the limited temperature range $360 \leq T \leq 425$ K, and derived a relative efficiency $k_{2.4}(N_2):k_{2.4}(Ar) = (1.64 \pm 0.12):1.0$. A collection of measured room temperature third body efficiencies relative to Ar can be found in Table 2.2. Table 2.2 also lists third body efficiencies corrected for relative collision velocities. With this correction, data for the rare gases behaves as expected, with the heavier rare gases being more efficient than the lighter ones. The same trend is seen in the data for polyatomic colliders; the exception, H_2O, probably owes its enhanced efficiency to its dipole moment.

Table 2.2. Relative third body efficiencies for H + NO + M → HNO + M; Ar = 1.0

ref.	Clyne and Thrush[65]		Hartley and Thrush[73]		Ahumada, et. al.[67]	
M	k/k_{Ar}	velocity corrected	k/k_{Ar}	velocity corrected	k/k_{Ar}	velocity corrected
H_2	1.7±0.3	0.19±0.04	1.9±0.3	0.20±0.03	1.60±0.09	0.18±0.01
He	0.8±0.2	0.15±0.04			1.10±0.04	0.23±0.01
Ne	0.8±0.2	0.6±0.1			0.50±0.03	0.35±0.02
Kr					1.40±0.09	1.81±0.11
CO_2			2.0±0.3	2.1±0.4		
N_2O			2.2±0.5	2.3±0.5		
SF_6			3.6±0.6	5.2±0.9		
H_2O			6.1±1.4	4.0±0.9		

Clyne and Thrush also measured the rate coefficient for production of HNO (A $\rightarrow$ X) chemiluminescence.[65] This was accomplished by first calibrating their detection system for absolute chemiluminescence detection efficiency. NO concentrations were obtained from flow rates, while H-atom concentrations were obtained by use of an isothermal calorimeter. The calorimetric determination of atom concentrations involves measurement of the heat released when atoms recombine on a suitable substrate, usually platinum foil or wire. These measurements sufficed to determine the chemiluminescent reaction rate coefficient, independent of the overall reaction rate coefficient. They found that the chemiluminescence rate coefficient was pressure independent, although it was dependent on the identity of the bath gas. They also determined the temperature dependence of the chemiluminescent channel, and found equally good fits to an Arrhenius form, with activation energy of -1.4 $\pm$ 0.3 kcal/mole, and a T^n form, with an n of -2.8 $\pm$ 0.4.

This rather striking behavior was explained by Clyne and Thrush using a simple reaction sequence:[65]

$$H + NO + M \rightarrow HNO\ (X) + M \tag{2.11}$$
$$H + NO + M \rightarrow HNO\ (A) + M \tag{2.12}$$
$$HNO\ (A) \rightarrow HNO\ (X) + h\nu \tag{2.13}$$
$$HNO\ (A) + M \rightarrow HNO\ (X) + M \tag{2.14}$$

By applying the steady-state assumption to [HNO (A)] they obtained an expression for the chemiluminescence intensity:

$$I = k_{2.13}[HNO\ (A)] = \{k_{2.12}k_{2.13}[M]/(k_{2.13} + k_{2.14}[M])\}[H][NO]$$

In the limit of $k_{2.13} \ll k_{2.14}[M]$, i.e. fluorescence quenching much faster than radiative decay, this expression becomes independent of pressure, but is still sensitive to the nature of the bath gas due to the M dependence of rates $k_{2.12}$ and $k_{2.14}$. The temperature dependence of the chemiluminescence also reflects temperature dependencies for both $k_{2.12}$ and $k_{2.14}$, and is thus not constrained to be equal to the temperature dependence of $k_{2.11}$. The radiative lifetime of HNO (A) has been measured to be τ = 22 - 29 μs,[75] and while the quenching rate with H_2 has not been measured, we would expect quenching to be at least competetive with radiative decay at Clyne and Thrush's pressures of 1-2 Torr.

Alternative mechanisms can be constructed in which the association of H and NO occurs only on the X^1A' surface, with HNO (A) formed by reverse internal conversion, and both X and A states undergoing collisional relaxation. For example, consider the scheme:

$$H + NO \leftrightarrow HNO\ (X)^{\dagger} \tag{2.15}$$
$$HNO\ (X)^{\dagger} \leftrightarrow HNO\ (A)^{\dagger} \tag{2.16}$$
$$HNO\ (X)^{\dagger} + M \rightarrow HNO\ (X) + M \tag{2.17}$$
$$HNO\ (A)^{\dagger} + M \rightarrow HNO\ (A) + M \tag{2.18}$$
$$HNO\ (A) + M \rightarrow HNO\ (X) + M \tag{2.19}$$
$$HNO\ (A) \rightarrow HNO\ (X) + h\nu \tag{2.20}$$

where $\dagger$ denotes vibrationally excited species. We can again solve for [HNO (A)] by use of the steady-state assumption, and obtain, in the low-pressure limit (which is appropriate for pressures below many atmospheres)

$$I = k_{2.20}[HNO\ (A)] = \{K_{2.15}K_{2.16}k_{2.18}k_{2.20}[M]/(k_{2.20} + k_{2.19}[M])\}[H][NO]$$

where the K's are equilibrium constants. Like Clyne and Thrush's equation, this equation ceases to be pressure dependent when quenching becomes dominant, retains the dependence on bath gas identity, and can have a temperature dependence different from that of the overall recombination reaction.

There has been one measurement of the limiting high pressure rate constant for the H + NO $\rightarrow$ HNO association, by Forte, who measured a rate of $k_\infty = (8 \pm 3)$ x 10^{-10} cm^3/molecule/s at pressures above 400 atm in N_2.[76]

2.4. H + NO (+M) $\leftrightarrow$ HON (+M)

H-atom can also add to the O-atom of NO to form HON. On the $^1A'$ surface this addition involves the formal promotion of an electron from O to N, and is expected to proceed with a barrier; the reactions on the $^{1,3}A''$ surfaces also involve breaking an NO π bond, and are also expected to proceed with barriers.[33] This analysis is borne out by the detailed calculations of Guadagnini et al.,[27] who find barriers of 3.9, 13.5 and 25.6 kcal/mole for the $^1A'$, $^3A''$ and $^1A''$ surfaces, respectively. Pauzat et al.[77] have also calculated the barriers on the $^1A'$ and $^3A''$ surfaces, and calculate values of 6.7 and 13.0 kcal/mole, using the SOCI method. Thus, one does not expect that direct formation of HON will be important in the recombination of H and NO. However, the lowest energy product channel in the H + NO reaction is N + OH. This channel is obviously not directly accessible from an HNO intermediate. In fact the ground state N ^5S and OH $^2\Pi$ products do not correlate with the singlet HNO/HON surfaces. However, they do correlate with the ground state $^3A''$ state of HON. One therefore expects that the isomerization process, reaction 2.8, may be important in the H + NO $\rightarrow$ N + OH reaction. Alternatively, the reaction may proceed directly through the $^3A'$ surface, although reaction on that surface presumably involves a barrier.

2.5. NH + O (+M) $\leftrightarrow$ HNO (+M) and N + OH (+M) $\leftrightarrow$ HON (+M)

The association reactions of NH with O and N with OH are very exothermic and possess thermodynamically open product channels. We therefore expect that they will have very short adduct lifetimes, and will not be stabilized under normal conditions. For example, Huang[39] calculated an RRK lifetime of 10^{-14} s for the adduct arising from NH adding to O, which would require pressures of $\sim 10^4$ bar before stabilization would be observable. We therefore expect these reactions to go on to yield bimolecular products, as discussed in 2.2 and 2.3.

2.6. HNO (+M) ↔ HON (+M)

The isomerization of HNO to HON is presumably a necessary step in reaction 2.3, assuming that the reaction goes via a stable intermediate. There is no experimental data on this isomerization. The most recent theoretical calculations[27] suggest that all the barriers to isomerization from HNO lie above the barriers to dissociation back to H plus NO, while for HON the barrier to isomerization on the 3A" surface is the only one which is below the barriers for dissociation to H plus NO. We thus expect that the major loss pathways for HNO and HON will be dissociation rather than isomerization.

3. THE H_2NO SYSTEM

We next consider the reaction of NH_2 with O, and, in general, reactions on the H_2NO potential energy surface.

3.1. The H_2NO Potential Energy Surface

Properties of dihydroxy nitrosyl radical, H_2NO, and its isomers have been the focus of several ab initio electronic structure studies.[77,80-91] The H_2NO radical is the simplest member of a series of nitrosyl-radical spin labels, R_1R_2NO. As such, accurate prediction of its geometry and ESR spectrum is important for a full understanding of the results of ESR investigations of more complex spin-labeled systems.[77,86,87] A common thread through various studies has been the concern about the planarity, or non-planarity, of H_2NO. Komaromi and

Table 3.1: Comparison of H_2NO Geometries

r_{NH}	r_{NO}	$\angle HNH$	$\angle HNOH$	Method	reference
0.99[b]	1.34	122.3[b]	180.0[b]	LMR	Davis et al.[78]
1.010 ± 0.010[b]	1.280 ± 0.004	122.7 ± 2.2	180.0	Microwave spectroscopy	Mikami et al.[79]
1.014	1.265	119.5	180.0	CISD/ cc-pVDZ	Soto et al.[80]
1.014	1.283	121.6	180.0	MRSDCI/ DZ + P	Cai [81]
1.013	1.291	120.3	180.0	QCISD/ D95(d,p)	Barone et al.[82]
1.037	1.298	114.0	180.0	CASSCF/VDZP	Walch[83]
1.015	1.268	119.4	180.0	CCD-ST/ 6-311+G(df,pd)	Komaromi and Tronchet[84]
1.017	1.273	119.4	180.0	G2	Yang et al.[85]

a) Bond lengths in Å; angles in degrees.
b) Fixed

Tronchet[84] have calculated geometries for H_2NO using upwards of 60 combinations of basis sets and methods, from HF/6-31G to CCD-ST/6-311+G(df,pd). Their results are representative of other investigators,[80,82,83,85,89] who find that the molecule has an almost flat bending potential out to about 30° from planarity, at which point the potential energy begins rapidly rising. There is consensus that the height of the inversion barrier is below the zero point energy in the bend.

There have been two experimental investigations of the H_2NO geometry, an LMR study by Davis, et al.,[78] and a microwave study by Mikami, Saito and Yamamoto.[79] Neither study allowed for a complete determination of the H_2NO geometry, and the data were analyzed using input from early theoretical calculations. Both experimental studies were able to establish that the H_2NO was planar. Their results are in reasonable agreement with recent theoretical calculations, as shown in Table 3.1.

Various stationary points on the H_2NO $^2A''$ potential energy surface have been characterized by Soto and Page,[80] Page, Soto and McKee,[92] Walch[83] and Yang, Koszykowski and Durant.[85] Table 3.2 lists the energies, in kcal/mole relative to the H + HNO asymptote, found by the various groups. The agreement between the various studies points out again the quality possible in modern electronic structure calculations. The potential energy surface is sketched out in Figure 3.1. It shows that the reaction of NH_2 with O is a highly exothermic, barrierless recombination producing an energized H_2NO species. This species can either fragment to products, or isomerize to HNOH. HNOH exists in both cis and trans conformations, which are linked by a low isomerization barrier. These species can fragment to H + HNO, or to NH + OH, or reisomerize to H_2NO.

Table 3.2. Comparison of Calculated H_2NO Energetics (kcal/mole, relative to H + HNO)

	Soto[80]	Page[92]	Walch[83]	Yang[85]	Experiment
NO + H_2	-55.7	--	--	-57.1	-56.2
NH + OH	--	--	--	18.4	16.9
H_2NO (planar)	-58.8	--	-57.2	-59.8	--
($H_2NO \rightarrow H_2 + NO$) ts	--	--	--	1.9	--
($H_2NO \rightarrow H + HNO$ (X)) ts	--	2.7	2.3	1.2	--
($H_2NO \rightarrow$ trans-HNOH) ts	--	--	-9.2	-8.3	--
trans-HNOH	-53.0	--	-52.1	-52.7	--
(trans-HNOH $\rightarrow$ H + HNO (X)) ts	--	9.0	8.3	7.6	--
(trans-HNOH $\rightarrow$ cis-HNOH) ts	--	--	--	-38.9	--
cis-HNOH	-47.9	--	-46.9	-47.5	--
(cis-HNOH $\rightarrow$ H + HNO (X)) ts	--	--	--	6.3	--

Lifshitz, et al. used tandem mass spectroscopy to study H_2NO and HNOH.[93] This work established the existence of stable H_2NO and HNOH isomers, and indicated that there was a substantial barrier to their interconversion, consistent with the results of Yang et al.[85]

Cai[81] also calculated the energies of the first two excited doublet states of H_2NO, which he placed 49.6 and 96.0 kcal/mole above the ground state. The lower of these, a $^2B_2$ state (in C_{2v} symmetry) correlates with reactants and is energetically accessible to $NH_2 + O$ reactants; its role in the overall reaction has not been examined in detail.

The last two sets of surfaces correlating to $NH_2 + O$ are two quartet surfaces, which connect $NH_2 + O$ with $NH + OH$ via a direct H-atom abstraction. The transition state on the lower of these surfaces has been characterized by Yang et al.,[85] they find a 6.7 kcal/mole barrier for the reaction.

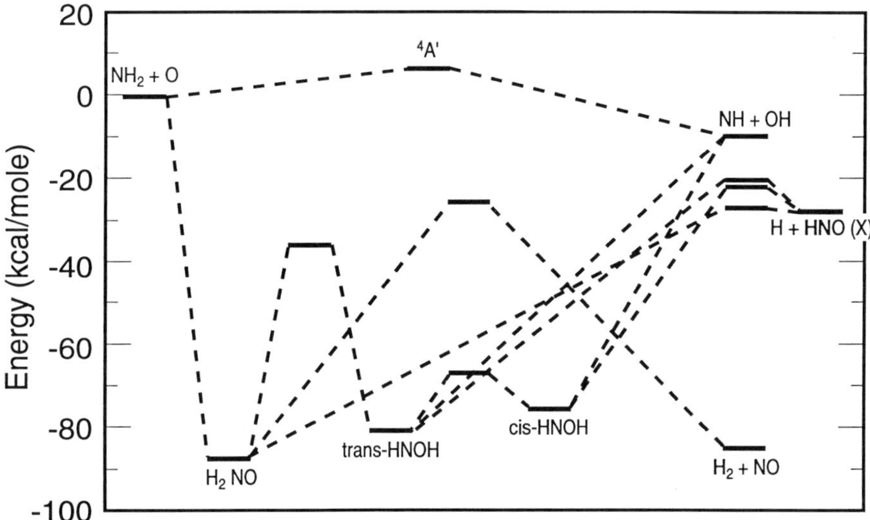

Figure 3.1: Potential energy surface for $^2A''$ and $^4A'$ states of H_2NO. Energies from G2/G2Q calculations of Yang, et al.[85]

3.2. The $NH_2 + O \rightarrow$ Products Reaction

There are four exothermic channels in the reaction of NH_2 with O:

$$NH_2 + O \rightarrow H + HNO \qquad \Delta H_{rxn} = -48.7 \text{ kcal/mole} \quad (3.1a)$$
$$\rightarrow NH + OH \qquad \Delta H_{rxn} = -10.2 \text{ kcal/mole} \quad (3.1b)$$
$$\rightarrow H_2 + NO \qquad \Delta H_{rxn} = -83.3 \text{ kcal/mole} \quad (3.1c)$$
$$\rightarrow H_2O + N \qquad \Delta H_{rxn} = -49.4 \text{ kcal/mole} \quad (3.1d)$$

There are three reported measurements of the overall reaction rate, $k_{3.1}$. Gehring and coworkers[94] used flash photolysis in a discharge-flow reactor to measure the reaction rate coefficient, obtaining $k_{3.1} = 3.5$ x 10^{-12} cm^3/molecule/s at 298 K. Dransfeld and coworkers[95] used the discharge-flow technique to study the reaction, and obtained a much larger rate coefficient, $k_{3.1} = (8.8 \pm 2.5)$ x 10^{-11} cm^3/molecule/s at 296 K. They also measured the channel-specific rates, finding $k_{3.1a} = (7.6 \pm 2.0)$ x 10^{-11} cm^3/molecule/s and $k_{3.1b} = (1.1 \pm 0.5)$ x 10^{-11} cm^3/molecule/s. They were unable to detect H$_2$O in their experiments, ruling out channel 3.1d, and from the time profiles of H$_2$ and NO products were also able to rule out channel 3.1c. Recently, Adamson and coworkers[41] used infrared kinetic spectroscopy to study the reaction. Their measurement of the overall reaction rate coefficient, $k_{3.1} = (6.5 \pm 1.3)$ x 10^{-11} cm^3/molecule/s, is in good agreement with Dransfeld, et al's rate. Adamson and coworkers established that $\Phi_{3.1b} <$ 8%, somewhat smaller, but still consistent, with Dransfeld, et al's results.

Dagdigian and coworkers have carried out a study of the HNO product from the NH$_2$ + O reaction,[96] and the OD product from the ND$_2$ + O reaction.[97] They found that the products were much colder rotationally than predicted statistically. They suggested that there are barriers present in both product channels. However, both G2Q[85] and MCSCF/CCI[92] calculations do not find significant barriers in these product channels. Recent work probing nascent rotational populations of fragments from the barrierless decomposition of H$_2$O$_2$,[98] HNO$_3$,[99] and CH$_2$CO[100] find that, for excess energies more than a few kcal/mole above thresholds, the product rotational distributions are considerably colder than phase space predictions. This suggests that rotationally cold fragments do not require the presence of barriers in the dissociation channel.

3.3. Other Reactions

Of the other reactions in the H$_2$NO system, only the reaction of H with HNO,
$$H + HNO \rightarrow products \tag{3.2}$$
has been directly measured. Dodonov and coworkers used the discharge-flow technique to obtain values of $k_{3.2} = (1.2 \pm 0.5)$ x 10^{-10} cm^3/molecule/s[101] and $k_{3.2} = (1.3 \pm 0.5)$ x 10^{-10} cm^3/molecule/s[102] at 295 K. This rate is consistent with work by Washida and coworkers,[103] who reported a lower limit of $k_{3.2} > 1.6$ x 10^{-12} cm^3/molecule/s at room temperature.

4. THE HNNO SYSTEM

The reactions of NH with NO and H with N$_2$O both access the HNNO potential energy surface. The NH + NO $\rightarrow$ N$_2$ + OH channel has been postulated as a source of OH in DeNOx systems.[104] Additionally, it has been claimed that the NH + NO $\rightarrow$ N$_2$O + H channel is a principle source of N$_2$O in combustion.[7] Of course other flame studies have suggested that the reaction produces very little

N_2O.[105] The $H + N_2O \rightarrow N_2 + OH$ reaction is one of the reactions converting N_2O, an undesirable combustion product, to N_2. The potential energy surface of HNNO is one of the better studied polyatomic potential energy surfaces, with a variety of theoretical calculations which have been checked by comparison to experiments.

The reaction of NH with NO has two thermodynamically open paths leading to ground state products:

$$NH + NO \quad \rightarrow H + N_2O \qquad \Delta H_{rxn} = -34.7 \text{ kcal/mole} \qquad (4.1a)$$
$$\rightarrow OH + N_2 \qquad \Delta H_{rxn} = -97.6 \text{ kcal/mole} \qquad (4.1b)$$

The reaction of H with N_2O has only one thermodynamically open path, although hot-atom studies have observed the endothermic channels (4.2b) and (4.2c):

$$H + N_2O \quad \rightarrow OH + N_2 \qquad \Delta H_{rxn} = -62.9 \text{ kcal/mole} \qquad (4.2a)$$
$$\rightarrow OH (A) + N_2 \qquad \Delta H_{rxn} = 29.7 \text{ kcal/mole} \qquad (4.2b)$$
$$\rightarrow NH + NO \qquad \Delta H_{rxn} = 34.7 \text{ kcal/mole} \qquad (4.2c).$$

4.1 HNNO Potential Energy Surface

Various portions of the HNNO potential energy surface have been calculated by a number of investigators.[106,108-112] The addition of NH $^3\Sigma$ to NO $^2\Pi$ gives rise to 12 electronic surfaces. In C_1 symmetry these are 2 doubly-degenerate ^2A surfaces and 2 4-fold-degenerate ^4A surfaces. The thermodynamically accessible products are N_2 $^1\Sigma$ + OH $^2\Pi$ and N_2O $^1\Sigma$ + H ^2S. These products will not correlate to the quartet states, and reaction to these products must take place on the doublet surfaces or violate spin conservation. Fueno et al.[109] have characterized the cis and trans HNNO isomers on both doublet surfaces and find them to be planar, of

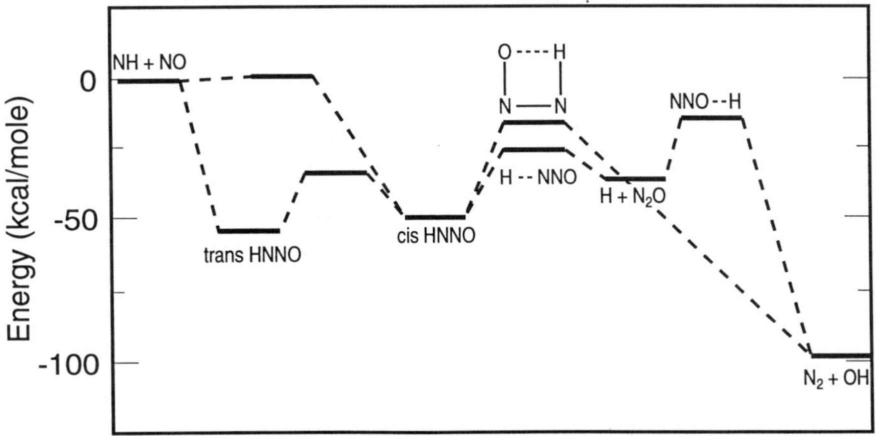

Figure 4.1: Potential energy surface of HNNO 2A' surface.[106,107]

A' and A" symmetry. Further calculations of the stationary points on the HNNO ground state surface predict that all of the stationary points are planar,[106,108-110,112] in which case the two ^2A states become a ground $^2A'$ surface and an excited $^2A''$ surface. Walch has considered the details of the correlations of the $^2A'$ and $^2A''$ surfaces, and has concluded that the $^2A''$ surface is not reactive for NH + NO reactants; the H + N_2O asymptote does not correlate with it, and reaching the N_2 + OH asymptote requires energetically unfavorable electron rearrangements.[112] As a result, the potential energy surface of interest is just that of the ground, $^2A'$ state. Results of the G2/G2Q calculation of the stationary points on the surface[106] are shown in Figure 4.1. Results of calculations for stationary points on the HNNO X^2A' potential energy surface are collected in Tables 4.1 and 4.2.

Table 4.1: Energetics of stable HNNO species[a]

cis-HNNO	trans-HNNO	N_2+OH	N_2O+H	reference
-48.9	-56.0	-96.9	-34.5	Durant[106]
-46.2	-51.5		-31.7	Walch[112]
-51.9	-48.8	-99.1	-37.6	Miller[111]
-30.9		-90.8	-28.9	Harrison[110, b]
-29.5		-91.6	-20.5	Harrison[110, c]
-60.3	-60.5	-92.4	-30.9	Fueno[109]
-50.2	-47.3	-97.5	-31.4	Melius[108]
-		-97.6	-34.7	experiment

a) all energies in kcal/mole, relative to NH + NO, 0 K
b) MP4/6-31G(d)//MP2/6-31G(d)
c) MP4/6-311G(d,p)//HF/6-311G(d,p)

Table 4.2: Energetics of HNNO transition-state species[a]

cis HN-NO	trans HN-NO	N_2-OH	H-N_2O	cis-trans ts	reference
2.9	0.1	-17.9	-25.3	-30.8	Durant[106]
6.3	3.2	-15.4	-21.4		Walch[112]
		-22.4	-28.2		Miller[111]
		3.0	-6.9		Harrison[110, b]
		3.6	-3.8		Harrison[110, c]
		-9.0	-12.2	-28.7	Fueno[109]
		-17.3	-27.7		Melius[108]

a) all energies in kcal/mole, relative to NH + NO, 0 K
b) MP4/6-31G(d)//MP2/6-31G(d)
c) MP4/6-311G(d,p)//HF/6-311G(d,p)

We turn our attention to a comparison of the recent calculations of Durant[106] with those of Walch.[112] The relative energies of the various stationary points from the two calculations are very close, with the only large error being in the location of the NH + NO asymptote. (Unfortunately, this is the reference point chosen by both authors, so the derived energetics relative to NH + NO, shown in Tables 4.1 and 4.2, systematically disagree by 2.5 - 4.5 kcal/mole.) The NH + NO → H + N_2O reaction represents a non-isodesmic process, meaning that there is a change in the number and type of bonds in the process. Traditionally it is harder to calculate accurate energies for non-isodesmic processes than for isodesmic processes, although the results of the G2 method of Pople and coworkers seems to be fairly accurate for both processes. The location of the N_2 + OH and H + N_2O asymptotes relative to the NH + NO asymptote can be calculated from experimental heats of formation, and indicates that the G2 surface of Durant is within 1 kcal/mole of the experimentally known energetics for asymptotic reactants and products. If Walch's surface is moved up to reproduce the experimentally known NH + NO → N_2O + H exothermicity then the two surfaces agree to within 1 kcal/mole, except for the trans-HNNO isomer, where the error increases to 1.5 kcal/mole. This suggests that these surfaces be considered as accurate to within 1.5 kcal/mole, and the other surfaces can be judged relative to them.

Experimental calibration of the HNNO potential energy surface has proceeded on several fronts. Most interesting would be the comparison of the properties calculated for the intermediate HNNO species with those experimentally determined. Unfortunately, there is no experimental thermochemistry or spectroscopic data for the intermediates. Kinetic information on the HNNO system includes measurements of the reaction rate coefficients for the NH + NO → Products and the H + N_2O → Products reactions, as well as measurement of the product branching fraction for the reaction of NH with NO. There are also a number of dynamics studies, where nascent product state distributions have been measured. These studies each contribute some knowledge concerning features of the potential energy surface.

4.2. NH + NO → Products

The reaction rate coefficient of NH with NO has been studied by a number of workers at temperatures from ambient[42,104,113-116] to shock tube and flame temperatures,[43,105,117-121] with recent work by Wolfrum and coworkers covering the range $293 \leq T \leq 1085$ K.[122] The directly measured rates are shown in Figure 4.2. There is general agreement as to the room temperature rate coefficient, with a weighted mean of $k_{4.1} = (5.2 \pm 0.2) \times 10^{-11}$ cm^3/molecule/s (1 σ). The recent shock tube measurements[43,119] are in relatively good agreement that $k_{4.1} \approx 1 - 4 \times 10^{-11}$ cm^3/molecule/s for $2000 \leq T \leq 3500$. The earlier study of Roose, Hanson and Kruger[118] relied on extensive modeling of shock heated $NO/NH_3/H_2/Ar$ mixtures; the disagreement between their results and the recent more direct determinations reflects on the quality of their model. Other temperature dependent studies include work of Harrison, Whyte and Phillips,[104] who report a temperature independent rate of $k_{4.1} = (5.8 \pm 0.6) \times 10^{-11}$ cm^3/molecule/s for $269 \leq$ T ≤ 377 K, and the results of Wolfrum, et al.,[122] who report a three parameter fit of $k_{4.1} = 2.5 \times 10^{-10}$ (T/298)$^{(-1.60 \pm 0.01)}$ exp(-0.91 $\pm$ 0.05 kcal/mole/RT) cm^3/molecule/s.

These results suggests that the reaction has a slight negative temperature dependence, as we would expect for a radical-radical association reaction occurring without a potential energy barrier. This result is consistent with the entrance channel of the potential energy surface calculated by Durant,[106] as well as a recalibration of the calculations of Walch,[112] as discussed above. The shock tube study of Mertens, et al.[43] is interesting, in that their results suggested that the reaction has a slight positive temperature dependence in the 2000 $\leq$ T $<$ 3000K range. Their experiments, utilizing shock heated HNCO/NO/Ar mixtures and cw, narrow-linewidth laser absorption monitoring of NH, are fairly direct, and although the observed increase in rate coefficient with temperature is within quoted uncertainties, it does appear real. One possible explanation of this is the possible opening of an additional channel in the reaction, possibly the endothermic production of $N_2H + O$ or the exothermic production of $N_2 + OH$ A$^2\Sigma$.

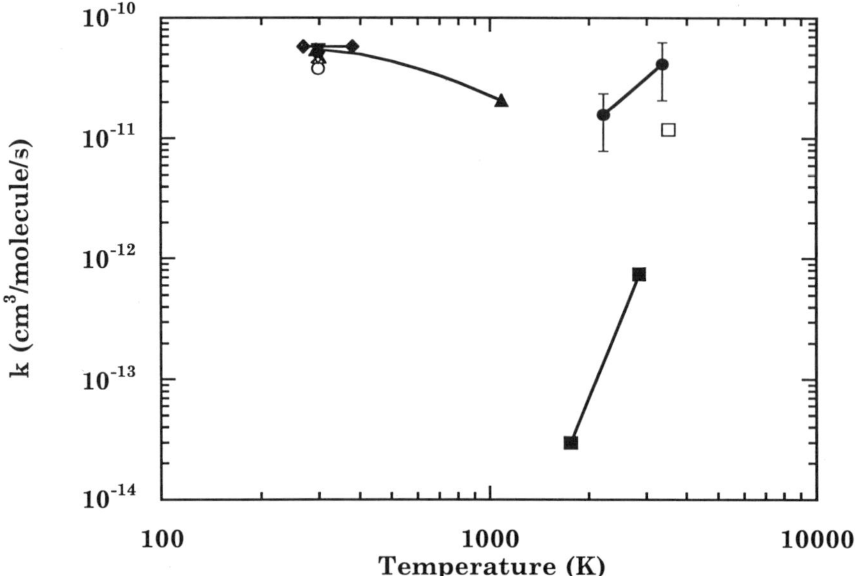

Figure 4.2: Experimentally measured rate coefficients for NH + NO $\rightarrow$ Products. O - ref. 113, ∇ - ref. 114, $\blacksquare$ - ref. 118 $\blacklozenge$ - ref. 104, $\bullet$ - ref. 43, $\triangle$ - ref. 116, $\square$ - ref. 119, X ref. 123, $\blacktriangle$ ref. 122, $\blacktriangledown$ ref. 42.

More work is clearly indicated here.

The product branching fraction for the reaction of NH with NO has been measured at room temperature by Matsui and coworkers[116] Yokoyami, et al.[119], Durant[106], Wolfrum and coworkers[122] and remeasured by Matsui and coworkers.[123] Matsui and coworkers used photolysis of HNCO to produce NH $a^1\Delta$. This was reacted with NO; a portion of the NH $a^1\Delta$ was quenched to NH $X^3\Sigma$, and a portion reacted, producing $N_2O + H$ or $N_2 + OH$ products. The NH $X^3\Sigma$ subsequently reacted with the NO, producing $N_2O + H$ and $N_2 + OH$ products. In their original study they observed OH profiles with growth rates different from the NH $a^1\Delta$ decay rates, and they assigned this OH to products of the reaction of NH $X^3\Sigma$. In their reinvestigation they added variable amounts of Xe, which efficiently quenches $NH\,a^1\Delta$, and were able to more cleanly separate contributions due to reaction of excited and ground state NH. They identified the growth rate in their observed OH profiles as arising from vibrational relaxation of OH, not reaction of NH $X^3\Sigma$, and concluded that the OH was produced by reactions of both NH electronic states, revising the interpretation in their earlier work. After careful calibration of their system they obtained a revised branching fraction of $k_{4.1a}/k_{4.1} = 0.80 \pm 0.18$. Yokoyami, et al., as part of their shock tube study of this reaction, also measured $k_{4.1a}/k_{4.1} = 0.81$ at 300 K. Durant used the discharge flow technique

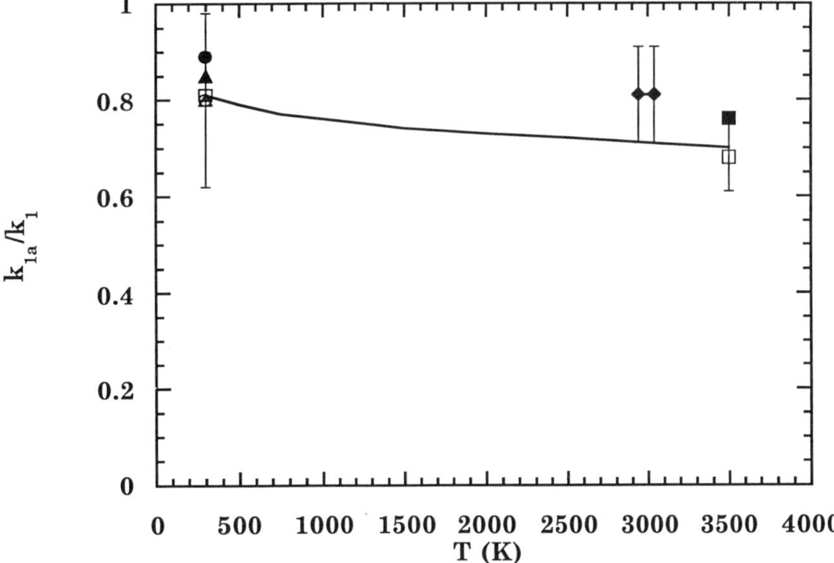

Figure 4.3: Product branching fraction for NH + NO $\rightarrow$ H + N$_2$O reaction. $\blacklozenge$ - ref. [124], $\square$ - ref. 119 (experiment), $\blacksquare$ - ref. 119 (theory), —— - ref. [111] $\bigcirc$ - ref. [106] (experimental), $\bullet$ - ref. [106] (theory), $\triangle$ - ref. 123, $\blacktriangle$ ref. 122.

with mass spectrometric detection of reaction products to determine the product branching fraction at room temperature. NH $X^3\Sigma$ was formed by reaction of NH_3 with excess F.[125,126] It was then reacted with excess $^{15}N^{18}O$, and the product $^{14}N^{15}N$ and $^{14}N^{15}N^{18}O$ were measured mass spectrometrically. After correction for $^{14}N^{15}N$ produced by reaction of residual NH_2 from the NH source he obtained values of $k_{4.1a}/k_{4.1} = 0.8 \pm 0.4$ for NH reactant, and $k_{4.1a}/k_{4.1} = 0.87 \pm 0.17$ for ND reactant. Wolfrum and coworkers also used the laser photolysis/LIF technique, photolyzing HNCO and using excess Xe to relax the NH $a^1\Delta$ to NH $X^3\Sigma$. They measured a room temperature branching fraction of $k_{4.1a}/k_{4.1} = 0.85 \pm 0.05$.

The product branching fraction for the reaction of NH with NO has also been measured in shock tube experiments by Yokoyami, et al.[119] and by Mertens, et al.[43] Both groups pyrolyzed mixtures of HNCO/NO/Ar and monitored NH and OH concentration profiles. Mertens, et al. operated in the temperature range from $2200 \leq T \leq 3350$ K, and used cw narrow-linewidth laser absorption to monitor the NH and OH. They obtained a branching fraction over the temperature range 2940 $\leq T \leq 3040$ K of $k_{4.1b}/k_{4.1} = 0.19 \pm 0.10$. Yokoyami, et al. used fluorescence from thermally excited OH $A^2\Sigma$ and NH $A^3\Pi$ to monitor ground state OH and NH populations. They derived a value for the other channel of $k_{4.1b}/k_{4.1} = 0.32 \pm 0.07$ at T = 3500 K. Thus one sees a gradual decrease in the importance of the H + N_2O channel, which is what we would expect to see given the ordering of the theoretically calculated barriers, i.e. E(NN-OH transition state) > E(H-NNO transition state). As the temperature increases the importance of the barrier heights for the two processes becomes less and less important, until at high temperature the product branching fraction should just reflect the relative Arrhenius A-factors for the two channels.

Yokoyami, et al.[119] also observed an intense OH (A $\rightarrow$ X) emission feature at short time in their experiments, which they attributed to a third product channel:

$$\text{NH} + \text{NO} \ \rightarrow \ \text{OH} (A^2\Pi) + N_2 \qquad \Delta H_{rxn} = \text{-5.0 kcal/mole} \qquad (4.1c)$$

They calculated an upper bound of $k_{4.1c}/k_{4.1} \leq 0.01$. This is the product channel expected for reaction on the excited, A^2A'' HNNO surface.[119] A final product channel which has been suggested at high temperature is the endothermic formation of the transient species N_2H:[127]

$$\text{NH} + \text{NO} \ \rightarrow \ N_2H + O \qquad \Delta H_{rxn} = \text{6.8 kcal/mole} \qquad (4.1d)$$

The opening up of these product channels at high temperature could give rise to the suggested positive temperature dependence of the rate coefficient observed by Mertens, et al.[111], and will also lead to a decrease in the branching fraction $k_{4.1a}/k_{4.1}$.

There have been calculations of the product branching fractions by Miller, using both a 1993 BAC/MP4 surface[111] and using the G2 surface of Durant[106] and by Yokoyami, et al.[119] using the MRDCI surface of Fueno, et al.[109] Miller calculated

$k_{4.1b}/(k_{4.1a}+k_{4.1b}) = 0.19$ at 300 K, increasing to 0.30 at 3500 K. Yokoyami obtained a value of $k_{4.1b}/(k_{4.1a}+k_{4.1b}) = 0.24$ at 3500 K. These values are plotted, together with the experimentally measured product branching fractions, in Figure 4.3. While the agreement between theory and experiment is not exact, it is reasonably good considering the experimental uncertainties in the measurements. The degree of agreement suggests that the theoretically calculated potential energy surfaces do a reasonable job of capturing the important features of the true surface, and that the measured product branching fraction is not overly sensitive to barrier heights (certainly not relative to the sensitivity of reaction rate coefficients on barrier heights!).

The measured product branching fractions confirm that the reaction of NH with NO will be a source of N_2O. It also establishes that it is unlikely to be the source of the OH product measured in several investigations of the $NH_2 + NO$ reaction.

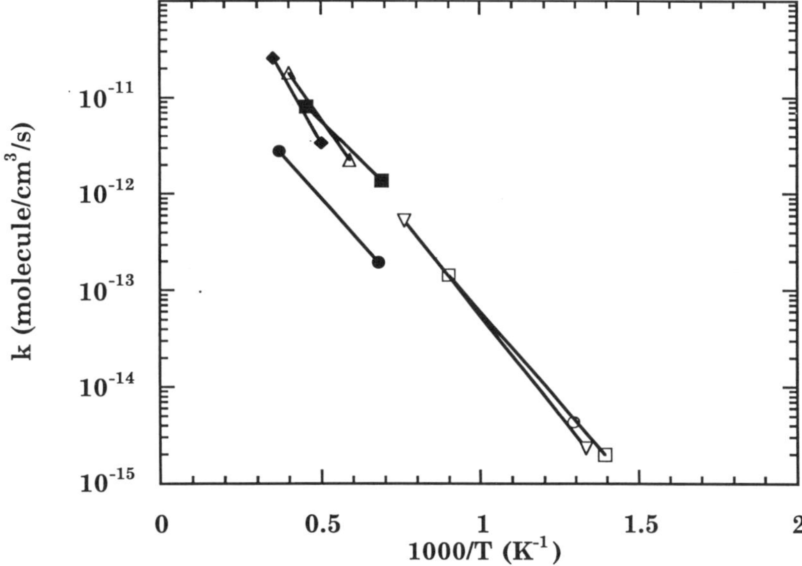

Figure 4.4: Experimentally measured rate coefficients for $H + N_2O \rightarrow$ Products. $\bigcirc$ - ref. [128], $\square$ - ref. [129], $\blacklozenge$ - ref. [130], $\bullet$ - ref. [131], $\triangle$ - ref. [132], $\blacksquare$ - ref. [133], $\triangledown$ - ref. [134].

4.3. $H + N_2O \rightarrow$ Products

As mentioned above, the HNNO potential energy surface can also be accessed from $H + N_2O$ reactants. This reaction has two channels: attack at the O-atom, which results in direct abstraction of the O, forming $N_2 + OH$ products; and attack at the terminal N-atom, which forms the HNNO intermediate before proceeding on to products. The reaction has been studied by a number of groups during the last 30 years,[135,136 128-134,137,138] and there is general agreement between the studies above ~700 K, as is evident in Figure 4.4. If we fit all of the direct determinations above 700K (with the exception of the somewhat low results of Glass and Quy[131] we obtain a consensus rate of $k_{4.2}$ = 5.3 x 10^{-10} exp(-18.1 kcal/mole/RT) cm^3/molecule/s.

We can use this activation energy to calculate a barrier height for each channel, assuming that it is the dominant channel, by first recognizing that[139]

$$E_a = E_0 + RT + <\Delta C_v^+>T.$$

We further recognize that

$$<\Delta C_v^+>T = [E(T)-E(0)]_{\text{transition state}} - [E(T)-E(0))]_{\text{reactants}}.$$

Calculation of the internal energy at temperature T for the reactants and transition state is straightforward, given knowledge of the frequencies of the species involved.[140] The frequencies of the transition state for OH elimination from HNNO[106] and for direct attack on the O-atom[107] have been calculated using the G2Q method. The midpoint, in terms of 1/T, of the interval $700 \leq T \leq 2850$ K is 1100 K, and we present the calculation of $<\Delta C_v^+>T$ for T=1100 K in Table 4.3.

From the experimental activation energy of 18.1 kcal/mole, the calculated values of $<\Delta C_v^+>T$, and knowing that RT = 2.2 kcal/mole at 1100 K we obtain values of E_0 = 16.5 kcal/mole for the barrier height for NN-OH reached via the HNNO intermediate, and E_0 = 16.2 kcal/mole for the barrier height for NN-OH reached via direct attack on the O-atom. These values are in good agreement with those calculated by the G2Q method, E_0 = 16.6 kcal/mole for the barrier height for NN-OH reached via the HNNO intermediate, and E_0 = 16.9 kcal/mole for the barrier height for NN-OH reached via direct attack on the O-atom. This suggests that both channels contribute to the thermal reaction.

The only measurements of the rate coefficient for reaction 4.2 below 700 are due to Fontijn and coworkers.[134,138] They found that their Arrhenius plots were strongly curved below T = 700 K, as shown in Figure 4.5. Their observed rates were best fit with a sum of Arrhenius rates. Their resulting rate expressions were:[138]

$$k(H + N_2O) = 4.2 \times 10^{-14} \exp(-4.55/RT) + 3.7 \times 10^{-10} \exp(-16.7/RT)$$
$$k(D + N_2O) = 3.5 \times 10^{-13} \exp(-7.15/RT) + 5.3 \times 10^{-10} \exp(-18.2/RT)$$

The kinetic isotope effect also appears to be the result of a sum of processes, decreasing sharply from a value of 2.2 at 450 K to ~1.3 at ~700 K, and remaining approximately constant to 1200 K. This was taken as a clear sign of a tunneling dominated reaction at temperatures below 700 K. The system was successfully modeled using a BAC-MP4 surface, assuming that the low temperature behavior corresponded to addition to form HNNO, followed by tunneling to form N_2 + OH.[134,138] Limitations in the accuracy of the BAC-MP4 energies for the various

Table 4.3: Calculation of $\langle \Delta C_v^+ \rangle T$

species		E-E$_0$ (kcal/mole)	frequency (cm^{-1})	E-E$_0$ (kcal/mole)
H				
	translational	3.28		
	total	3.28		
N$_2$O				
	translational	3.28		
	rotational	2.19		
	vibrational	3.71		
			2633	0.25
			1392	0.77
			689	1.35
			689	1.35
	total	9.18		
NN-OH transition state via HNNO				
	translational	3.28		
	rotational	3.28		
	vibrational	5.31		
			1765	0.56
			1280	0.84
			958	1.10
			805	1.23
			471	1.58
	total	11.87		
NN-OH transition state via direct attack				
	translational	3.28		
	rotational	3.28		
	vibrational	5.56		
			2146	0.39
			1192	0.91
			836	1.20
			567	1.47
			475	1.58
	total	12.12		
$\langle \Delta C_v^+ \rangle T$ (H + N$_2$O $\rightarrow$ NN-OH via HNNO) =				-0.6 kcal/mole
$\langle \Delta C_v^+ \rangle T$ (H + N$_2$O $\rightarrow$ NN-OH via direct attack) =				-0.3 kcal/mole

barriers precluded definitive assignment of the microscopic mechanism for the high temperature reaction, although the BAC-MP4 surface places the barrier to the direct NNOH transition state 3 kcal/mole above that for the indirect NNOH transition state, which lead to a prediction that the indirect channel dominated reactivity for T < 2000 K.[134] If one instead uses the energetics calculated using G2Q[106,107] one would expect that the activation energies of the two processes would be equal, and the branching at high temperature into the two channels would be solely determined by the ratio of the Arrhenius A-factors.

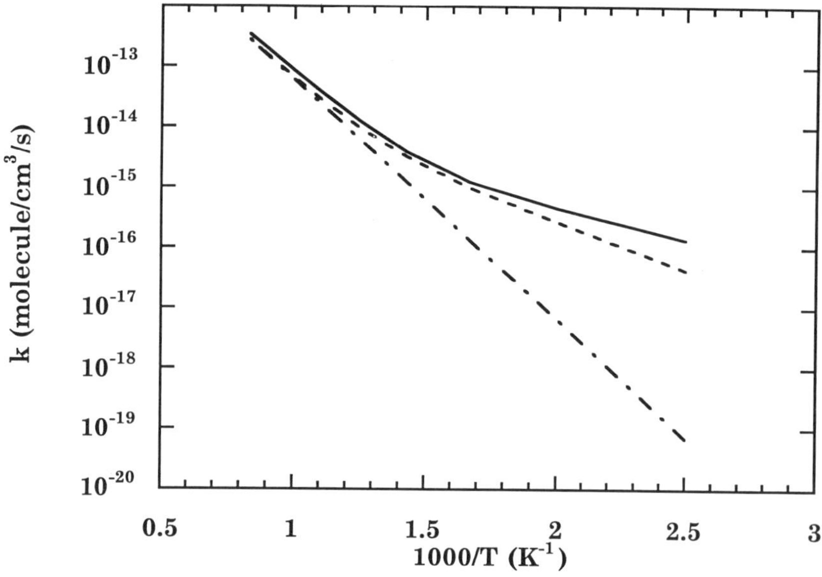

Figure 4.5: Arrhenius plot of rates of: —— H + N$_2$O → Products,[138] - - - - D + N$_2$O → Products,[138] and — · — fit of data above 700 K (see text)

4.4. H + N$_2$O Dynamics Studies

The reaction of translationally hot H-atoms with N$_2$O has been the focus of considerable interest by the dynamics community, due, in part, to the ease of LIF detection of the OH product. However, the earliest of these studies, by Oldershaw and Porter[141] predated the pre-eminence of laser methods, and used mass spectrometry and gas chromatography to measure H$_2$, N$_2$ and NO products from the photolysis of HI/N$_2$O mixtures. By varying the photolysis wavelength they were able to map out the reaction cross section as a function of translational energy. By extrapolating these back to zero cross sections they were able to derive values for the reaction thresholds. They obtained values of 10 ± 6 kcal/mole for the threshold for reaction forming OH + N$_2$, 4.2a, and 38 ± 4 kcal/mole for reaction forming NH + NO, 4.2c. These values compare favorably with later determinations of the barrier for reaction 4.2a (vide supra) and the experimental endothermicity of reaction 4.2c.

Nascent OH rotational populations from the H + N$_2$O $\rightarrow$ OH + N$_2$ reaction have been measured by a number of workers. By varying the H atom precursor and the photolysis wavelength they have probed the reaction from its threshold up to several eV above threshold. The measured rotational distributions are typically made up of two Boltzmann distributions, which has been taken as evidence for the two channels available for reaction. Measurement of the relative lambda-doublet populations in the OH show a preponderance of the $\Pi(A')$ state. This is the state which is expected to arise from dissociation from a planar complex, as is the case for the reaction channel which proceeds through HNNO.

Additional information can be gained by carrying out the reaction with a HX-N$_2$O van der Waals cluster as the reactant. As in the case of the bulk reaction, the process is started by photolysis of the HX molecule. The H-atom approach to the N$_2$O is now dependent on the (presently unknown) geometries of the van der Waals clusters. Ohoyami et al.[142] carried out an early series of experiments, using 266 nm photolysis of HI-N$_2$O clusters. They found OH rotational populations which consisted of two Boltzmann distributions, with temperatures of 120 K and 4000 K, in bulk samples and in beams with low stagnation pressures. A third distribution, with a temperature of 1500 K, was found to grow in at higher stagnation pressures. They found that the relative population of the 120 K component increased with increasing stagnation pressure, which they took as evidence that it corresponded to a reaction channel associated with a more stable van der Waals cluster geometry. There is general consensus that the more stable geometry is the one with the H-atom located on the O end of the N$_2$O molecule. The low rotational temperature is taken as evidence of the reactive collisions only having small impact parameters for the O-atom attack. They also observed the relative lambda-doublet population, $\Pi(A')/\Pi(A'')$, as a function of stagnation pressure, and saw that it declined from a value of ~2 in the bulk sample to ~1 at stagnation pressures of 2.5 atm. They take this as evidence that the reaction through the direct attack on the O-atom does not preferentially produce one of the lambda doublets. This would be expected for a reaction which is not constrained to take place in a plane; presumably the H-atom can attack the O-atom to leave the unpaired electron either in the plane or out of the plane of the OH rotation. Finally, they observed the ratio of the $^2\Pi_{1/2}$ to $^2\Pi_{3/2}$ spin states as a function of rotational level. It was found to be near unity for most rotational lines, consistent with results observed by others in bulk samples.[143] However, they find a preponderance of the $^2\Pi_{3/2}$ level for very low rotational quantum numbers. They attribute this to angular momentum coupling between the nascent OH and the I-atom. They theorize that this is more important in the case of the direct O-atom attack because the I-atom would not have had time to move away from the OH, whereas it would have had more opportunity to move away in the case of the indirect reaction as it proceeded through the HNNO intermediate.

Wittig and coworkers have also carried out a number of studies on the H + N_2O system, both in bulk samples and in van der Waals clusters.[144-149] They find that both OH and NH are formed in these reactions, with the branching ratio, $k_{4.2c}/k_{4.2a}$ approaching 0.5 for highly exothermic systems.[144] This ratio was found to be environment sensitive, decreasing in clusters of HI-N_2O relative to bulk samples. This behavior is taken to indicate an entrance channel specificity, i.e. that attack on the N end of N_2O is more likely to lead to NH product, and that the H-atom produced by photolysis of HI-N_2O is more likely to be directed to the O end of N_2O.[149] This behavior is in keeping with our expectations based on our knowledge of the HNNO potential energy surface, and suppositions about the geometry of the HX-N_2O van der Waals clusters. Hoffman, et al.[144] studied the highly exothermic reaction initiated by ArF photolysis of HBr. In bulk samples they observed intense OH A $\rightarrow$ X chemiluminescence, which was absent in reactions in the clusters. The chemiluminescence was also present, at much lower levels, in reactions initiated by photolysis of HI at $240 \leq l \leq 260$ nm, but has not been observed in reactions with hot D-atoms.[149] This is taken as evidence for the non-adiabaticity of the reaction forming OH (A).

4.5. NH + NO Dynamics Studies

The OH product channel in the reaction of NH with NO has been also been examined in experiments of Patel-Misra and Dagdigian.[150] They studied the reaction in a crossed beam apparatus, using ArF photolysis of NH_3 as their NH source. They observed the nascent OH product from the reaction using pulsed LIF, allowing them to measure the rovibronic distribution of the OH product. They found OH rotational distributions which could be fit, at high J, to temperatures of ~4400 K, with enhanced populations of low J levels. Their observations are reminiscent of those of Hollingsworth, et al.[143] and Hoffmann, et al.[144,145] on products from the H + N_2O reaction (vide supra). They also observed the preferential population of the $\Pi(A')$, pointing to a planar transition state. One point of divergence with the observations from the H + N_2O reaction is Patel-Misra and Dagdigian's observation of preferential formation of the $^2\Pi_{3/2}$ levels for high J levels. The relative populations of the spin-orbit levels is determined as the system is evolving from the polyatomic transition state, where the spin-orbit splitting is quenched, to the diatomic OH product which is spin-orbit split. However, what these observations are telling us is still uncertain.

5. THE H_2N_2O SYSTEM

We will next turn our attention to the reaction of NH_2 with NO, which accesses the ^1A surface of H_2N_2O. There are several proposed channels for this reaction:

$$NH_2 + NO \rightarrow N_2 + H_2O \qquad \Delta H_{rxn} = -124.4 \pm 0.3 \text{ kcal/mole} \qquad (5.1a)$$
$$NH_2 + NO \rightarrow N_2H + OH \qquad \Delta H_{rxn} = 3 \pm 2 \text{ kcal/mole} \qquad (5.1b)$$
$$NH_2 + NO \rightarrow N_2 + H + OH \qquad \Delta H_{rxn} = -6.5 \pm 0.4 \text{ kcal/mole} \qquad (5.1c)$$
$$NH_2 + NO \rightarrow H_2 + N_2O \qquad \Delta H_{rxn} = -46.9 \pm 0.3 \text{ kcal/mole} \qquad (5.1d).$$

(We have used the Gaussian 2 heat of formation for N_2H of 61 ± 2 kcal/mole.[151]) The reverse reactions -5.1a to -5.1d are mechanistically improbable, and will not be considered further.

Recent years have seen the introduction of various DeNOx schemes which involve injection of NHx containing species into exhaust gas from a combustion process.[1-4] If the temperature is in the appropriate range there will be a conversion of NO to N_2. The reaction chiefly responsible for this reduction is reaction 5.1a. Modeling of the complex reaction system suggests that reactions 5.1b and/or 5.1c are also important as radical sources, allowing the DeNOx chemistry to be self-sustaining. The obvious commercial impact of such processes has resulted in a great deal of study of this reaction in the last twenty years.

5.1. The H_2NNO Potential Energy Surface

Portions of the potential energy surface for the NH_2 + NO reaction have been calculated by a number of workers in the last fifteen years. Calculated energies for a number of stationary points important in reaction 5.1 are listed in Table 5.1. Casewit and Goddard[152] used Generalized Valence Bond (GVB) calculations to evaluate energies of various isomers of H_2N_2O, including species such as NN(H)OH and HNN(O)H. Melius and Binkley[108] used BAC-MP4 to characterize not only the isomers, but also the transition states linking them. Their calculated transition state energies for the various 1,2 H-atom transfers were all energetically inaccessible to NH_2 + NO reactants. As a result, NN(H)OH and HNN(O)H species are not considered important intermediates in reaction 5.1. Abou-Rachid and coworkers used SCF and CI calculations with the very small 4-31G basis set to examine stationary points along a proposed reaction path.[153,154] Harrison, MacIagan and Whyte carried out a more extensive study, performing MP4SDQ/6-31G* calculations on the thermally accessible stationary points on the NH_2 + NO $\rightarrow$ Products potential energy surface.[155] However, neither their study, nor the earlier work of Abou-Rachid and coworkers was of sufficiently high level of theory to quantitatively capture features of the surface. Recently, Walch has used CASSCF/ICCI with large basis sets to characterize a number of points on the potential energy surface.[156] His calculations agree well with the more extensive study of Wolf and Durant,[157] using the Gaussian-2 method, as proposed by Pople, and extended for transition states by Durant and Rohlfing.[17] The convergence of the theoretical methods can be seen by examining the agreement between these last two studies, as listed in Table 5.1, and by comparing the overall reaction exothermicities calculated using G2 with the experimental values. Overall the agreement is quite satisfactory, with errors generally <2 kcal/mole. The potential energy surface of Wolf and Durant[157] is shown in Figure 5.1.

Table 5.1: Energetics of H_2NNO Species

reference	Casewit[152]	Melius[108]	Abou-Rachid[153]	Harrison[155]	Walch[156]	Yang[158]	experiment
I	-29.8	-48.1	-39.9	-37.0	-44.0	-45.8	
II	-32.0	-47.5		-36.2	-45.7	-45.6	
III	-31.2	-45.5	-38.3	-34.4	-44.5	-45.1	
IV	-24.0	-41.1		-29.2		-39.2	
V	-32.4	-47.2		-36.1	-45.5	-45. 4	
VI		-19.7	0.5	-2.8	-14.4	-14.0 (-14.4[b])	
VII				-26.2		-36.7[b]	
VIII				7.8		-6.8 (-7.2[b])	
IX				-25.9		-36.2[b]	
X		-7.0		6.5	-7.4	-9.8 (-10.3[b])	
XI		-25.2		-7.6	-21.0	-23.4 (-24.6[b])	
N_2-H ts + OH		8.7		11.5		10.0 (9.9[b])	
N_2H + OH		2.4	-3.5	6.3		4.5	
N_2 + H + OH		-8.8	-14.7	-13.0		-4.7	-6.5
N_2 + H_2O		-134.2	-113.1	-113.2		-122.8	-124.4

a) All energies are in kcal/mole, relative to NH_2 + NO, 0 K, including zero point energies. Energies are evaluated at G2 level for equilibrium species and G2Q level for transition states, unless otherwise noted.
b) Energy is evaluated at G2 level for transition state

The qualitative mechanism emerging from these calculations starts with the barrierless association of NH_2 with NO to form a highly energized H_2NNO adduct. This adduct can then undergo a 1,3 H-atom migration, yielding an HNNOH species, again with considerable excess internal energy. This species, which has a trans conformation about the NN bond and a cis conformation about the NO bond, must isomerize to a cis conformation about the NN bond and a trans conformation about the NO bond before it can undergo a second 1,3 H-atom migration, yielding N_2 + H_2O products. These isomerizations can take place in either order. Isomerization about the NO bond is effected by simple torsional rotation, while isomerization about the NN bond occurs by passage through a linear HNN transition state. The transition state for formation of N_2 + H_2O products has a non-planar geometry, with a very flat bending potential. All of the HNNOH species are connected to N_2H + OH products by a simple bond fission, which is slightly endothermic relative to the NH_2 + NO reactants. The N_2H species, and the transition state linking it with N_2 + H, have been characterized

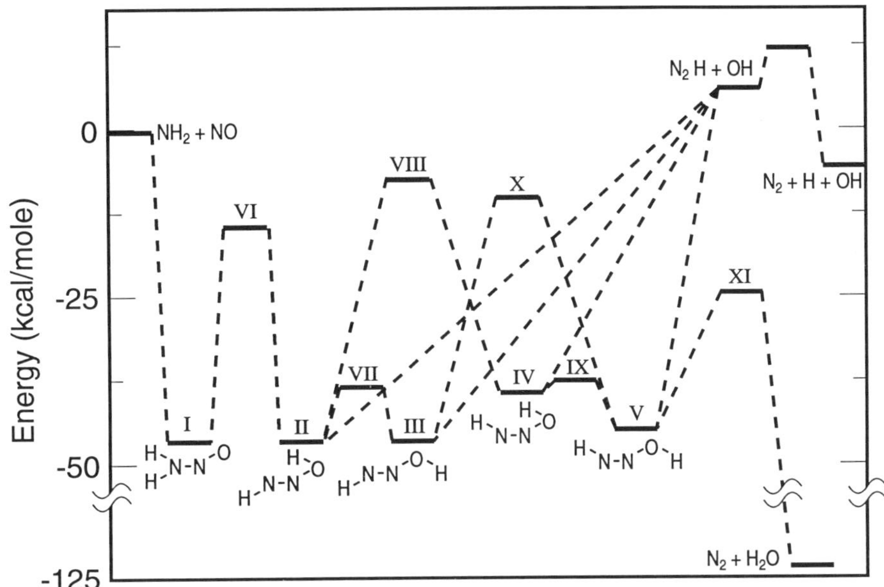

Figure 5.1: H_2NNO Potential Energy Surface

by a number of high level ab initio calculations.[151,159-161] The small barrier to dissociation (5.5 - 8.5 kcal/mole) suggests that N_2H will have a very short lifetime with respect to dissociation. Indeed, calculations by Koizumi, Schatz and Walch[162] predict that only the lowest two N_2H states will have lifetimes longer than 1 ps. Lifetimes for the (000) and (001) states were calculated to be 3 ns and 0.2 ns, respectively. The N_2H radical has been produced in a series of neutralized ion beam experiments in which N_2H was formed by neutralization of a N_2H^+ beam by metal vapor.[163] They established an upper limit on the N_2H lifetime in their apparatus of <0.5 µs, based on their inability to detect undissociated N_2H from the neutralization of the ion beam. We therefore expect that 5.1b will be rapidly followed by 5.2,

$$N_2H \rightarrow N_2 + H \tag{5.2}$$

with 5.1c as the net reaction. Mechanistically, channel 5.1c will be entropically disfavored, since it requires simultaneous fission of two bonds, and we expect that most of the OH-formation will proceed through channel 5.1b.

5.2. The NH_2 + NO $\rightarrow$ Products Reaction Rate Coefficient

There has been a flurry of experimental studies of the reaction rate coefficient in the last 25 years, utilizing discharge-flow,[5,94,164-166] flash photolysis,[157,167-176] pulsed radiolysis,[177-179] and shock tube techniques.[118] The results of these various studies are plotted in Figure 5.2. One of the more provocative features of this plot is the factor of ~2 difference between the room temperature discharge flow rate coefficients and rate coefficients measured using other techniques. This discrepancy has been noted in the past, but never satisfactorily explained. Although there are fewer studies, similar behavior is apparent in the reaction of NH_2 with NO_2, where the flow tube results of Hack, et al.[5] are approximately a factor of 2 slower than the results of other workers.[173,179-182]

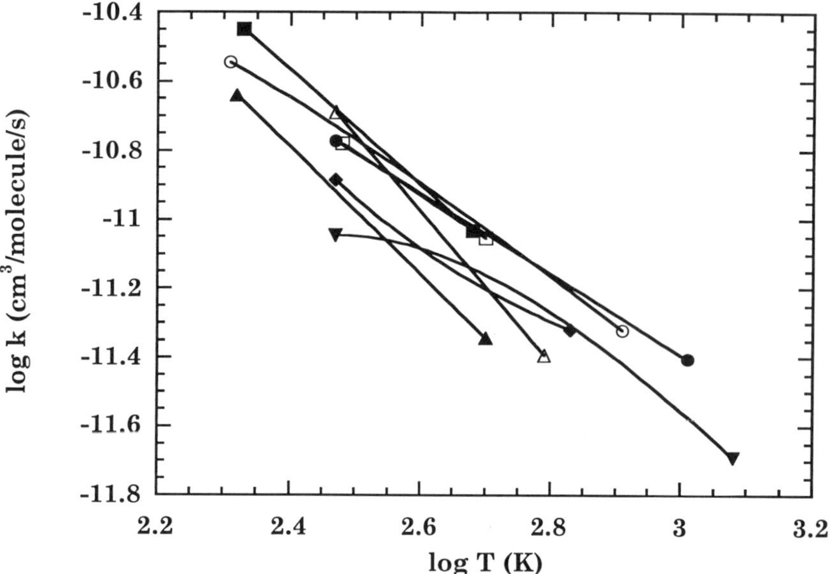

Figure 5.2: Temperature Dependent Studies of the NH_2 + NO $\rightarrow$ Products Reaction. $\square$ - ref. 167, $\blacktriangle$ - ref. 5, $\blacktriangledown$ - ref. 183, $\blacksquare$ - ref. 172, $\triangle$ - ref. 184, $\bullet$ - ref. 175, $\blacklozenge$ - ref. 176, $\bigcirc$ - ref. 157.

The wide variety of NH_2 sources and detection methods utilized in measuring the reaction rate coefficient $k_{5.1}$ makes explaining the discrepancy in rate coefficients difficult. Discharge-flow experiments have used reaction of OH-radicals with NH_3,[94] reaction of F-atoms with NH_3,[5,165] and SF_6-sensitized IRMPD of N_2H_4[166] as NH_2 sources, while other techniques have used flash lamp photolysis of NH_3,[167-170,172,174] laser photolysis of NH_3[157,171,173,175,176] or pulsed radiolysis of NH_3.[177-179] Detection schemes include mass spectrometry,[5,94] resonance fluorescence,[172] laser-induced fluorescence,[157,164-166,168,171,173,175] long-path laser absorption,[167] intracavity laser absorption[169,170,174] and cavity-ring down.[176]

The one shock tube study, by Roose, Hanson and Kruger,[118] consisted of shock heating mixtures of NO, NH_3, H_2 and Ar. NH_2 concentrations were measured by observing emission from electronically excited NH_2 at 538 nm. NH_3, NH, N_2O and NO were also monitored by emission, and NO by IR laser absorption. Their precision was limited by uncertainties in calibrating their detection systems; as a result, their data is consistent with a temperature independent rate, although they reported a positive temperature dependence for reaction $k_{5.1}$. They also observed N_2O products, although again their error bars were large enough to preclude any definitive statements about the importance of the channel. Despite its lack of precision, the work of Roose, et al. is consistent with a simple extrapolation of lower temperature results.

A number of investigators have also examined the effect of pressure on the NH_2 + NO $\rightarrow$ Products reaction. They all found that the reaction is pressure independent for pressures below 2 atm.[5,94,157,165-167,169,173,174,177,185,186] Particularly noteworthy are studies by Gordon and Mulac,[177] who measured $k_{5.1}$ in NH_3 for pressures from 250 to 1520 Torr, Lesclaux, et al.,[167] who used N_2 as the bath gas in the pressure range $2 \leq P \leq 700$ Torr and Sarkisov, et al.,[185] who also used N_2 in the pressure range $1 \leq P \leq 700$ Torr. These measurements are in keeping with our expectations for an association-type reaction with open product channels below the energy of the reactants.

5.3. Products from the NH_2 + NO $\rightarrow$ Products Reaction

Experimental interest in the products of the reaction of NH_2 with NO has a long history. In 1939 Bamford[187] photolyzed NH_3 in the presence of NO at 373 K in a series of stationary photolysis experiments . He identified H_2O and N_2 as photolysis products, and proposed a mechanism in which reaction 5.1a was the loss channel for NH_2. Later work by Serewicz and Noyes[188] in the vicinity of room temperature also identified N_2O as a product of the photolysis; presumably it was destroyed by mercury-sensitized photolysis in Bamford's work, since he made extensive use of mercury seals in his vacuum apparatus. Srinivasan[189] used $^{15}NH_3$ to establish that the N_2O came solely from NO, while the N_2 incorporated one N-atom from ammonia and one from nitric oxide, thus ruling out reaction 5.1d as a significant process. The N_2O was assumed to be formed by the reaction of two HNO radicals, producing N_2O and H_2O. The final "classical" photochemical study was that of Jayanty, Simonaltis and Heicklen[190] whose actinometric measurements established that the quantum yield for formation of N_2 as 1.0 ± 0.1 at 298 K, making N_2 + H_2O the major channel in the reaction of NH_2 with NO at room temperature.

The reaction of NH_2 with NO presumably proceeds through formation of a vibrationally excited H_2NNO adduct. This adduct can be expected to have a very short lifetime with respect to isomerization to HNNOH; RRKM calculations by Phillips[191] suggested that the lifetime was of the order of 10 ps. This result is consistent with the lack of any observed pressure dependence of the reaction rate coefficient. Gehring, et al.[94] observed m/e = 45 and m/e = 46 peaks in the mass spectrum of the products of the reaction of NH_2 with NO. Similarly, Farber and Harris[192] observed m/e = 45 and m/e = 46 mass peaks from the products of the reaction of NH_3 with NO over a vanadium oxide catalyst. These observations would imply that the H_2NNO adduct lifetime was orders of magnitude larger than 10 ps. However, neither group performed isotopic substitution experiments to confirm their assignments of m/e = 46 to H_2NNO^+ and m/e = 45 to $HNNO^+$. Foner and Hudson[193] observed $N_3H_3^+$ and $N_4H_4^+$ in the products of electrodeless discharge dissociated hydrazine which had been condensed in a liquid nitrogen trap; presumably other hydrides of N_3 and N_4 are present in the gas phase following condensation reactions of NHx species, and could have given rise to the peaks observed by Gehring, et al. and Farber and Harris.

Table 5.2: Room Temperature Product Branching Fractions for NH_2 + NO $\rightarrow$ Products Reaction

$\Phi(N_2 + H_2O)$	$\Phi(OH)$	$\Phi(H)$	$\Phi(HNO)$	reference
1.0 ± 0.1				Jayanty, et al.[190]
	0.4 ± 0.1			Silver and Kolb[165]
	(<0.33)			(Silver and Kolb[194])
	≥0.65	≤0.05		Andresen, et al.[171]
	<0.22			Stief, et al.[172]
	0.13 ± 0.02			Hall, et al.[195]
	<0.13			Dolson[196]
	0.10 ± 0.02			Bulatov, et al.[174]
	0.10 ± 0.025			Atakan, et al.[175]
			≤0.01	Unfried, et al.[197]
	0.07			Pagsberg, et al.[179]
0.85	0.10			Stephens, et al.[198]
	0.09 ± 0.02			Yang and Durant[158]

One might expect to be able to stabilize the H_2NNO intermediate using matrix isolation techniques, and Crowley and Sodeau[199] observed IR absorption features which they attributed to the H_2NNO intermediate formed by photolysis of NH_3/NO or NH_3/NOCl mixtures in Ar or N_2 matrices. The two sets of bands observed are in reasonable agreement with predictions based on scaled HF/6-31G* calculations,[200] both in their location and in their response to ^{15}N or D isotopic substitution. Interestingly enough, their experiments[201] suggest that N_2O + H_2 are the primary products of the NH_2 + NO reaction at 4.2 K, with the yield of H_2O + N_2 being considerably surpressed. The barrier for reaction 5.1d has been calculated by Melius and Binkley[108] to be 32 kcal/mole above the reactants, making it thermally inaccessible. However, quantum mechanical tunneling may be responsible for the reaction at 4.2 K.

Because of its important role in various DeNOx processes, the product branching fraction for the NH_2 + NO reaction has been the focus of a number of direct determinations. Table 5.2 summarizes room temperature measurements, while temperature dependent product branching fractions are presented in Figure 5.3.

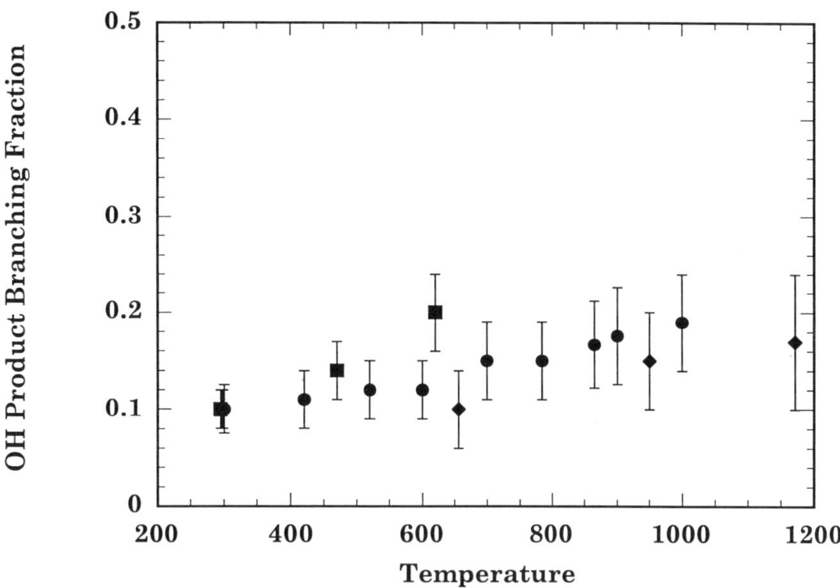

Figure 5.3: Temperature dependence of $(k_{5.1b} + k_{5.1c})/k_{5.1}$. ● - ref. 175, ■ - ref. 174, ◆ - ref. 198.

One of the interesting conclusions drawn from modeling DeNOx schemes is that reactions 5.1b and/or 5.1c must have non-negligible rates at temperatures in excess of 1000 K.[7] This is a result of the requirement that the overall DeNOx process be self-sustaining. Reaction 5.1a is chain terminating, removing one radical center, NH_2. By contrast, reactions 5.1b and 5.1c are strongly chain branching. They directly produce two radicals for every one consumed. Under conditions encountered in most DeNOx schemes (excess O_2 and H_2O) the H-atom reacts rapidly with O_2, producing another OH-radical and an O-atom. The O-atom is rapidly converted to two more OH-radicals by reaction with H_2O, resulting in production of four radical centers for every one consumed.

$$NH_2 + NO \rightarrow N_2H + OH \rightarrow N_2 + H + OH$$
$$H + O_2 \rightarrow OH + O$$
$$O + H_2O \rightarrow 2OH$$

net: $NH_2 + NO + O_2 + H_2O \rightarrow N_2 + 4OH$

In the absence of other important chain branching or chain terminating reactions the models require $(k_{5.1b} + k_{5.1c})/k_{5.1} > 0.25$ for the DeNOx process to be self-sustaining. Of course, the models may well be wrong; extrapolation of lower temperature data on the product branching fraction is presently too uncertain to say definitively.

6. CONCLUSION

At this point we conclude our brief survey of NHx chemistry. It is hoped that this examination of a facet of nitrogen chemistry has been useful. We have tried to demonstrate the state of experimental studies of reaction rate coefficients and products, and of theoretical studies of structures and energies of stable and transient species encountered in these reactions. The combination of theoretical and experimental studies has allowed us to learn things about these systems that would not be possible without the integration of both approaches. It should also be apparent that the richness of these systems leaves much still left to learn. Questions range from questions about reaction rate coefficients and products to questions about energy disposal in these reactions and the affect of multiple electronic surfaces on the reaction mechanisms.

ACKNOWLEDGEMENTS

The author would like to acknowledge the support of the U.S. Department of Energy, Office of Basic Energy Sciences, Division of Chemical Sciences.

REFERENCES

(1)   Lyon, R. K. *Int. J. Chem. Kinetics* **1976**, *8*, 315.

(2)   Muzio, L. J.; Arand, J. K.; Teixeira, D. P. *Symp. (Int.) Combust. [Proc.]* **1977**, *16*, 1977.

(3)   Perry, R. A.; Siebers, D. L. *Nature* **1986**, *324*, 657.

(4)   Lyon, R. K. *Environ. Sci. Technol.* **1987**, *21*, 231.

(5)   Hack, W.; Schacke, H.; Schroter, M.; Wagner, H. G. *Symp. (Int.) Combust. [Proc.]* **1979**, *17*, 505.

(6)   Lesclaux, R. *Reviews of Chemical Intermediates* **1984**, *5*, 347.

(7)   Miller, J. A.; Bowman, C. T. *Prog. Energy Combust. Sci.* **1989**, *15*, 287.

(8)   Hanson, R. K.; Salimian, S. In *Combustion Chemistry*; J. W.C. Gardiner, Ed.; Springer-Verlag: NY, 1984.

(9)   Mallard, W. G.; Westley, F.; Herron, J. T.; Hampson, R. F.; Frizzell, D. H. *NIST Standard Reference Data Database 17: Chemical Kinetics*: Gaithersburg, MD, 1993.

(10)   Baulch, D. L.; Cobos, C. J.; Cox, R. A.; Esser, C.; Frank, P.; Just, T.; Kerr, J. A.; Pilling, M. J.; Troe, J.; Walker, R. W.; Warnatz, J. *J. Phys. Chem. Ref. Data* **1992**, *21*, 411.

(11)   Cohen, N.; Westberg, K. R. *J. Phys. Chem. Ref. Data* **1991**, *20*, 1211.

(12)   Tsang, W.; Herron, J. T. *J. Phys. Chem. Ref. Data* **1991**, *20*, 609.

(13)   Tsang, W.; Herron, J. T. *J. Phys. Chem. Ref. Data* **1992**, *21*, 753.

(14)   Atkinson, R.; Baulch, D. L.; Cox, R. A.; Hampson, R. F., Jr.; Kerr, J. A.; Troe, J. *J. Phys. Chem. Ref. Data* **1992**, *21*, 1125.

(15)   DeMore, W. B.; Sander, S. P.; Golden, D. M.; Molina, M. J.; Hampson, R. F.; Kurylo, M. J.; Howard, C. J.; Ravishankara, A. R. In *JPL Publication 90-1, Pasadena, CA*.1992; pp 1.

(16)   Curtiss, L. A.; Raghavachari, K.; Trucks, G. W.; Pople, J. A. *J. Chem. Phys.* **1991**, *94*, 7221.

(17)   Durant, J. L.; Rohlfing, C. M. *J. Chem. Phys.* **1993**, *98*, 8031.

(18)   Bauschlicher, C. W.; Langhoff, S. R. *Sci.* **1991**, *254*, 394.

(19)   Chase, M. W.; Davies, C. A.; Downey, J. R.; Frurip, D. J.; McDonald, R. A.; Syverud, A. N. *J. Phys. Chem. Ref. Data* **1985**, *14 (Suppl. No. 1)*,

(20)   Anderson, W. R. *J. Phys. Chem.* **1989**, *93*, 530.

(21)   Berkowitz, J.; Ellison, G. B.; Gutman, D. *J. Phys. Chem.* **1994**, *98*, 2744.

(22)   Dixon, R. N.; Jones, K. B.; Noble, M.; Carter, S. *Molecular Physics* **1981**, *42*, 455.

(23)   Huber, K. P.; Herzberg, G. *Molecular Spectra and Molecular Structure Constants of Diatomic Molecules*; Van Nostrand Reinhold Company: New York, 1979.

(24)   Ogilvie, J. F. *Nature* **1973**, *243*, 210.

(25)   Ellis, H. B.; Ellison, G. B. *J. Chem. Phys.* **1983**, *78*, 6541.

(26)   Bancroft, J. L.; Hollas, J. M.; Ramsay, D. A. *Canadian Journal of Physics* **1962**, *40*, 322.

(27)   Guadagnini, R.; Schatz, G. C.; Walch, S. P. *J. Chem. Phys.* **1994**, *submitted*,

(28)   Bruna, P. J.; Marian, C. M. *Chem. Phys. Lett.* **1979**, *67*, 109.

(29)   Heiberg, A.; Almlöf, J. *Chem. Phys. Lett.* **1982**, *85*, 542.

(30)   Melius, C. F.; Binkley, J. S. In *The Chemistry of Combustion Processes*; A. M. Dean; M.-S. Chou and D. Stern, Ed.; ACS Symposium Series: 1984; Vol. 249.

(31)   Adams, G. F.; Bent, G. D.; Barlett, R.; Purvis, G. D. In *Potential Energy Surfaces and Dynamics Calculations*; D. G. Truhlar, Ed.; Plenum: New York, 1981; pp 133.

(32)   Nomura, O.; Ikuta, S.; Igawa, A. In *Applied Quantum Chemistry*; V. H. Smith Jr.; H. F. Schaefer III and K. Morokuma, Ed.; Reidel: Dordrecht, 1984; pp 243.

(33)   Walch, J. P.; Rohlfing, C. M. *J. Chem. Phys.* **1989**, *91*, 2939.

(34)   Walch, S. P. *J. Chem. Phys.* **1990**, *93*, 8036.

(35)   Pauzat, F.; Ellinger, Y.; Berthier, G.; G'rin, M.; Viala, Y. *Chem. Phys.* **1993**, *174*, 71.

(36)   Lee, T. J. *J. Chem. Phys.* **1993**, *99*, 9783.

(37)   Carter, S.; Mills, I. M.; Murrell, J. N. *J. Chem. Soc. Faraday Trans. 2* **1979**, *1*, 148.

(38)   Colton, M. C.; Schatz, G. C. *J. Chem. Phys.* **1985**, *83*, 4823.

(39)   Huang, Y.-L.; Dagdigian, P. J. *J. Chem. Phys.* **1992**, *97*, 180.

(40)   Temps, F. *Ber. MPI Strömungsforschung* **1983**, *4*, 140.

(41)   Adamson, J. D.; Farhat, S. K.; Morter, C. L.; Glass, G. P.; Curl, R. F.; Phillips, L. F. *J. Phys. Chem.* **1994**, *98*, 5665.

(42)   Hack, W.; Wagner, H. G.; Zasypkin, A. *Ber. Bunsenges. Phys. Chem.* **1994**, *98*, 156.

(43)   Mertens, J. D.; Chang, A. Y.; Hanson, R. K.; Bowman, C. T. *Int. J. Chem. Kinet.* **1991**, *23*, 173.

(44)   Zeldovich, J. *Acta Physiocochim. URSS.* **1946**, *21*, 577.

(45)   Howard, M. J.; Smith, I. W. M. *Chem. Phys. Lett.* **1980**, *69*, 40.

(46)   Howard, M. J.; Smith, I. W. M. *J. Chem. Soc. Faraday Trans. 2* **1981**, *77*, 997.

(47)   Brune, W. H.; Schwab, J. J.; Anderson, J. G. *J. Phys. Chem.* **1983**, *87*, 4503.

(48)   Rowe, B. R.; Sims, I. R.; Smith, I. W. M., *private communication*, **1994**.

(49)   Bradley, J. N.; Craggs, P. *Proceedings of the International Combustion Symposium* **1975**, *15*, 833.

(50)   Duxbury, J.; Pratt, N. H. *Proceedings of the International Combustion Symposium* **1975**, *15*, 843.

(51)   Flower, W. L.; Hanson, R. K.; Kruger, C. H. *Proceedings of the International Combustion Symposium* **1975**, *15*, 823.

(52)   Koshi, M.; Ando, H.; Oya, M.; Asaba, T. *Proceedings of the International Combustion Symposium* **1975**, *15*, 909.

(53)   Ando, H.; Asaba, T. *Int. J. Chem. Kinetics* **1976**, *8*, 259.

(54)   Flower, W. L.; Hanson, R. K.; Kruger, C. H. *Combustion Science and Technology* **1977**, *15*, 115.

(55)   McCullough, R. W.; Kruger, C. H.; Hanson, R. K. *Combustion Science and Technology* **1977**, *15*, 213.

(56)   Smallwood, H. M. *J. Am. Chem. Soc.* **1929**, *51*, 1985.

(57)   Harteck, P. *Ber.* **1933**, *66*, 423.

(58)   Cashion, J. K.; Polanyi, J. C. *J. Chem. Phys.* **1959**, *30*, 317.

(59)   Clement, M. J. Y.; Ramsay, D. A. *Canadian Journal of Physics* **1961**, *39*, 205.

(60)   Dalby, F. W. *Canadian Journal of Physics* **1958**, *36*, 1336.

(61)   Dixon, R. N.; Rosser, C. A. *Chem. Phys. Lett.* **1984**, *108*, 323.

(62)   Petersen, J. C. *J. Mol. Spect.* **1985**, *110*, 277.

(63)   Dixon, R. N.; Noble, M.; Taylor, C. A.; Delhoume, M. *Discussions of the Faraday Society* **1981**, *71*, 1.

(64)   Bruna, P. J. *Chem. Phys.* **1980**, *49*, 39.

(65)   Clyne, M. A. A.; Thrush, B. A. *Discussions of the Faraday Society* **1962**, *33*, 139.

(66)   Hikida, T.; Eyre, J. A.; Dorfman, L. M. *J. Chem. Phys.* **1971**, *54*, 3422.

(67)   Ahumada, J. J.; Michael, J. V.; Osborne, D. T. *J. Chem. Phys.* **1972**, *57*, 3736.

(68)   Atkinson, R.; Cvetanovic, R. J. *Canadian Journal of Chemistry* **1973**, *51*, 370.

(69)   Oka, K.; Singleton, D. L.; Cvetanovic, R. J. *J. Chem. Phys.* **1977**, *67*, 4681.

(70)   Oka, K.; Singleton, D. L.; Cvetanovic, R. J. *J. Chem. Phys.* **1977**, *66*, 713.

(71)   Ishikawa, Y.; Sugawara, K.; Sato, S. *Bull. Chem. Soc. Jpn.* **1979**, *52*, 3503.

110

(72)   Clyne, M. A. A.; Thrush, B. A. *Trans. Faraday Soc.* **1962**, *57*, 1305.

(73)   Hartley, D. B.; Thrush, B. A. *Roy. Soc. London Proc. A* **1967**, *297*, 520.

(74)   Campell, I. M.; Handy, B. J. *J. Chem. Soc. Faraday Trans.* **1975**, *71*, 2097.

(75)   Obi, K.; Matsumi, Y.; Takeda, Y.; Mayama, S. *Chem. Phys. Lett.* **1983**, *95*, 520.

(76)   Forte, E. N. *J. Photochem.* **1981**, *17*, 13.

(77)   Pauzat, F.; Gritli, H.; Ellinger, Y.; Subra, R. *J. Phys. Chem.* **1984**, *88*, 4581.

(78)   Davis, P. B.; Dransfeld, P.; Temps, F.; Wagner, H. G. *J. Chem. Phys.* **1984**, *81*, 3763.

(79)   Mikami, H.; Saito, S.; Yamamoto, S. *J. Chem. Phys.* **1991**, *94*, 3415.

(80)   Soto, M. R.; Page, M. *Chem. Phys. Lett.* **1991**, *187*, 335.

(81)   Cai, Z. L. *Chem. Phys.* **1993**, *169*, 75.

(82)   Barone, V.; Grand, A.; Minichino, C.; Subra, R. *J. Phys. Chem.* **1993**, *97*, 6355.

(83)   Walch, S. P. *J. Chem. Phys.* **1993**, *99*, 3804.

(84)   Komaromi, I.; Tronchet, J. M. *J. Chem. Phys. Lett.* **1993**, *215*, 444.

(85)   Yang, D. L.; Koszykowski, M. L.; Durant, J. L., Jr. *J. Chem. Phys.* **1994**, *101*, 1361.

(86)   Salotto, A. W.; Burnelle, L. *J. Chem. Phys.* **1970**, *53*, 333.

(87)   Burton, B.; Claxton, T. A.; Ellinger, Y. *Computer Physics Communications* **1979**, *17*, 27.

(88)   Barone, W.; Cristinziano, P. L.; Lelj, F.; Pastore, A.; Russo, N. *J. Mol. Struct.* **1982**, *90*, 59.

(89)   Kysel, O.; Mach, P.; Haring, M. *J. Mol. Struct.* **1986**, *138*, 299.

(90)   Sana, M.; Leroy, G.; Peeters, D.; Younang, E. *J. Mol. Struct.* **1987**, *151*, 325.

(91)   So, S. P. *J. Phys. Chem.* **1990**, *94*, 2344.

(92)   Page, M.; Soto, M. R. *J. Chem. Phys.* **1993**, *99*, 7709.

(93)   Lifshitz, C.; Ruttink, P. J. A.; Schaftenaar, G.; Terlouw, J. K. *Rapid Commun. Mass Spectrom.* **1987**, *1*, 61.

(94)   Gehring, M.; Hoyermann, K.; Schacke, H.; Wolfrum, J. *Symp. (Int.) Combust., [Proc.]* **1973**, *14*, 99.

(95)   Dransfeld, P.; Hack, W.; Kurzke, H.; Temps, F.; Wagner, H. G. *Symp. (Int.) Combust. [Proc.]* **1985**, *20*, 655.

(96)   Patel-Misra, D.; Dagdigian, P. J. *Chem. Phys. Lett.* **1991**, *185,*387.

(97)   Patel-Misra, D.; Sauder, D.; Dagdigian, P. *J. Chem. Phys.* **1991**, *95*, 955.

(98)   Luo, X.; Rizzo, T. R. *J. Chem. Phys.* **1992**, *96*, 5129.

(99)   Sinha, A.; Vander Wal, R.; Crim, F. F. *J. Chem. Phys.* **1990**, *92*, 401.

(100)   Garcia-Moreno, I.; Lovejoy, E. R.; Moore, C. B. *J. Chem. Phys.* **1994**, *100*, 8890.

(101)   Dodonov, A. F.; Zelenov, V. V.; Tal'rose, V. L. *Dokl. Phys. Chem.* **1980**, *252*, 642.

(102)   Dodonov, A. F.; Zelenov, V. V.; Strunin, V. P.; Tal'rose, V. L. *Kinet. Catal.* **1981**, *22*, 689.

(103)   Washida, N.; Akimoto, H.; Okuda, M. *J. Phys. Chem.* **1978**, *82*, 2293.

(104)   Harrison, J. A.; Whyte, A. R.; Phillips, L. F. *Chem. Phys. Lett.* **1986**, *129*, 346.

(105)   Dean, A. M.; Chou, M. S.; Stern, D. *Int. J. Chem. Kinetics* **1984**, *16*, 633.

(106)   Durant, J. L. *J. Phys. Chem.* **1994**, *98*, 518.

(107)   Durant, J. L. *unpublished results* **1994**.

(108)   Melius, C. F.; Binkley, J. S. *Symp. (Int.) Combust. [Proc.]* **1984**, *20*, 575.

(109)   Fueno, T.; Fukuda, M.; Yokoyama, K. *Chem. Phys.* **1988**, *124*, 265.

(110)   Harrison, J. A.; Maclagan, R. G. A. R. *J. Chem. Soc. Faraday Trans.* **1990**, *86*, 3519.

(111)   Miller, J. A.; Melius, C. F. *Symp. (Int.) Combust. [Proc.]* **1993**, *24*, 719.

(112)   Walch, S. P. *J. Chem. Phys.* **1993**, *98*, 1170.

(113)   Gordon , S.; Mulac, W.; Nangia, P. *J. Phys. Chem.* **1971**, *75*, 2087.

(114)   Hansen, I.; Hoeinghaus, K.; Zetzsch, C.; Stuhl, F. *Chem. Phys. Lett.* **1976**, *42*, 370.

(115)   Cox, J. W.; Nelson, H. H.; McDonald, J. R. *Chem. Phys.* **1985**, *96*, 175.

(116)   Yamasaki, K.; Okada, S.; Koshi, M.; Matsui, H. *J. Chem. Phys.* **1991**, *95*, 5087.

(117)   Morley, C. *Symp. (Int.) Combust. [Proc.]* **1981**, *18*, 23.

(118)   Roose, T. R.; Hanson, R. K.; Kruger, C. H. *Symp. (Int.) Combust., [Proc.]* **1981**, *18*, 853.

(119)   Yokoyama, K.; Sakane, Y.; Fueno, T. *Bull. Chem. Soc. Jpn.* **1991**, *64*, 1738.

(120)   Vandooren, J.; Sarkisov, O. M.; Balakhnin, V. P.; Van Tiggelen, P. J. *Chem. Phys. Lett.* **1991**, *184*, 294.

(121)   Sausa, R. C.; Anderson, W. R.; Dayton, D. C.; Faust, C. M.; Howard, S. L. *Combust. Flame* **1994**, *in press*,

(122)   Lillich, H.; Schuck, A.; Volpp, H.; Wolfrum, J. *Symp. (Int.) Combust. [Proc.]* **1994**, *25*, in press.

(123)   Okada, S.; Miyoshi, A.; Matsui, H. *J. Chem. Phys.* **1994**, *submitted.*

(124)   Mertens, J. D.; Kohse-Hoinghaus, K.; Hanson, R. K.; Bowman, C. T. *Int. J. Chem. Kinet.* **1991**, *23*, 655.

(125)   Walther, C. D.; Wagner, H. G. *Ber. Bunsenges. Phys. Chem.* **1983**, *87*, 403.

(126)   Hack, W.; Kurzke, H.; Wagner, H. G. *J. Chem. Soc. Faraday Trans. 2* **1985**, *81*, 949.

(127)   Walch, S. P.; Duchovic, R. J.; Rohlfing, C. M. *J. Chem. Phys.* **1989**, *90*, 3230.

(128)   Baldwin, R. R.; Gethin, A.; Walker, R. W. *Journal of the Chemical Society, Faraday Transactions 1* **1973**, *69*, 352.

(129)   Albers, E. A.; Hoyermann, K.; Schacke, H.; Schmatjko, K. J.; Wagner, H. G. *Symp. (Int.) Combust. [Proc.]* **1975**, *15*, 765.

(130)   Dean, A. M.; Steiner, D. C.; Wande, E. E. *Combust. Flame* **1978**, *32*, 73.

(131)   Glass, G. P.; Quy, R. B. *J. Phys. Chem.* **1979**, *83*, 30.

(132)   Dean, A. M.; Johnson, R. L.; Steiner, D. C. *Combust. Flame* **1980**, *37*, 41.

(133)   Hidaka, Y.; Takuma, H.; Suga, M. *Bull. Chem. Soc. Jpn.* **1985**, *58*, 2911.

(134)   Marshall, P.; Fontijn, A.; Melius, C. F. *J. Chem. Phys.* **1987**, *86*, 5540.

(135)   Fenimore, C. P.; Jones, G. W. *J. Phys. Chem.* **1959**, *63*, 1154.

(136)   Dixon-Lewis, G.; Sutton, M. M.; Williams, A. *Symp. (Int.) Combust. [Proc.]* **1965**, *10*, 495.

(137)   Balakhnine, V. P.; Vandooren, J.; Van Tiggelen, P. J. *Combust. Flame* **1977**, *28*, 165.

(138)   Marshall, P.; Ko, T.; Fontijn, A. *J. Phys. Chem.* **1989**, *93*, 1922.

(139) Benson, S. W.; O'Neal, H. E. *Kinetic Data on Gas Phase Unimolecular Reactions*; National Bureau of Standards: Washington, 1970; Vol. 21.

(140) Mayer, J. E.; Mayer, M. G. *Statistical Mechanics*; Wiley: New York, 1957.

(141) Oldershaw, G. A.; Porter, D. A. *Nature* **1969**, *223*, 490.

(142) Ohoyama, H.; Takayangi, M.; Nishiya, T.; Hanazaki, I. *Chem. Phys. Lett.* **1989**, *162*, 1.

(143) Hollingsworth, W. E.; Subbiah, J.; Flynn, G. W.; Weston, R. E. *J. Chem. Phys.* **1985**, *82*, 2295.

(144) Hoffmann, G.; Oh, D.; Iams, H.; Wittig, C. *Chem. Phys. Lett.* **1989**, *155*, 356.

(145) Hoffmann, G.; Oh, D.; Wittig, C. *J. Chem. Soc. Faraday Trans. 2* **1989**, *85*, 1141.

(146) Hoffmann, G.; Oh, D.; Chen, Y.; Engel, Y. M.; Wittig, C. *Isr. J. Chem.* **1990**, *30*, 115.

(147) Shin, S. K.; Chen, Y.; Nickolaisen, S.; Sharpe, S. W.; Beaudet, R. A.; Wittig, C. *Adv. Photochem.* **1991**, *16*, 249.

(148) Shin, S. K.; Chen, Y.; Oh, D.; Wittig, C. *Philos. Trans. R. Soc. London Ser. (A)* **1990**, *332*, 361.

(149) Boehmer, E.; Shin, S. K.; Chen, Y.; Wittig, C. *J. Chem. Phys.* **1992**, *97*, 2536.

(150) Patel-Misra, D.; Dagdigian, P. J. *J. Phys. Chem.* **1992**, *96*, 3232.

(151) Pople, J. A.; Curtiss, L. A. *J. Chem. Phys.* **1991**, *95*, 4385.

(152) Casewit, C. J.; Goddard, W. A., III *J. Am. Chem. Soc.* **1982**, *104*, 3280.

(153) Abou-Rachid, H.; Pouchan, C.; Chaillet, M. *Chem. Phys.* **1984**, *90*, 243.

(154) Abou-Rachid, H.; Pouchan, C. *J. Mol. Struct.* **1985**, *121*, 299.

(155) Harrison, J. A.; Maclagan, G. A. R.; Whyte, A. R. *J. Phys. Chem.* **1987**, *91*, 6683.

(156) Walch, S. P. *J. Chem. Phys.* **1993**, *99*, 5295.

(157) Wolf, M.; Durant, J. L. *unpublished results* **1994**,

(158) Yang, D. L.; Durant, J. L. *unpublished results* **1994**,

(159) Curtiss, L. A.; Drapcho, D. L.; Pople, J. A. *Chem. Phys. Lett.* **1984**, *103*, 437.

(160) Walch, S. P.; Duchovic, R. J.; Rohlfing, C. M. *J. Chem. Phys.* **1989**, *90*, 3230.

(161)   Walch, S. P. *J. Chem. Phys.* **1990**, *93*, 2384.

(162)   Koizumi, H.; Schatz, G. C.; Walch, S. P. *J. Chem. Phys.* **1991**, *95*, 4130.

(163)   Selgren, S. F.; McLoughlin, P. W.; Gellene, G. I. *J. Chem. Phys.* **1989**, *90*, 1624.

(164)   Gersh, M. E.; Silver, J. A.; Zahniser, M. S.; Kolb, C. E.; Brown, R. G.; Gozewski, C. M.; Kallelis, S.; Wormhoudt, J. C. *Rev. Sci. Instrum.* **1981**, *52*, 1213.

(165)   Silver, J. A.; Kolb, C. E. *J. Phys. Chem.* **1982**, *86*, 3240.

(166)   Jeffries, J. B.; McCaulley, J. A.; Kaufman, F. *Chem. Phys. Lett.* **1984**, *106*, 111.

(167)   Lesclaux, R.; Khe, P. V.; Dezauzier, P.; Soulignac, J. C. *Chem. Phys. Lett.* **1975**, *35*, 493.

(168)   Hancock, G.; Lange, W.; Lenzi, M.; Welge, K. H. *Chem. Phys. Lett.* **1975**, *33*, 168.

(169)   Sarkisov, O. M.; Cheskis, S. G.; Sviridenkov, E. A. *Bull. Acad. Sci. USSR, Div. Chem. Sci.* **1978**, *27*, 2612.

(170)   Nadtochenko, V. L.; Sarkisov, O. M.; Cheskis, S. G. *Fiz.-Khim. Protsessy Gazov. Kondens. Fazakh* **1979**, 21.

(171)   Andresen, P.; Jacobs, A.; Kleinermanns, C.; Wolfrum, J. *Symp. (Int.) Combust. [Proc.]* **1982**, *19*, 11.

(172)   Stief, L. J.; Brobst, W. D.; Nava, D. F.; Borkowski, R. P.; Michael, J. V. *J. Chem. Soc. Faraday Trans. 2* **1982**, *78*, 1391.

(173)   Whyte, A. R.; Phillips, L. F. *Chem. Phys. Lett.* **1983**, *102*, 451.

(174)   Bulatov, V. P.; Ioffe, A. A.; Lozovsky, V. A.; Sarkisov, O. M. *Chem. Phys. Lett.* **1989**, *161,141-146*,

(175)   Atakan, B.; Jacobs, A.; Wahl, M.; Weller, R.; Wolfrum, J. *Chem. Phys. Lett.* **1989**, *155*, 609.

(176)   Diau, E. W.; Yu, T.; Wagner, M. A. G.; Lin, M. C. *J. Phys. Chem.* **1994**, *in press*,

(177)   Gordon, S.; Mulac, w.; Nangia, P. *J. Phys. Chem.* **1971**, *75*, 2087.

(178)   Gordon, S.; Mulac, W. A. *Int. J. Chem. Kinet. (Symp. 1)* **1975**, 289.

(179)   Pagsberg, P.; Sztuba, B.; Ratajczak, E.; Sillesen, A. *Acta Chem. Scandinavica* **1991**, *45,329-334*,

(180)   Kurasawa, H.; Lesclaux, R. *Chem. Phys. Lett.* **1979**, *66*, 602.

(181)   Xiang, T.-X.; Torres, L. M.; Guillory, W. A. *J. Chem. Phys.* **1985**, *83*, 1623.

(182)   Bulatov, V. P.; Ioffe, A. A.; Lozovsky, V. A.; Sarkisov, O. M. *Chem. Phys. Lett.* **1989**, *159*, 171.

(183)   Silver, J. A.; Kolb, C. E. *Chem. Phys. Lett.* **1980**, *75*, 191.

(184)   Bulatov, V. P.; Buloyan, A. A.; Iogansen, A. A.; Sarkosov, O. M.; Cheskis, C. G. *Sov. J. Chem. Phys.* **1982**, 878.

(185)   Sarkisov, O. M.; Cheskis, S. G.; Nadtochenko, V. A.; Sviridenkov, E. A.; ; Vedeneev, V. I. *Arch. Combust.* **1984**, *4*, 111.

(186)   Diau, E. W.-G.; Tso, T.-L.; Lee, Y.-P. *J. Phys. Chem.* **1990**, *94*, 5261.

(187)   Bamford, C. H. *Trans. Faraday Soc.* **1939**, *35*, 568.

(188)   Serewicz, A.; Noyes, W. A., Jr. *J. Phys. Chem.* **1959**, *63*, 843.

(189)   Srinivasan, R. *J. Phys. Chem.* **1960**, *64*, 679.

(190)   Jayanty, R. K. M.; Simonaltis, R.; Heicklen, J. *J. Phys. Chem.* **1976**, *80*, 433.

(191)   Phillips, L. F. *Chem. Phys. Lett.* **1987**, *135*, 269.

(192)   Farber, M.; Harris, S. P. *J. Phys. Chem.* **1984**, *88*, 680.

(193)   Foner, S. N.; Hudson, R. L. *J. Chem. Phys.* **1958**, *29*, 442.

(194)   Silver, J. A.; Kolb, C. E. *J. Phys. Chem.* **1987**, *91*, 3713.

(195)   Hall, J. L.; Zeitz, D.; Stephens, J. W.; Kasper, J. V. V.; Glass, G. P.; Curl, R. F.; Tittel, F. K. *J. Phys. Chem.* **1986**, *90*, 2501.

(196)   Dolson, D. A. *J. Phys. Chem.* **1986**, *90*, 6714.

(197)   Unfried, K. G.; Glass, G. P.; Curl, R. F. *Chem. Phys. Lett.* **1990**, *173*, 337.

(198)   Stephens, J. W.; Morter, C. L.; Farhat, S. K.; Glass, G. P.; Curl, R. F. *J. Phys. Chem.* **1993**, *97*, 8944.

(199)   Crowley, J. N.; Sodeau, J. R. *J. Phys. Chem.* **1990**, *94*, 8103.

(200)   Harrison, J. A.; Maclagan, R. G. A. R.; Whyte, A. R. *Chem. Phys. Lett.* **1986**, *130*, 98.

(201)   Crowley, J. N.; Sodeau, J. R. *J. Phys. Chem.* **1987**, *91*, 2024.

Research in Chemical Kinetics, Volume 3
R.G. Compton and G. Hancock (editors)
© 1995 Elsevier Science B.V. All rights reserved.

Recent Advances in the Kinetics of Radiolytic Processes

Simon M. Pimblott[a] and Nicholas J. B. Green[b]

[a]Radiation Laboratory, University of Notre Dame, Notre Dame, Indiana 46556, U.S.A.

[b]Department of Chemistry, King's College London, Strand, London WC2R 2LS, U.K.

Theoretical methods for elucidating the ultra-fast kinetics in the fast electron radiolysis of liquids are described and discussed.

1 INTRODUCTION

As a high energy radiation particle, such as a fast electron, passes through a liquid it leaves in its wake a track of events where energy has been transferred from the particle to the liquid.[1,2] Each energy loss event gives rise to a discrete ionization or excitation, or to a correlated group of ionizations and excitations. The molecular ions, low energy electrons and excited molecules undergo very rapid thermalization, fragmentation, and ion-molecule reactions, and on a picosecond time scale these processes result in a spatially nonhomogeneous cluster of highly reactive radicals and ions.[3,4] The radiation-induced species diffuse rapidly and undergo near diffusion-controlled reaction.[5] Consequently, irradiation is followed by a period of ultra-fast chemistry which reflects not only the reaction mechanisms, but also the initial spatial distribution. The time scales of importance in radiolysis are summarized in Figure 1.[6]

The short-time chemical kinetics following irradiation contain the most direct experimentally available information about the primary processes of radiolysis, and an accurate theory is necessary to extract this information. Constructing a framework for modeling poses a number of problems: (i) the determination of the relative disposition of the primary energy loss events, (ii) the description of the physico-chemical processes which determine the number of reactants in each radiation-induced cluster, (iii) the calculation of the spatial distribution of the reactants within each cluster, and (iv) the diffusion-kinetic modeling of the chemical reactions. Any complete analysis must contain an acceptable treatment of all these problems.

This article describes recent progress in the understanding of fast electron radiolysis. The following section discusses the energy loss of a fast electron in water and the consequences of the energy transfer. The third section of the review describes different theoretical treatments of radiation-chemical kinetics. The development of the kinetic models and their application to problems of particular interest are detailed, and studies addressing some of the important unresolved questions are reviewed. Finally, the different models are critiqued.

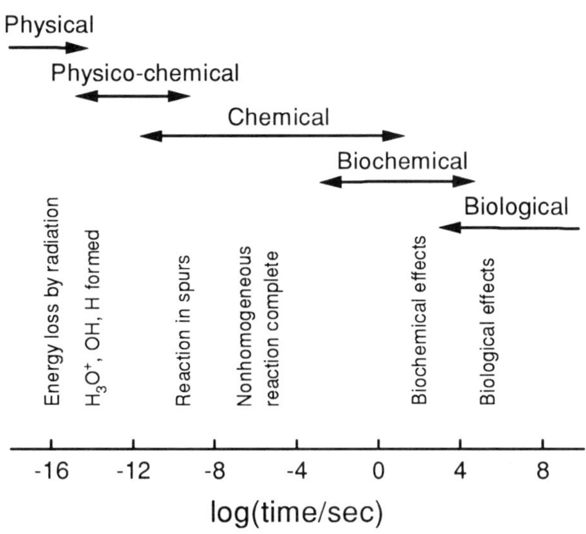

Figure 1. Time scales of the various processes of importance in radiolysis.

2 FAST ELECTRON RADIOLYSIS

2.1 Energy loss and spurs

The spatial distribution of the primary energy loss events along a radiation track is determined by the properties of the particle and of the medium[7]. The effects of radiation quality on chemistry or biology are usually parameterized in terms of the stopping power or linear energy transfer (LET) of the radiation, $S = - dE/dx$, which is a measure of the average rate at which energy is lost per unit length of track.[8]

Recently, accurate theoretical formulations for the stopping power, S, and for the inelastic mean free path, Λ, have been developed.[9,10] The theory is based upon an extrapolation of the dielectric response function of the medium in the energy-momentum plane,[11] and relates the energy loss properties of the electron to the dipole oscillator strength distribution of the medium, $f(\varepsilon)$. The resulting equations for the stopping power and for the mean free path have the form[12,13]

$$S = 2\chi n \int_{\varepsilon_o}^{\varepsilon_{max}} f(\varepsilon)G(\varepsilon / E)d\varepsilon \tag{1}$$

$$\Lambda^{-1} = \chi n \int_{\varepsilon_o}^{\varepsilon_{max}} f(\varepsilon)L(\varepsilon / E) / \varepsilon d\varepsilon \tag{2}$$

where G and L are functions of ε/E, $\chi = \pi e^4 / E$, with e and m being the charge and the rest mass of an electron, n is the number density of molecules, and ε_{max} and ε_0 are the maximum and minimum permitted energy losses, respectively.

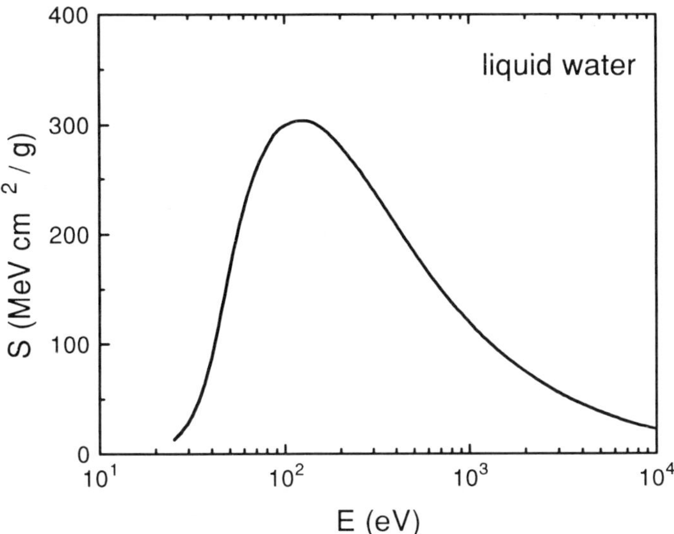

Figure 2. Effect of incident energy on the density-normalized stopping power of liquid water for electrons.

The energy dependence of the density-normalized stopping power of an electron in liquid water is shown in Figure 2, and the corresponding dependence of the density-normalized inelastic mean free path is presented in Figure 3. The figures demonstrate that for fast electrons the primary energy loss events are well-separated. Given a typical inter-cluster separation of 100 nm and a radical diffusion coefficient of 10^{-8} m^{2}s^{-1} this means that clusters of reactants

120

(commonly known as spurs) are effectively isolated from one another on a microsecond time scale.

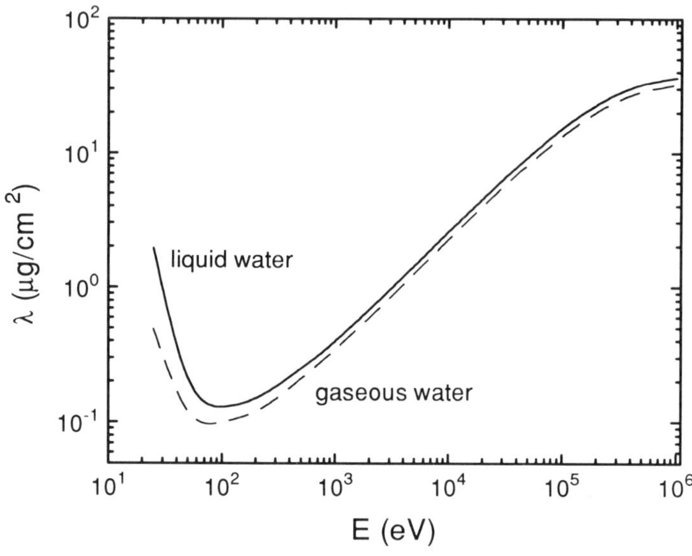

Figure 3. Density-normalized inelastic mean free path of electrons in water.

Stopping power is an average quantity which does not make any allowance for the spatial integrity of the energy loss events comprising a track. A variety of different methods have been developed for modeling the structure of electron tracks.[13-19] The simplest of these calculations give only the probability distribution of the energy losses [13-15], while more complex calculations attempt to model the detailed spatial structure of the track.[16-19] The accuracy of all the results is limited by the cross-sections employed. Accurate inelastic cross-sections are available for many gases.[13,15,20] Some cross-sections have been inferred from studies of condensed films[21], but information for condensed media is **very** limited.[15,22]

Using the same approximation employed for the stopping power and for the mean free path, the inelastic cross-section for events with energy less than ε can also be expressed in terms of the dipole oscillator strength density,[13,15]

$$\sigma(\varepsilon, E) = \left(\Lambda^{-1}(E) + (\chi / E) \int_{\varepsilon_o}^{\varepsilon} f(\gamma) M(\gamma / E, \varepsilon / E) d\gamma \right) / n \qquad (3)$$

where M is a known function of γ, ε and E. Figure 4 compares the normalized cumulative inelastic cross-sections,[15]

$$Y = \sigma(\varepsilon, E) / \sigma(\varepsilon_{max}, E), \tag{4}$$

for 1 MeV electrons in gaseous and in liquid water. Energy losses are typically larger in liquid water than in water vapor. Taken in conjunction with the differences shown in Figure 3, this means that any attempt to model a track in liquid water using gas-phase cross-sections[23,24] will give energy losses which are too small and too frequent, and therefore the model of the track will not be accurate.

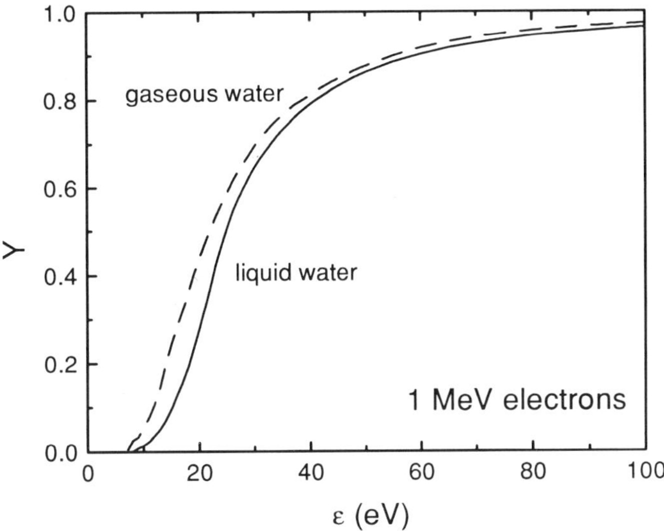

Figure 4. Normalized cumulative cross-section for inelastic collisions for a 1 MeV electron in gaseous water (- - -) and in liquid water (———).

Straightforward Monte Carlo simulation techniques employing electron energy-dependent inelastic cross-sections, like those shown in Figure 4, have been used to model the energy degradation of fast electrons and the probability distribution function of the energy losses.[13,15,22] By convention, primary energy loss events of energetic electrons are divided into three classes : spurs, blobs and short tracks.[14] Spurs are events of less than 100 eV, blobs are between 100 eV and 500 eV and short tracks are between 500 eV and 5 keV. Energy loss

122

events greater than 5 keV are assumed to produce energetic electrons that can be treated in the same way as the primary electron. This categorization of energy loss events is arbitrary and was originally introduced to reflect the spatial nonhomogeneity of the clusters of reactants produced.[14]

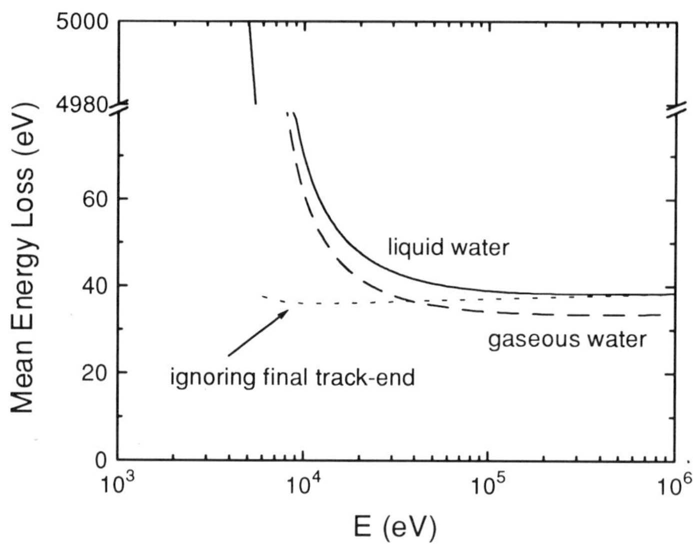

Figure 5. Effect of incident electron energy on the mean energy loss in gaseous water (- - -) and in liquid water (———).

The calculated mean energy loss per event in water as a function of incident electron energy is shown in Figure 5.[15] The increase in the mean energy loss as the energy of the primary electron decreases is simply a result of the convention that the final 5 keV track-end of any track is regarded as a single entity. If the final track end is ignored, the mean energy loss is almost independent of primary electron energy. The mean energy loss for a 1 MeV electron is 34 eV in the gas and 38 eV in the liquid. However, these means are influenced by a long tail towards high energy, as shown in Figure 6.[15] The energy loss distribution for liquid water is clearly shifted to higher energies than that for gaseous water. The most probable energy loss in the gas is 14 eV, while in the liquid it is 22 eV. For energies larger than 10 keV, the incident energy has very little effect on the energy loss distribution in both phases.[15] Clearly differences between the two phases are significant and must be recognized in any acceptable analysis of the radiation chemistry of aqueous systems or in radiation biology.

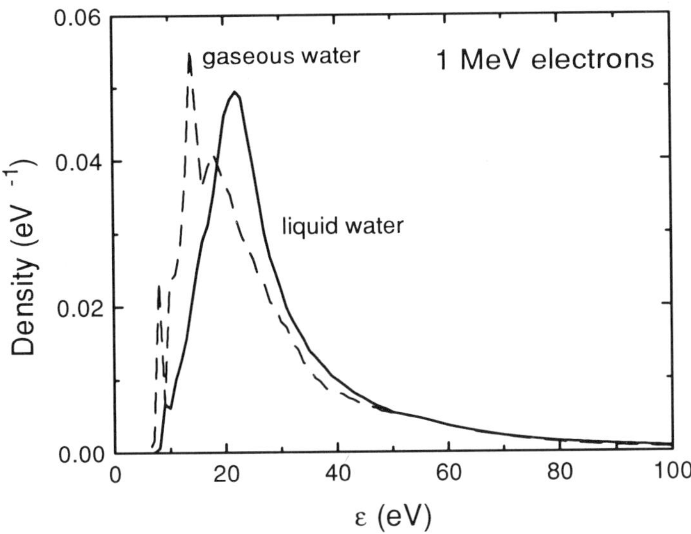

Figure 6. Probability density of energy losses along the track of a 1 MeV electron in gaseous water (- - -) and in liquid water (———).

Similar calculations have been performed for ice,[15] gaseous hydrocarbons[13], ethers and alcohols[20] and for dry DNA.[22] There are large differences between the predictions, reflecting differences in the UV spectra of the media, and hence the dipole oscillator strength distributions, $f(\varepsilon)$, used in the calculation of the cross-sections. Results obtained for the series of gaseous alkanes C_nH_{2n+2} are all very similar, but they differ from those for gaseous water. The largest differences seem to be between the gas phase and the condensed phases; predictions for liquid water and ice and for dry DNA show some similarity, but are very different from all the gas phase media studied, including gaseous water. The phase effects observed are not simply due to density, as the energy loss spectrum shifts. Since the energy loss properties of fast electrons depend upon the irradiated medium, it is important to use the correct model system in any study.

More sophisticated Monte Carlo calculations have been used to provide idealized track structures in gaseous and liquid water, which means in the first instance the spatial positions of the energy losses.[16-19] While the spectrum of energy losses in liquid water, and in numerous gases, is now well-characterized, other important features which are necessary to generate a track in 3 dimensions, such as the angle-dependence of differential cross-sections and the behavior of low-energy electrons, are not well-understood. These calculations rely heavily upon the cross-sections employed. While most of the necessary cross-sections are available for gaseous water, the most studied medium, there is little information for condensed media. The differential cross-sections necessary to describe many of the elastic and inelastic processes are simply not available for condensed media and have to be extrapolated from the gas phase.[19]

2.2 Ionization and excitation distribution

The modeling of the radiation chemical kinetics following irradiation requires a description of the initial spatial distribution of the radiation-induced reactants.[5,25-30] This requires more information than simply the energy losses which comprise the track, since kinetic calculations require a distribution describing the chemical contents of the spurs and their spatial distributions.[5,28,30] The first step towards this end is to analyze the radiation physics of the energy loss: at the simplest level, given a particular energy loss, is the event an ionization or an excitation?

There is a great deal of knowledge about radiation physics in the gas phase, where the structure in the ultra-violet optical spectrum gives cross-sections and thresholds for well-defined excitations and ionizations. However, experiments on amorphous condensed matter are much more difficult, and the ultra-violet spectrum is less well understood. For many years, for example, there has been controversy about the assignment of the 22 eV maximum in the UV spectrum of liquid water.[31]

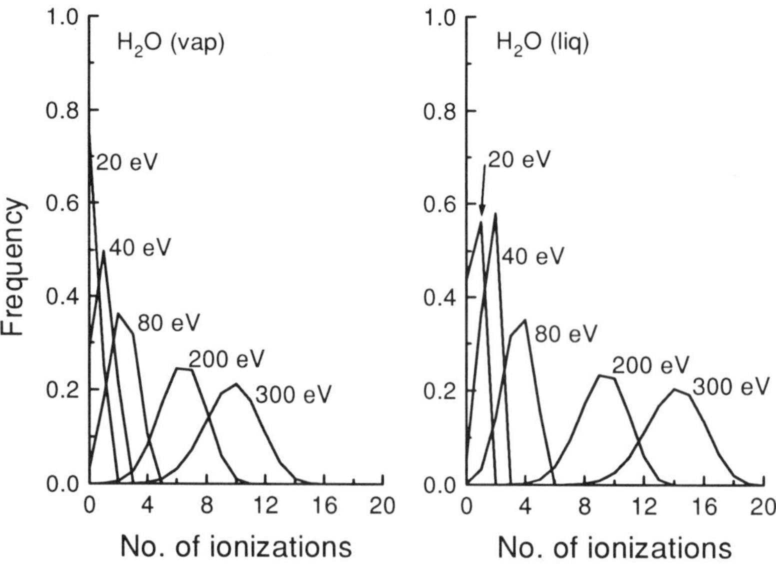

Figure 7. Effect of secondary electron energy on the distribution of the number of ionizations.

Recently a new approximate, analytic theoretical approach has been developed for the calculation of the entity-specific ionization and excitation probabilities for the fast electron radiolysis of water.[32] This method incorporates existing data on the ionization threshold and the energy dependence of the ionization efficiency. When applied to water vapor these analytic calculations are in close agreement with the predictions of Monte Carlo track structure calculations.[17]

Figure 7 shows the effect of secondary (progenitor) electron energy on the probability distribution of the number of ionizations occurring in gaseous[32] and in liquid water. The numbers of ionizations and excitations are not exactly proportional to the spur energy. For a given energy loss the yield of (secondary) ionizations is considerably higher in the liquid than in the gas. This difference primarily reflects the smaller amount of energy needed to create an ionization in the liquid compared to the gas.

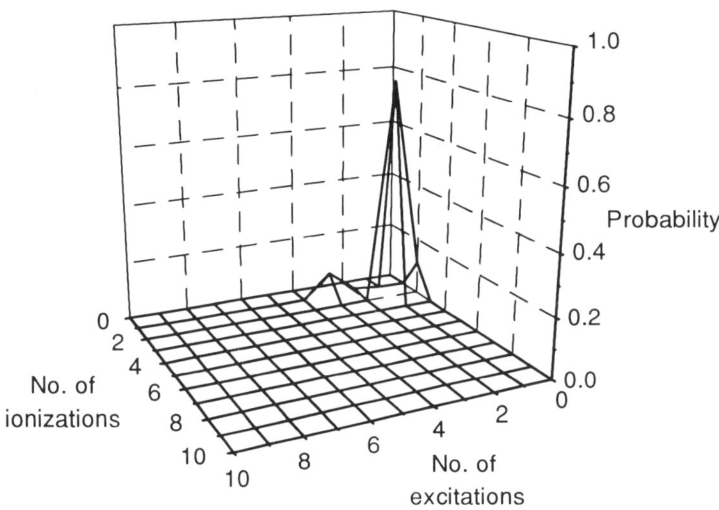

Figure 8. Distribution of numbers of ionizations and excitations for a 40 eV spur in liquid water.

The joint distribution of the numbers of ionization and excitation events for a 40 eV spur in liquid water is shown in Figure 8. Figure 9 shows the corresponding distribution per event in a 1 MeV track.[32] The distribution for the "typical" 40 eV energy loss event is very different from that for the whole track. The most probable spurs in a 1 MeV track contain one

126

ionization, two ionizations or one ionization and one excitation, while a 40 eV spur usually contains an excitation and one or two ionizations. The differences between Figures 8 and 9 show that it is necessary to incorporate an acceptable model for the whole track in a full kinetic model of radiolysis. Since the chemistry of a spur depends strongly on the numbers of ionizations and excitations it contains, the "typical" spur model is not likely to give accurate insights into the relationship between track structure and chemistry.

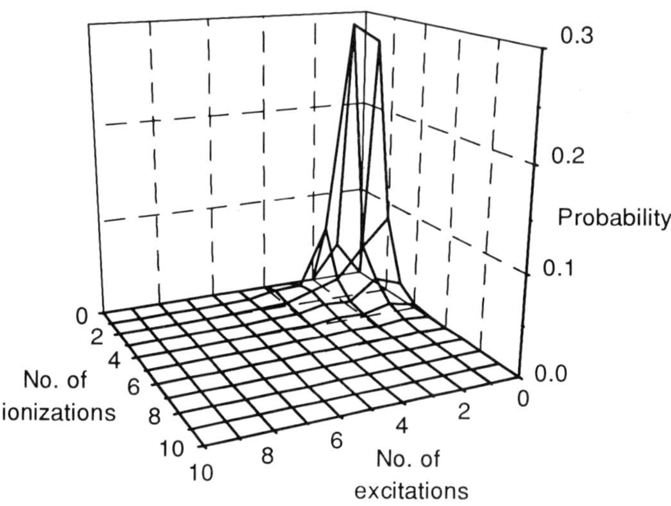

Figure 9. Distribution of ionizations and excitations for the fast electron radiolysis of liquid water.

The radiation physics of liquid hydrocarbons has not been studied in the same detail as water, and the distributions of the numbers of ionizations and excitations in spurs are frequently assumed to be the same as in water.[33] As there are significant differences between the energy loss distributions and the ionization thresholds in water and in hydrocarbons, this assumption is doubtful.

Although the problem of excitation and ionization is a major gap in our knowledge at present, the theoretical advances described here seem to provide a promising way forward. It is likely that studies using synchrotron radiation, where the energy deposited in an event can be carefully controlled, will provide new experimental data, which will enable this approach to be tested properly.

2.3 Thermalization and fragmentation

The excited states and ions produced in radiolysis only have very short lifetimes in liquid water. They rapidly fragment or react with surrounding molecules. The electrons produced in ionization events thermalize and are solvated very rapidly. Recent photoionization experiments have given considerable insight into the formation of solvated electrons and OH radicals.[34-38] Additional information about these processes has been provided by low-energy electron scattering experiments,[21] and synchrotron studies. However, the current state of knowledge is still very incomplete, even in a qualitative sense. It seems likely that not all the important processes have been identified yet; for example, a resonance in the low-energy electron energy loss spectrum of amorphous ice has recently been found, which coincides with the direct production of molecular hydrogen.[21] This could be dissociative electron attachment to a water molecule, and has not been included in any published code for modeling radiation chemistry to date.

At present, therefore, there is no detailed quantitative understanding of fragmentation and thermalization processes of the molecular ions, low energy electrons or excited states produced, either in liquid water or in other condensed media. As a result, it is necessary to make simple parameterizations to convert a given energy loss to the spatial distribution of reactive species that is necessary to model chemistry. These parameters do not usually have any theoretical foundation, but are manipulated to give acceptable initial yields of the radiation-induced species.[25,29] This is one of the major unsolved problems of radiation chemistry.

3 RADIATION CHEMICAL KINETICS

In gaseous water, gaseous hydrocarbons, liquid water and solid DNA, well over 70 percent of the energy of a fast electron is lost in primary events smaller than 100 eV,[22] which result in fewer than six or seven ionization/excitation events.[32] The initial distribution of the reactants in these small clusters is spatially nonhomogeneous and consequently the diffusion of the reactants as well as the reaction rate parameters influences the fast chemistry following radiolysis.

A number of different theoretical treatments have been developed for studying the nonhomogeneous radiation chemical kinetics of solutions. These techniques can be classified into two broad groups: deterministic[5,39-42] and stochastic.[26,28-30] The former use conventional macroscopic treatments of concentration, diffusion and reaction to model the chemistry of "typical, average" spurs. In contrast, stochastic analyses consider realistic clusters of reactants and use probabilistic methods to model the kinetics. Each approach has advantages and limitations. At present both treatments have a valuable role to play in the modeling of short time radiation chemistry.

3.1 Deterministic kinetic models

Conventional deterministic treatments of spur kinetics model the fast chemistry using a typical spur[41,42] or distribution of spurs.[5] The spur is characterized by a space-dependent concentration c_i for each species i and the spur chemistry is described by a set of coupled diffusion-reaction equations. Each equation comprises terms which describe the diffusion of species i, its reaction with other radiation-induced species and with homogeneously distributed solutes (scavengers), and its production by reaction,

$$\frac{\partial c_i}{\partial t} = D_i \nabla^2 c_i - \sum k_{ij} c_i c_j + \sum k_{mn} c_m c_n. \tag{5}$$

In the equation D_i is the diffusion coefficient of species i. The initial spatial distributions of the reactants are represented by spherical concentration profiles, which are usually assumed to be Gaussian,

$$c_i(r) = \frac{N_i}{(2\pi\sigma_i^2)^{3/2}} exp(-r^2 / 2\sigma_i^2), \tag{6}$$

where N_i is the number of particles of species i in the spur, although alternative distributions have also been used.[43] The parameter σ_i is the standard deviation of the Gaussian distribution, while N_i is calculated from the initial G-value, $N_i = G_i^0 \varepsilon_{spur} / 100\ eV$ (where ε_{spur} is the energy of the spur). The set of diffusion-reaction equations is solved by numerical methods[41,42] or by using the prescribed diffusion approximation,[5,39] which assumes that the concentration profile remains Gaussian and analyzes the way in which the Gaussian spreads by diffusion and reaction.

3.1.1 FACSIMILE model

Several different numerical treatments have been developed for solving the set of coupled diffusion-reaction equations (5).[41-43] Most recent studies have been performed using a technique in which a "typical, average" spur is used as a representation of the electron track. The spur is divided into concentric shells as shown in Figure 10.

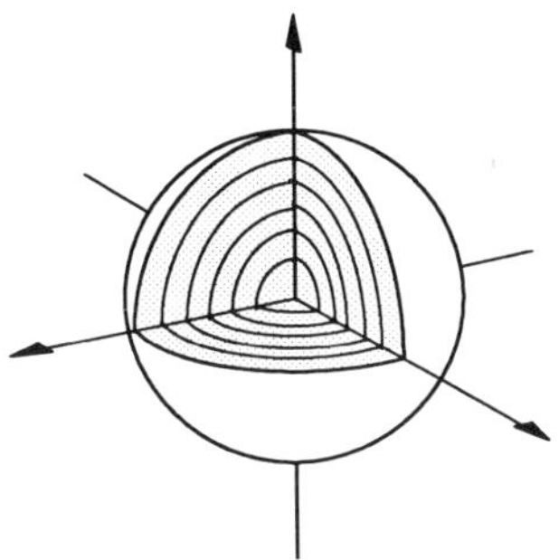

Figure 10. Division of the typical spur into concentric shells in the deterministic treatment of spur kinetics using the FACSIMILE algorithm.

Each shell is sufficiently thin that the concentration of reactants within the shell can be regarded as constant. Diffusion takes place between adjacent shells and reactions within a shell are modeled using deterministic rates. The resulting set of coupled equations describing the diffusion and the kinetics is solved using the FACSIMILE implementation of the Gear algorithm.[44] The parameters defining the spur for aqueous solutions have been optimized to match measured experimental yields in simple scavenger studies.[42] The optimum standard deviations for the initial concentration profiles, σ_i, are given in Table 1 and the energy of the spur, ε_{spur}, is 62.5 eV.

3.1.2 Prescribed diffusion

The prescribed diffusion approximation employed for solving the coupled diffusion-reaction equations (5) assumes that reaction affects only the density of the reacting particles and not the form of their nonhomogeneous concentration profile which is only modified by diffusion. The prescribed diffusion approximation is suggested by the Green's function for the diffusion equation being a Gaussian whose variance increases linearly with time so that

$$\sigma(t)^2 = \sigma(t=0)^2 + 2Dt. \tag{7}$$

The prescribed diffusion methodology was originally used to study the kinetics of prototype single species systems[4,45,46] and has been extended to more general reaction schemes.[5]

Table 1. Diffusion coefficients and spur parameters for the radiation-induced species in aqueous solutions.[42]

Reactant	$D \times 10^9$, m^{2}s^{-1}	σ, nm	G^0, molec./ 100eV	G^{esc}, molec./ 100eV	α, ns
e_{aq}^-	4.5	2.23	4.8	2.55	0.91
H_{aq}^+	9.0	0.85	4.8	-	-
OH	2.8	0.85	5.50	2.56[a] 2.66[b]	0.30[a] 0.26[b]
H	7.0	0.85	0.42	-	-
H_2	5.0	0.85	0.15	0.44	0.96
OH^-	5.0˙	-	-	-	-
H_2O_2	2.2	-	0	0.72	0.38

[a] deaerated solution, [b] aerated solution

The assumption of a linear time dependence for σ is incorrect if the reaction kinetics are nonlinear, as in radiolysis.[6,47,48] Several attempts have been made to modify the approximation to account for faster reaction close to the center of a spur, due to second order kinetics.[5,39] These modifications introduce a nonlinear time dependence for the width of the concentration profile, i.e.

$$\frac{d\sigma^2}{dt} = 2D + f(t).$$

(8)

The form of $f(t)$, depends on the approximation by which it is derived.

3.2 Stochastic methods

A variety of different stochastic treatments are being developed for modeling fast radiation-chemical kinetics.[49] By far the most commonly used approach involves the Monte Carlo random flights simulation of the trajectories of the radiation-induced radicals and ions.[29,50-52] This type of calculation is elaborate and computer intensive, but provides detailed information about both the diffusion and the reaction processes. As the technique is vulnerable to discretization (time step) errors, it is important to validate the simulation methodology to demonstrate that the predictions are quantitatively reliable.[49]

Recently, a more efficient simulation method, the independent reaction times (IRT) model, has been developed which requires only the generation of random reaction times for the radiation-induced reactants.[30,53-55] This technique is based on the independent-pairs approximation that is implicit in the Smoluchowski-Noyes treatment of diffusion-limited reactions.[56] An elegant, analytic master equation (ME) formulation for spur kinetics, which relies on the independent pairs approximation, has also been described.[28,57-59]

3.2.1 Random flights simulation

The random flights modeling of radiation chemical kinetics involves the simulation of the diffusion and the reaction of the particles in an isolated track entity or a "representative section of track". Several simulation methodologies at different levels of sophistication have been developed, and have been described in detail.[29,50-52] Each simulation begins from a realizable configuration, which may be generated at random from the assumed spatial profile of a spur[50-52] or from an idealized track structure.[29] The diffusive trajectories of all the surviving particles are simulated using a discretization of the appropriate stochastic differential equations.[49] Reaction is modeled by testing for overlap at the end of a time step or, as it is quite possible for two particles to diffuse together and then separate during a time step, by estimating the probability of encounter during a time step using a bridging process.[60] Each realization is continued until all reaction is complete or a predefined cut-off time is attained. The kinetics are obtained by repeated simulation. Typically 10^5 to 10^6 realizations are required to obtain statistical accuracy.

The methodology for random flights simulation has been described in detail.[51,52,54,61] In the calculations reported here the diffusive jump of a particle i is given by an Euler-type discretization of the stochastic differential equation[62,63]. Each diffusive jump has two components,

$$\delta \mathbf{r}_i = \sqrt{2D_i \delta t}\, \mathbf{N}_3(0,1) + \frac{D_i \mathbf{F}_i}{k_B T} \delta t \tag{9}$$

where D_i is the diffusion coefficient of the particle, δt is the length of the time step and $\mathbf{N}_3$ is a three dimensional normal random vector of unit variance.[51,52] The first term in the expression (9) represents the motion caused by the buffeting of the solvent and the second reflects the drift caused by any force $\mathbf{F}$. Simulations have been reported, and the methodology validated, for several types of idealized system.[51,52,54,59,61]

For neutral reactants the force $\mathbf{F} = 0$, and therefore[52]

$$\delta\mathbf{r}_i = \sqrt{2D_i\delta t}\,\mathbf{N}_3(0,1) \tag{10}$$

In this case, the simulation provides exact samples from the diffusive trajectories. When the particles are charged, the strong Coulomb inter-ion forces modify the diffusion. The force on ion i is then described by[51,52]

$$\mathbf{F}_i = -k_BT\sum_j \frac{z_iz_jr_c}{r_{ij}^3}\mathbf{r}_{ij} \tag{11}$$

where z_i is the charge on ion i, and $\mathbf{r}_{ij}$ is the inter-ion vector. The Onsager distance, r_c, is the distance at which the potential energy due to the Coulomb interaction of two unit charges has magnitude k_BT. In the presence of an applied electric field, $\mathbf{E}$, the expression for the force on an ion is further modified, giving[59]

$$\mathbf{F}_i = -k_BT\sum_j \frac{z_iz_jr_c}{r_{ij}^3}\mathbf{r}_{ij} + z_ie\mathbf{E}\,. \tag{12}$$

where e is the electron charge.

In some simulations in the literature, the length of the time step, δt, is fixed.[50] This fixed time step treatment is computationally inefficient: when the particles are well-separated the probability of encounter is small. Sophisticated methods have been developed which allow more efficient computation by incorporating variable time steps.[52,54] In these treatments, the time step, δt, is determined by the proximity of the particles. The time step is selected by one of two methods: (i) so that the pair with the shortest interparticle separation has an encounter probability of less than 1 percent for the time step[54], or alternatively (ii) so that the interparticle drift of no pair changes by more than 10 percent during the time step.[52] In both treatments, the minimum time step becomes increasingly small as two particles approach and a minimum time step of at least 0.1 ps is usually employed to ensure that the simulation does not become prohibitively slow.[64]

If a reaction is diffusion-controlled then reaction occurs with certainty on encounter. In a simulation employing finite time steps, it is feasible for the simulated trajectory of a pair to encounter during a time step and then "separate" before the end of the time step. The pair should react, but the reaction would not be registered in a simple model which only tests for overlap at the end of the time step. The modeled chemistry then underestimates both the rate and the amount of reaction. As long as the time step in which the encounter occurs is sufficiently short that no third particle interferes, the problem of encounter during a time step can be overcome using the conditional encounter probability for an interpolating "bridging process".[60] Two forms of bridging process have been used : the Bessel bridge, which assumes that the separation behaves as a Bessel process, and the Brownian bridge in which the separation between the two particles is approximated by a Wiener process.

In the absence of any interparticle forces, the separation between two diffusing particles is a Bessel process, and the probability of two particles with relative diffusion coefficient D' encountering at a distance a while diffusing from a separation x to a separation y in a time step δt is

$$W_{Bes} = \frac{exp(-(x-a)(y-a)/D'\delta t) - exp(-xy/D'\delta t)}{1 - exp(-xy/D'\delta t)}. \tag{13}$$

The Bessel bridge is applicable to any size time step for a pair of uncharged reactants. The Brownian bridge is based on the assumption that the radial drift of the pair does not change significantly during the time step. This limitation is equivalent to method (ii) for determining the length of the diffusive time step. Under these circumstances, the probability of encounter is

$$W_{Bro} = exp(-(x-a)(y-a)/D'\delta t). \tag{14}$$

The Brownian bridge can be applied to both charged and neutral pairs.

The geminate recombination of a pair of neutrals and of a pair of ions in solvents of differing permittivity (water, alcohols, ethers and hydrocarbons) have been simulated using the random flights method described above, incorporating the Coulomb and applied forces. The simulations match the corresponding exact solutions of the Debye-Smoluchowski equation accurately,[51,52,54,59,61,64] as demonstrated in Figures 11, 12 and 13. The time dependence of the diffusion-controlled reaction of a geminate pair of neutrals is shown in Figure 11. Included in the figure are the results of simulations ignoring the possibility of encounter and separation during a time step, but using the same reaction distance. The consequences of this error are very significant; the time dependence of the modeled kinetics is too slow and the escape probability is overestimated. Clearly, neglecting the possibility of encounter during a diffusive time step is equivalent to using a reaction distance smaller than the distance at which encounter occurs.

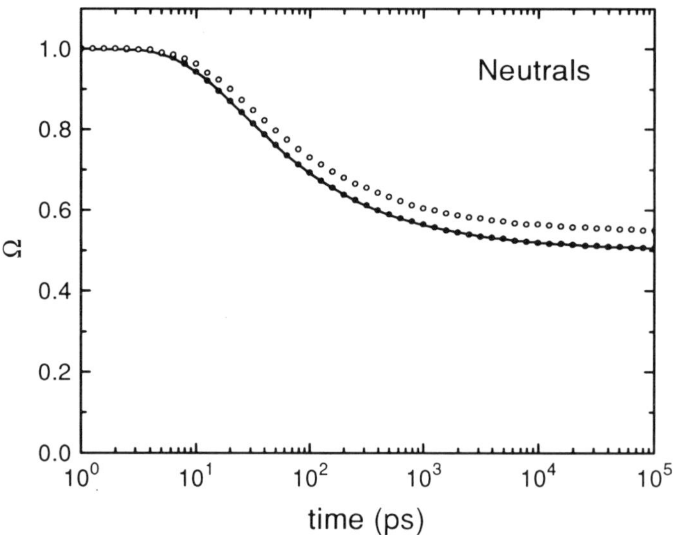

Figure 11. Geminate recombination kinetics of a pair of uncharged radicals initially 1 nm apart with relative diffusion coefficient 5×10^{-9} m^2 s^{-1}. The full line (———) is the exact analytic solution, the filled points (●) are the predictions of random flights simulation incorporating the Brownian bridge and the open points (O) are the predictions of random flights simulation with the possibility of encounter and separation during a time step neglected.

The simulated reaction kinetics of a molecular cation - molecular anion pair in a low permittivity solvent like n-hexane are presented in Figure 12. Random flights simulation accurately reproduces the numerical solution of the Debye-Smoluchowski equation. The results of simulations examining the effect of weak applied electric fields on the yield of geminate ion-pair escape are given in Figure 13.[59] The escape probability predicted by random flights simulation is independent of the relative diffusion coefficient of the ion pair and is in good agreement with that predicted by an analytic Onsager analysis.

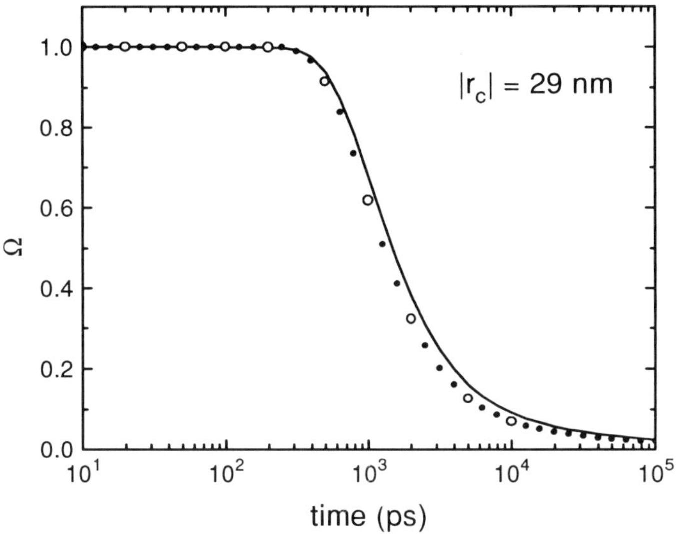

Figure 12. Geminate recombination kinetics of an ion pair initially 6 nm apart with relative diffusion coefficient 2.5×10^{-9} m^2 s^{-1} in a n-hexane-like low permittivity solvent with $|r_c| = 29$ nm. The open points (O) are the numerical solution of the Debye-Smoluchowski equation, the filled points (●) are the predictions of random flights simulation and the solid line (———) is the extrapolated solution from the grid constructed for the generation of reaction times in the IRT model (see next section).

The random flights simulation technique has been extended to include systems where separation and re-encounter occur, i.e. where reaction is only partially diffusion-controlled [61] or spin-controlled. When re-separation is possible following an encounter, it is necessary to modify the simulated positions of the encountering particles, as they are not permitted to penetrate one another. The technique for calculating the positions of the particles following a reflecting encounter is described in detail in reference [61], but in the simplest case, where the two particles have the same diffusion coefficient, it is essentially as follows.

If the positions of the two particles are denoted X_1 and X_2 then the two vectors $R = X_2 - X_1$ and $S = X_2 + X_1$ are considered. Following a diffusive jump R and S become $R+\delta R$ and $S+\delta S$, and the latter is unaffected by a reflective encounter. The new vector $R+\delta R$ is resolved into components, one parallel and one perpendicular to R. Of these only the parallel component is altered by an encounter. The distance of closest approach, which is prescribed by the nature of the collision, is calculated and this distance is used to modify the perpendicular component of the vector $\mathbf{R}+\delta\mathbf{R}$. A new vector is constructed to replace $\mathbf{R}+\delta\mathbf{R}$ and the new positions $\mathbf{X}_1+\delta\mathbf{X}_1$ and $\mathbf{X}_2+\delta\mathbf{X}_2$ evaluated from this new vector and from $\mathbf{S}+\delta\mathbf{S}$.

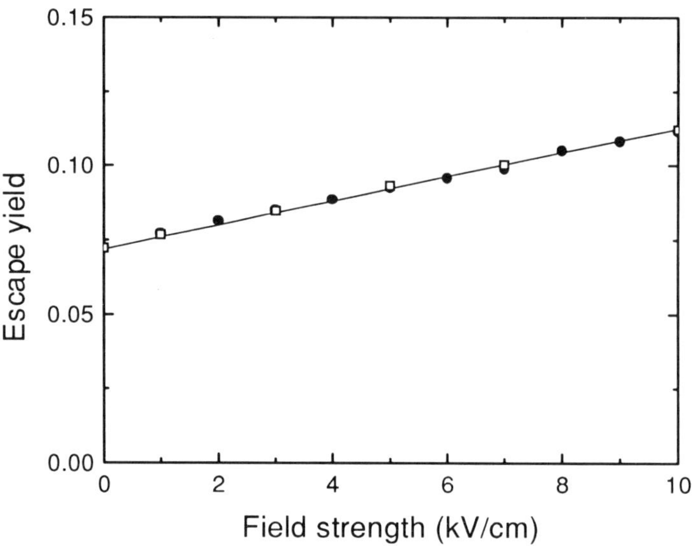

Figure 13. Effect of applied field strength on the escape probability of an ion pair in n-hexane. The initial relative distribution is a Gaussian of standard deviation 6.5 nm. The solid line is the analytic Onsager solution and the points refer to the predictions of random flights simulation for a molecular cation - molecular anion pair ($\bullet$) and for a molecular cation-electron pair ($\square$).

No additional approximations are introduced in the simulation of the kinetics of multiple particle clusters, therefore the observed agreement for a geminate pair confirms random flights simulation as an acceptable model with which to test more approximate modeling techniques.

3.2.2 Independent reaction times model

Full random flights treatment of diffusion and reaction provides a great deal more information than is required for kinetic studies, which ultimately use only the reaction times of the reacting radicals and ions. The IRT model is a Monte Carlo simulation technique which employs the independent pairs approximation to allow the calculation of reaction times without having to chart the diffusive trajectories of the reactants.[53] As in the random flights method, the calculation starts from a set of coordinates generated from a suitable spatial distribution and deals appropriately with any particles formed in a reactive configuration. The inter-particle separations between surviving particles are then used to generate random reaction times from the reaction time probability distribution functions appropriate for isolated pairs with the same separations. It is here that the independent pairs approximation is employed - each reaction time is generated independently of any times already generated for other pairs. The minimum of the resulting ensemble of times is taken to be the reaction time of the first pair. This pair is allowed to react, and new reaction times are generated for any reactive products formed. The minimum of the new ensemble of times is the next reaction time. The simulation proceeds in this manner until the cut-off time is reached, or all the potentially reactive pairs have reacted. Kinetics for a given spur are obtained by repeated simulation. The methodology of the IRT model has been described in detail elsewhere.[30,51-54,61]

The success of the IRT model is largely dependent on using the correct probability distribution functions for generating the reaction times. This reaction time distribution function is determined by the nature of the pair and by its initial separation. For a diffusion-limited reaction, the function is assumed to be the time-dependent reaction probability, $W(r,a;t)$, for the pair as if it were in isolation. The reaction probability is obtained by solution of the backward, or adjoint, diffusion equation for the pair,[65]

$$\frac{\partial W}{\partial t} = D'\frac{\partial^2 W}{\partial r^2} + D'\left(\frac{2}{r} - \frac{dU}{dr}\right)\frac{\partial W}{\partial r} \tag{15}$$

Here r and a are the initial separation and the encounter distance of the pair, respectively, U is the potential energy of interaction between the pair in units of k_BT and D' is the relative diffusion coefficient.

For neutral particles, in the first approximation, there is no potential energy of interaction between the pair, i.e. $U = 0$, and exact analytic solution of the equation is possible both when reaction is fully diffusion-controlled,[66]

$$W(r,a;t) = \frac{a}{r}\,erfc\left(\frac{r-a}{2(D't)^{1/2}}\right) \tag{16}$$

and when it is partially diffusion-controlled (approximated using the radiation boundary condition)[67]

$$W(r,a;t) = \frac{a}{r(1+\delta)}\left(erfc(\beta) - exp(2\beta\gamma + \gamma^2)erfc(\beta+\gamma)\right) \tag{17}$$

where $\beta = (r-a)/(4D't)^{1/2}$, $\delta = D'/va$ and $\gamma = (D't)(1+\delta^{-1})/a$, with v being the reaction velocity defined by the radiation boundary condition.[68]

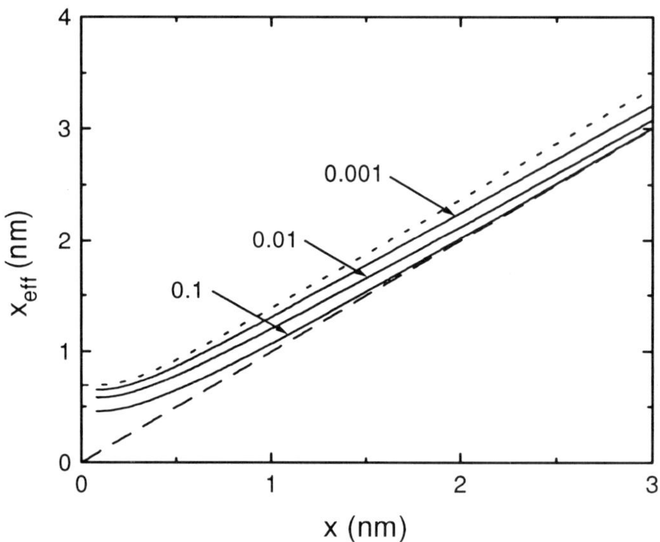

Figure 14. Comparison of effective distance scales for pure Coulomb potential (dotted line), Debye-Hückel screened Coulomb potential for solutions of ionic strength 0.001 M, 0.01 M and 0.1 M (solid lines) and zero potential (dashed line).

A number of approximate analytic solutions are also available for ionic reactions in high permittivity solvents.[69-76] It can be shown that for the Coulomb and for the screened Coulomb potential the approximations are equivalent to the use of an effective distance scaling[67], which represents the natural distance scale in the diffusion equation,[77] and governs both the ion-pair escape probability and the asymptotic approach to this escape probability. The transformed distance scale is[67]

$$x_{eff} = \left(-\int_{\infty}^{x} exp(U(y))/y^2 dy\right)^{-1} \tag{18}$$

where $U(y)$ is the distance-dependent inter-particle potential in units of k_BT. Figure 14 compares the distance scaling for a cation-anion pair in water. It shows the results for a pure Coulomb potential (dotted line) and for the screened potential arising from Debye-Hückel theory for solutions of ionic strength 0.001 M, 0.01 M and 0.1 M. Acceptable approximate solutions for $W(r,a;t)$ for an ion-pair in a high permittivity solvent, are obtained by substituting a_{eff} and r_{eff} for a and r in equations (16) and (17).[67]

Reaction times are generated from the (approximate) reaction time distribution functions by the inversion method.[78]

When the Coulomb force between an ion pair is strong, as in low permittivity solvents such as hydrocarbons, there is no suitable simple approximation for $W(r,a;t)$. The exact Laplace transform of the geminate reaction probability has been obtained,[79] however, the only exact solutions available in the time domain are numerical, either by numerical inversion of the Laplace transform[80] or by direct numerical solution of the backward equation.[76] Consequently, a method has been developed for generating reaction times for the IRT model by sampling and interpolation from a grid of numerical solutions. The technique has been described in detail elsewhere.[51] The important steps are the construction of the grid of solutions and the interpolation between the grid points for the reaction time.

The grid of numerical solutions was constructed in a coordinate space obtained by scaling the backward equation to a dimensionless form[81] using $\rho = 2r/r_c$ and $\tau = 4D't/r_c^2$, and then making a transformation,

$$W(r,a;t) = W(a,r;\infty)\, W^*(r,a;t), \tag{19}$$

to separate out the infinite time asymptote, $W(r,a;\infty)$, which is known analytically[67],

$$W(r,a;\infty) = a_{eff} / r_{eff}. \tag{20}$$

The resulting transformed backward equation[76]

$$\frac{\partial W^*}{\partial t} = \frac{\partial^2 W^*}{\partial \rho^2} + \frac{2}{\rho^2}(\rho - coth(\rho^{-1}))\frac{\partial W^*}{\partial \rho} \tag{21}$$

was solved using the Crank-Nicolson method over the range $-3 \leq ln(\rho) \leq 2$ and $-3 \leq ln(\zeta) \leq 1.5$, where ρ and ζ are the transformed inter-ion separation and reaction distance, respectively. Outside the range of this grid the approximate solutions of the backward equation detailed in references [76] and [82] are acceptable.

The random reaction time for an ion pair is generated by transforming the reaction distance and interparticle separation for the pair. The three closest lattice points in the solution grid are determined and a linear interpolation in the $ln(\rho)$-$ln(\zeta)$ plane gives the reaction time for a given reaction probability. The geminate pair reaction probability obtained by the sampling method is also included in Figure 12 and is in good agreement with the numerical solution and with the random flights simulation.[51,64]

The random flights and IRT simulation techniques have been widely used to study stochastic effects in idealized systems and to examine radiation chemical problems in water, in intermediate permittivity solvents like alcohols and in hydrocarbons.[64] Examples of the use of these techniques are presented in Section 5. Some attempts have also been made to apply the IRT method to complete radiation tracks or track sections, modeled by Monte Carlo methods.[19,26,83] These methods are still in a developmental stage, but it seems likely that the IRT method will be a very useful tool in analyzing realistic tracks in the near future.

3.2.3 Master equation analysis

The random flights simulation and the IRT method both treat particles individually, so that each realization of the simulation gives a typical life history of the system modeled. In a given realization the number of particles of each species is an integer-valued function of time. Many independent realizations are required to obtain statistically significant kinetics. The master equation method attempts to avoid the need for many repetitive simulations by analyzing the probability distribution of the spur contents. This can be done in the following way. A state space is set up in which the states are defined by the number of particles of each species present. Every time a reaction occurs some reactants are removed and the appropriate products added; this corresponds to a transition from one state to another. The state space is the collection of all possible such states, and a probability distribution for the spur contents assigns a probability to each possible state.

In the master equation method all the possible transitions between states are given model rates, which are calculated from the independent pairs approximation. This procedure gives a set of coupled linear differential equations.[28,49,57-59,84,85] Each equation describes the time dependence of the probability that the spur occupies a specific state , i.e. that it contains the appropriate particles. For instance, the probability that a spur is in state j, P_j, obeys an equation of the form

$$\frac{dP_j}{dt} = \sum_i \omega_{i,j} \ell_{i,j} P_i - \sum_k \omega_{j,k} \ell_{j,k} P_j \tag{22}$$

where $\omega_{i,j}$ and $\omega_{j,k}$ are the numbers of independent routes leading from state $i \rightarrow j$ and from $j \rightarrow k$, respectively, in one transition. The first summation describes the formation of state j from all other possible states i and the second summation, the decay of state j to the states k. The rate coefficients describing the transition from one state to another, the ℓ, are time dependent. For reaction between two spur particles, the rate coefficient is dependent upon the nonhomogeneous distribution of the reactants and the independent pairs approximation is used to express the rate coefficient in terms of the time-dependent pair survival probability, Ω, of the reacting pair if they were in isolation,[57]

$$\ell = -\frac{1}{\Omega}\frac{d\Omega}{dt}. \tag{23}$$

The survival probability is given by

$$\Omega(t) = \int_r g(\mathbf{r})(1 - W(r,a;t))d\mathbf{r} \tag{24}$$

where $g(\mathbf{r})$ is the probability density of the interparticle vector. If the reaction involves one of the spur particles with a homogeneously distributed scavenger, the appropriate rate coefficient is the time-dependent scavenging capacity,[47] i.e.

$$\ell = k(t)[S] \tag{25}$$

where $k(t)$ is the time-dependent second-order rate coefficient for the reacting species[56,65] and $[S]$ is the concentration of the scavenger.

This type of approach can be applied to all systems, however, in practice it is useful for only a limited number of schemes where the reaction mechanism and the nonhomogeneous spatial distribution are simple.[49] This is because the dimensionality of the state space increases by one for every new chemical species added, and because no method has yet been developed for dealing with reactive products in the master equation. Work on the latter problem is in progress.

Table 2 : Reaction scheme for the radiolysis of water and aqueous solutions.[91]

	Reaction			$k\ (/\ 10^{10}\ \mathrm{M^{-1}s^{-1}})$
R1	$e_{aq}^- + e_{aq}^-$	$\rightarrow$	$H_2 + OH + OH^-$	0.55
R2	$e_{aq}^- + H_{aq}^+$	$\rightarrow$	H	2.3
R3	$e_{aq}^- + H$	$\rightarrow$	$H_2 + OH^-$	2.5
R4	$e_{aq}^- + OH$	$\rightarrow$	OH^-	3.0
R5	$e_{aq}^- + H_2O_2$	$\rightarrow$	$OH + OH^-$	1.1
R6	$H_{aq}^+ + OH^-$	$\rightarrow$	H_2O	14.3
R7	$H + H$	$\rightarrow$	H_2	0.78
R8	$H + OH$	$\rightarrow$	H_2O	2.0
R9	$H + H_2O_2$	$\rightarrow$	$OH + H_2O$	0.009
R10	$OH + OH$	$\rightarrow$	H_2O_2	0.55
R11	$e_{aq}^- + O_2$	$\rightarrow$	O_2^-	1.9
R12	$H + O_2$	$\rightarrow$	HO_2	2.1
R13	$H_{aq}^+ + O_2^-$	$\rightarrow$	HO_2	3.8

4 RADIATION CHEMISTRY OF AQUEOUS SYSTEMS

On the picosecond time scale, fast electron radiolysis of water produces groups of very reactive radicals and ions,[6]

$$H_2O^+ + e^- \longrightarrow H_{aq}^+ + OH + e_{aq}^-$$
$$H_2O \xrightarrow{radn} H_2O^* \longrightarrow H + OH$$
$$H_2O^* \longrightarrow H_2 + O \longrightarrow H_2 + 2OH$$

The radiation-induced reactants undergo fast reactions[86,87], which are close to diffusion control,[88] and the observed chemistry therefore depends not only on the reaction mechanism, but also on the spatial distribution of the reactants and the relative diffusion of the particles.[5] The diffusion coefficients of the most important reactive species are given in Table 1 (earlier) and a simple reaction scheme is given in Table 2.

Two types of experiment are commonly used to probe the nonhomogeneous kinetics of water : pulse radiolysis[89] and scavenger studies.[90] In the former a short pulse of radiation is used to produce a non-equilibrium chemical system whose behavior is then monitored using an optical, electrochemical or magnetic resonance technique. Scavenger experiments study the influence of an added solute, known as a scavenger, on an observable yield, which may be the product of the reaction of a scavenger with a radiation-induced radical or one of the molecular products, H_2 and H_2O_2. The effect of scavenger concentration on the yield is then examined.

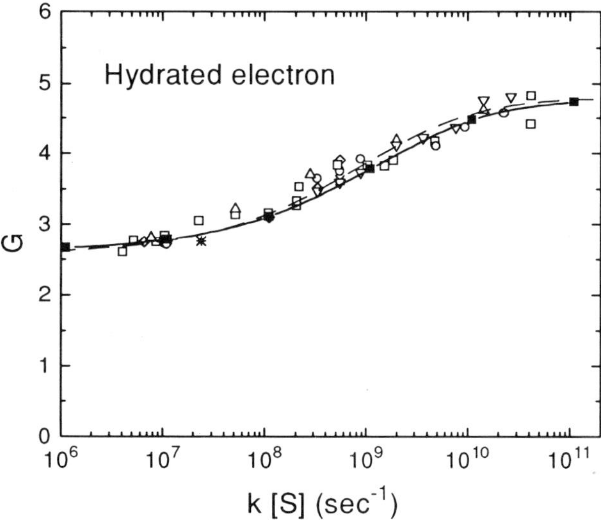

Figure 15. Effect of hydrated electron scavenger on the yield of scavenged hydrated electrons. Solid lines and filled points refer to the predictions of deterministic diffusion-kinetic calculations. The open points are a compilation of experimental measurements[42,92] and the dashed line is a best fit to this data.

4.1 Scavenger studies

Deterministic calculations have been used to help elucidate the short time radiation chemistry in the fast electron radiolysis of aqueous solutions and to quantify the effects of scavengers.[42] Figure 15 shows the results of calculations examining the effects of a hydrated electron scavenger on the scavenged yield of hydrated electron, $G_{e^-}(s)$. The results are presented as a function of the steady state scavenging capacity, $s = k[S]$, and are compared with a compilation of the available experimental data for a variety of different scavengers.[42,92] Agreement between the modeled chemistry and the yields measured experimentally is good.

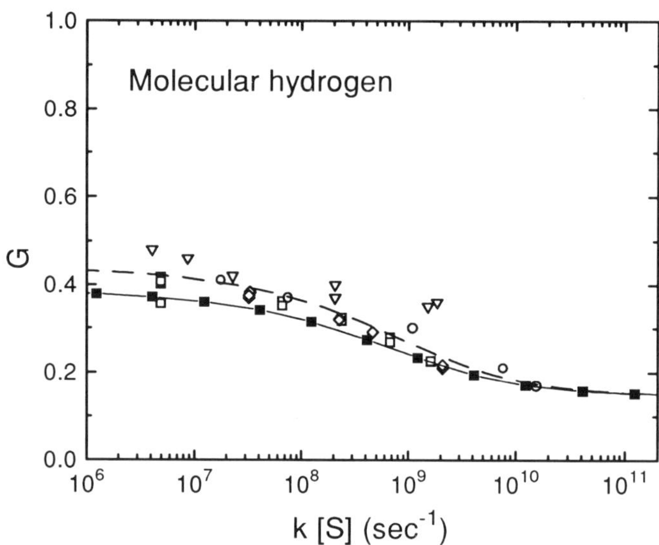

Figure 16. Effect of hydrated electron scavenger on the yield of molecular hydrogen. Key as in Figure 15.

In the radiolysis of an aqueous solution, the yield of scavenged hydrated electrons clearly depends on the scavenging capacity of the solution. This dependence has been parameterized using a number of different empirical equations.[42,93] The best fit of the experimental data in Figure 15 to the function

$$G(s) = G^{esc} + (G^0 - G^{esc}) \frac{(\alpha s)^{1/2} + \alpha s/2}{1 + (\alpha s)^{1/2} + \alpha s/2} \tag{26}$$

suggested by LaVerne and Pimblott[42] is also shown in the figure. The scavenged yield of hydrated electron increases from an escape yield, G^{esc}, of 2.55 molecules / 100 eV to a (theoretical) upper limit, G^0, of 4.8 molecules / 100eV. The time constant obtained is 0.9 ns.

Figure 16 examines the effects of hydrated electron scavenger on the yield of molecular hydrogen, $G_{H_2}(s)$.[42,94] Again there is good agreement between calculation and experiment. The yield of molecular hydrogen decreases as the scavenging capacity of the solution for the hydrated electron increases. For calculation to match experiment a non-scavengeable, initial yield of hydrogen of 0.15 molecules / 100eV is necessary.[42]

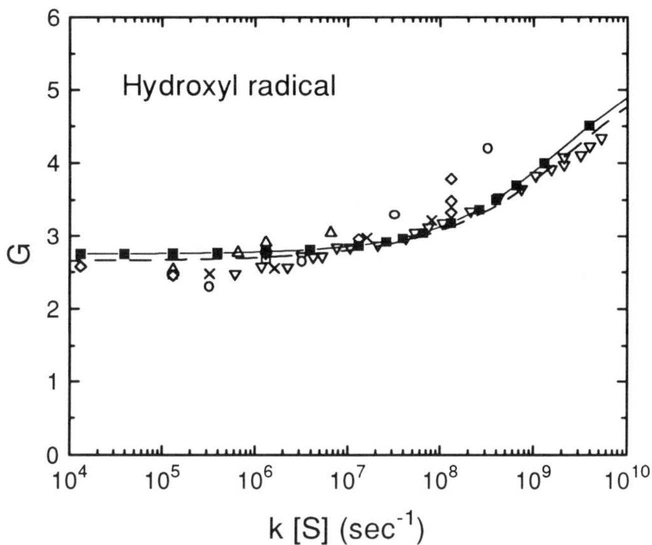

Figure 17. Effect of hydroxyl radical scavenger on the scavenged yield of hydroxyl radical in aqueous solutions. Key as in Figure 15.

Similar calculations for hydroxyl radical scavengers are shown in Figures 17 and 18.[42,94] The agreement with experiment is of the same order as found for the hydrated electron and molecular hydrogen. Notice that there is no initial yield of hydrogen peroxide. Best fit parameters for the four sets of data are given in Table 1 above. The time constants for the molecular products were obtained from fitting equation (26) with $s = k[S]$, however, for future reference it should be remembered that scavenging either of the precursor reactants prevents formation of the molecular product.

The calculations presented in Figure 15-18 suggest that the deterministic treatment provides an acceptable way of parameterizing the chemistry in the fast electron radiolysis of water and for investigating chemical problems of interest.

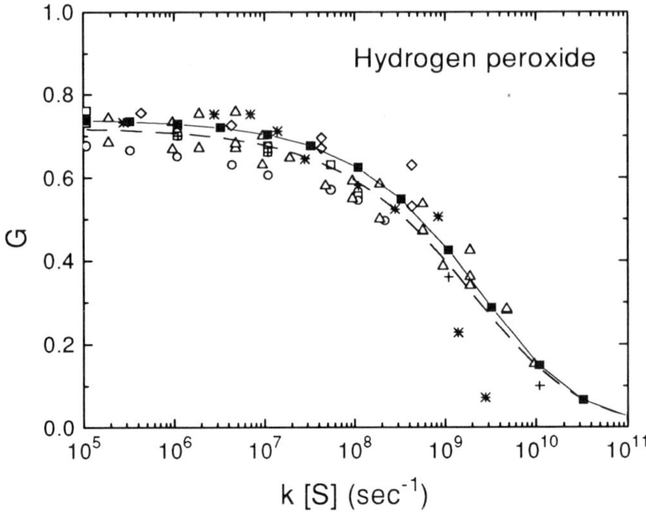

Figure 18. Effect of hydroxyl radical scavenger on the yield of hydrogen peroxide in aqueous solutions. Key as in Figure 15.

4.2 Laplace transform kinetics

While many experimental data about the influence of scavengers on the radiolysis of water are available,[42,92] there are very few direct measurements of time-dependent yields, $G(t)$,[86,87] and so an alternative source of information is necessary. The dependence of the scavenged yield of a geminate pair, $\Omega(s)$, on the steady-state scavenging capacity, $s = k[S]$, and the decay kinetics, $\Omega(t)$, in the absence of the scavenger, are linked mathematically by a Laplace transform relation

$$\Omega(s) = s\int_0^\infty \Omega(t)\,exp(-st)\,dt. \tag{27}$$

Radiation-induced spurs frequently result from more than one ionization event, and typically contain more than two reactant particles.[32] Deterministic modeling studies have been used to investigate whether a similar relationship can be used for the observed G-values, in spite of the fact that the spur contains more than two particles and that the kinetics are basically second-order,

$$G(s) = s\int_0^\infty G(t)\,exp(-st)\,dt \tag{28}$$

If this type of relation is accurate, it can be used to describe the radiolysis of water and to predict the time-dependent $G(t)$ from the scavenged yields, $G(s)$.[42,94]

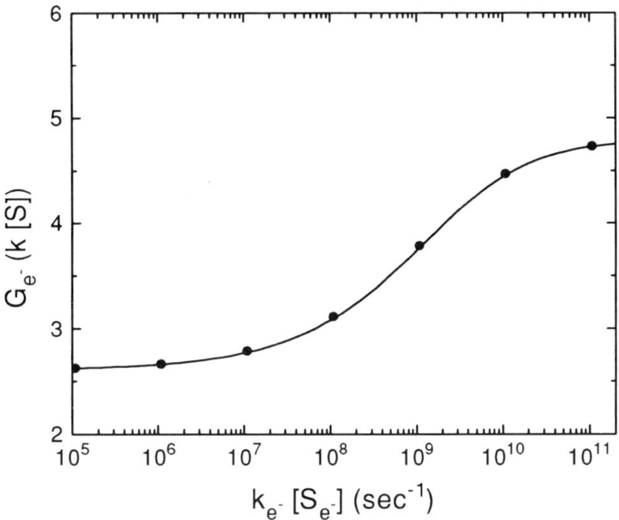

Figure 19. The relation between the yield of scavenged hydrated electrons and the underlying decay kinetics of the hydrated electron in the absence of scavenger. The solid points are the scavenged yields predicted by deterministic calculations, c.f. Figure 15, and the solid line is the Laplace transform of the modeled decay kinetics in the absence of scavenger.

In Figure 19 the method is tested by comparing the Laplace transform of the modeled hydrated electron decay kinetics in the absence of scavenger with the scavenged yields presented in Figure 15.[42,92] The Laplace relation is not exact because the kinetics are not first order,[6,47,48] however, calculations of the type reported here show it to be very accurate for the chemistry of the hydrated electron (and for the hydroxyl radical).[42]

A Laplace transform relationship has also been found to link the yields of the molecular products, H_2 and H_2O_2, in aqueous scavenger solutions and their formation kinetics in the radiolysis of water in the absence of scavenger.[42,94] The appropriate transform variable is $2k[S]$ rather than $k[S]$, as either of two reacting particles may be scavenged.[94] A comparison of the calculated scavenger concentration-dependent hydrogen yield and the Laplace transform estimate is shown in Figure 20. The Laplace relations for the molecular products provide access to their "experimental" formation kinetics from the experimental data summarized in Figures 16 and 18.

148

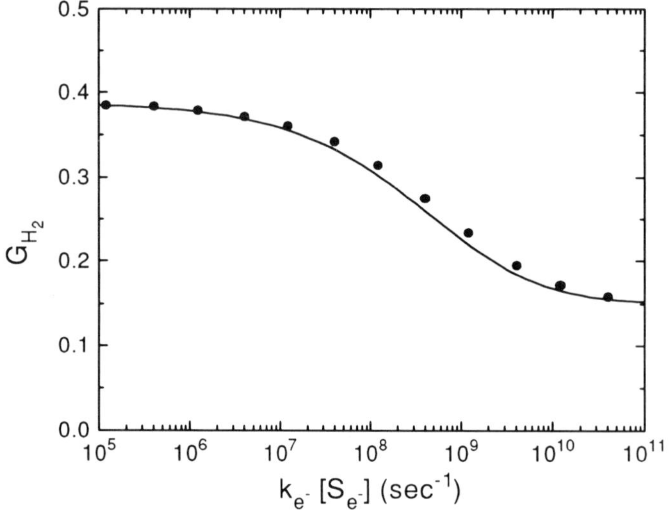

Figure 20. The relation between the yield of molecular hydrogen and the underlying formation kinetics of hydrogen in the absence of scavenger. The solid points are the yields predicted by deterministic calculations, c.f. Figure 16, and the solid line is the Laplace transform of the formation kinetics in the absence of scavenger.

The inverse transform of equation (26) is

$$G\,(t) = G^{esc} + 2(G^0 - G^{esc})F_f(\sqrt{4t/\alpha}\,).\tag{29}$$

where F_f is the auxiliary function for the Fresnel integrals.[95] Simple substitution of the best fit parameters given in Table 1 gives an accurate estimate of the desired decay/formation kinetics. (The best fit parameters in Table 1 refer to the variation in the yields as a function of $k[S]$. As the Laplace variable is $2k[S]$ for the molecular products, the α value in Table 1 has to be divided by a factor of two before substitution in equation (29), i.e. replaced by $\alpha/2$.) This information on the formation of the molecular products is not directly experimentally accessible, and cannot be obtained in any other way.

4.3 Cooperative scavenging

Deterministic kinetic calculations show that the radiation chemistry of aqueous solutions is dominated by the reaction of the hydrated electron with the hydroxyl radical,[96,97]

$$e_{aq}^{-} + OH \xrightarrow{3\times10^{10}\,\mathrm{M^{-1}s^{-1}}} OH^{-} \tag{R4}$$

The modeled yield of this reaction is about 1.2 molec. / 100eV, which corresponds to the value estimated from experimental yields,

$$G_{OH}^{0} - G_{OH}^{esc} - 2G_{H_2O_2}^{esc} \approx 1.4 \text{ molec.} / 100 \text{ eV}$$

when the amount of H + OH (R8) reaction, 0.2 molecules / 100eV, is subtracted.

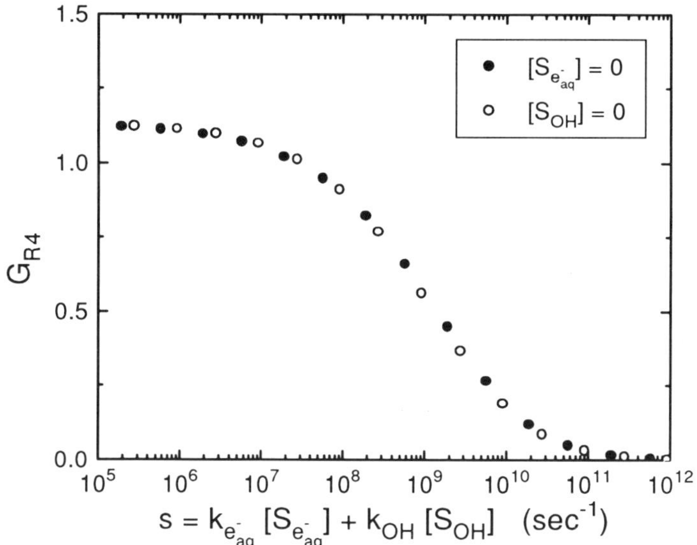

Figure 21. Effect of scavenging capacity on the yield of e_{aq}^{-} + OH reaction. The filled and open points are for calculations with a hydroxyl radical scavenger and with a hydrated electron scavenger respectively.

Scavenging either the hydrated electron or the hydroxyl radical will affect the subsequent chemistry of the other, and the spur model has been used to examine the cooperative effects of scavengers.[96,97] Calculations shown in Figure 21 demonstrate that the yield of reaction (R4) does not depend on whether the hydrated electron or the hydroxyl

150

radical is scavenged, but only on the total scavenging capacity of the solution for the hydrated electron and the hydroxyl radical,

$$s = k_{e^-} [S_{e^-}] + k_{OH} [S_{OH}].$$ (29)

As a result, the scavenged yield of the hydrated electron, and of the hydroxyl radical, depend on both the scavenging capacity of hydrated electron scavenger and that of hydroxyl radical scavenger.

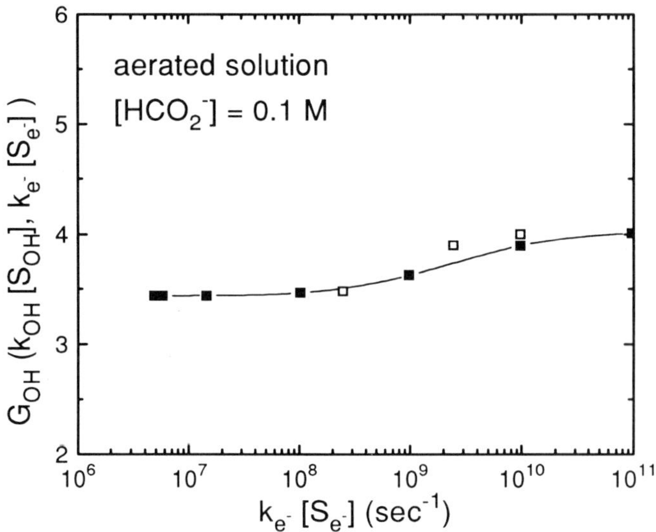

Figure 22. Cooperative effect of a hydrated electron scavenger on the scavenged yield of hydroxyl radical from aerated 0.1 M formate solution. The open points are experimental data[98] and the solid line and filled points are theoretical predictions.

Figure 22 shows how the yield of CO_2 due to OH radicals in the radiolysis of 0.1 M formate solution is affected by the presence of a hydrated electron scavenger. The predictions of the spur model[97] are in good agreement with experimental measurements[98]. The scavenged yield increases from 3.3 molec./100 eV in aerated solution to an upper (theoretical) limit of 4.1 molec./100 eV. The difference represents the amount of reaction (R4) in aerated 0.1 M formate solution. A surface plot calculated to show how the scavenging capacity of hydrated electron scavenger and of hydroxyl radical scavenger influences the yield of the hydrated electron scavenged is presented in Figure 23. Only the valley of the surface is accessible in experimental studies.

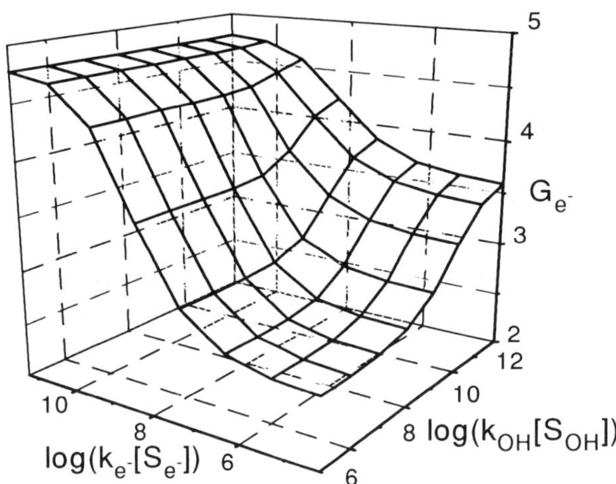

Figure 23. Cooperative effects of scavengers on the scavenged yield of hydrated electron.

4.4 Nitrous oxide solution

Studies of radiation chemical kinetics have not been restricted to the primary radiation-induced radicals (e_{aq}^-, OH, and H) and the molecular products (H_2 and H_2O_2). Deterministic calculations have also been employed to describe the kinetics of the ionic species in water and in nitrous oxide solution.[99] Nitrous oxide is an efficient scavenger of the hydrated electron,[91]

$$e_{aq}^- + N_2O \xrightarrow{9.1 \times 10^9 \, M^{-1} s^{-1}} N_2 + O^-$$

$$O^- + H_2O \xrightarrow{1.8 \times 10^6 \, M^{-1} s^{-1}} OH + OH^-$$

While most radiation chemical studies have focused on the conversion of e_{aq}^- to OH,[100] the scavenging reaction also provides a convenient way of producing non-equilibrium concentrations of OH^-.

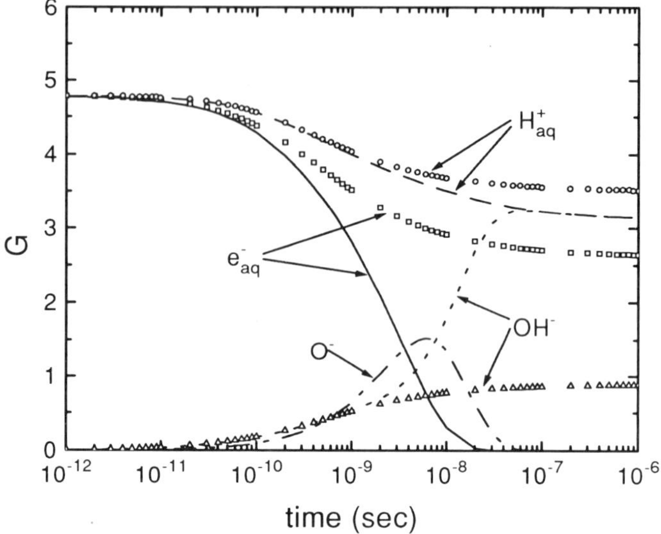

Figure 24. Time dependences of the ionic species produced by the fast electron radiolysis of water (open points) and saturated nitrous oxide solution (lines).

Figure 24 compares the predictions of calculations for the kinetics of the ionic species in the radiolysis of water and of saturated nitrous oxide solution.[99] The calculations show a considerable difference in the time dependences of the yields of the cationic (H_{aq}^{+}) and the anionic (e_{aq}^{-} + OH + O^{-}) species. The difference primarily reflects an increase in charge annihilation resulting because the reaction

$$H_{aq}^{+} + OH^{-} \xrightarrow{1.4 \times 10^{11} M^{-1} s^{-1}} H_2O \tag{R6}$$

is considerably faster than reaction (R4). The difference in the yields is very apparent in the predicted time dependence of the conductivity expected for the solutions, Figure 25.[99] The dip in the conductivity of the nitrous oxide solution at about 7 ns is due to the transient O^{-} which has a smaller equivalent conductance than either its parent, e_{aq}^{-}, or its daughter, OH^{-}.

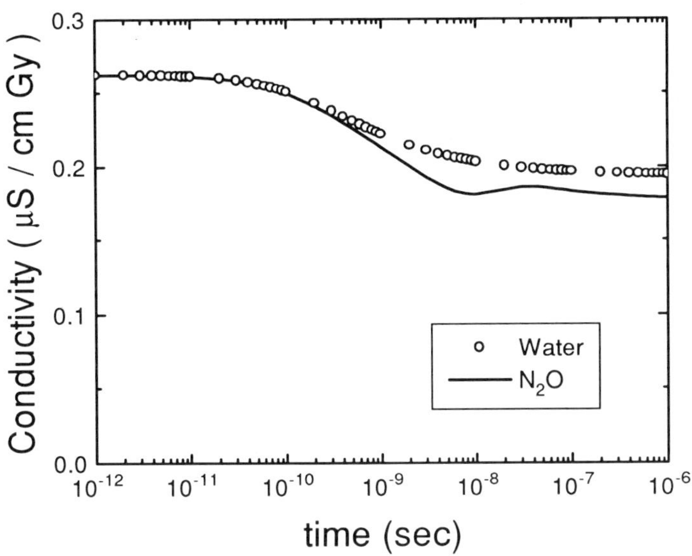

Figure 25. Time dependences of the transient conductivity produced by the fast electron radiolysis of water (open points) and saturated nitrous oxide solution (line).

4.5 Temperature dependence of yields

Recent experiments, prompted by the need for information for nuclear plant design, have probed the temperature dependence of radiation chemical kinetics.[88,101-103] Analysis of these experiments also provides information about diffusion and reaction processes influencing the basic radiation chemical kinetics. The temperature dependence of the fast electron radiolysis of aqueous solutions has been modeled using deterministic diffusion-kinetic methods.[104] The best fit of model to experiment occurs if the width of the spatial distribution of the hydrated electron has an Arrhenius-like temperature dependence with an activation energy similar to that for the movement of an electron between traps in water. The spatial distribution of the other primary radiation-induced species appears to be unaffected by temperature.

Figure 26 shows the influence of temperature on the radiation chemical yields of hydrated electron, hydroxyl radical, molecular hydrogen and hydrogen peroxide in neutral solution. The calculations shown in the figure predict an increase in the yields of the hydrated electron and the hydroxyl radical and a decrease in the yields of molecular hydrogen and hydrogen peroxide with increasing temperature. These results are in reasonable agreement with the available experimental data. The good match between calculation and experiment suggests that the initial yields of the primary radiation-induced species are independent of temperature and there is probably no need for a thermally activated route to molecular hydrogen production.[104]

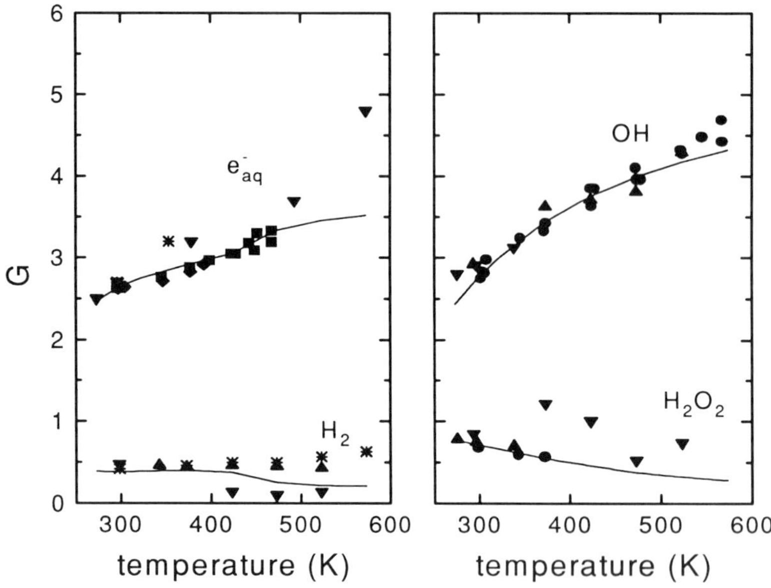

Figure 26. Temperature dependences of radiation chemical yields. The solid lines are the predictions of deterministic diffusion-kinetic calculations and the points are values obtained experimentally. (Key as for Figures 5 and 6 of reference [104])

4.6 Critique of deterministic methods

The information presented demonstrates that the deterministic treatment of the radiation chemical kinetics of water and of aqueous solutions provides a good method for analyzing experimental data. The deterministic approach is reasonably simple to understand and to use and is based on familiar concepts from traditional macroscopic kinetics. Deterministic methods are very flexible and can be used to investigate complex reaction mechanisms in a simple manner. The studies described show that a single set of consistent parameters gives wide agreement with an extensive set of experiments using scavengers. In addition to being easily used, the deterministic models provide a useful method for testing the approximations employed in the analysis of experimental data, c.f. the Laplace transform relationship between the yields in scavenger experiments and the underlying decay and formation kinetics.

While deterministic methods provide a simple and useful model for predicting the radiation chemical kinetics of aqueous systems, the track of a fast electron is condensed into an averaged spur with several characteristic parameters. In consequence the treatment is insensitive to the detailed structure of the track, and to the spur size distribution. The typical average spur has no real physical significance: the optimized spur which reproduces the experimentally observed chemistry of aqueous solutions in electron radiolysis has an energy of 62.5 eV, while the average energy loss of a fast electron in liquid water is slightly less than 40 eV. Furthermore, as the deterministic models of spur chemistry employ a typical average spur with a nonhomogeneous concentration profile for each reactant, the calculations cannot provide any insight into the effects of the spatial correlations of the reactants on the chemical kinetics. This type of model is very useful for considering problems of a mechanistic nature, but does not provide information about the details of the spatial distributions in a spur.

5 STOCHASTIC ANALYSES

Spurs produced in fast electron radiolysis are comprised of only small numbers of particles.[32] Conventional models for the kinetics of these systems employ deterministic methods to model reaction, which do not consider the individual, and discrete, nature of the particles.[57] This fault is best demonstrated by considering a cluster of two (A/B) pairs. The reaction distances for the three reactions A+A, A+B and B+B are the same, and the initial coordinates of the four reactants are obtained from identical and independent distributions. As there is one A-A distance, four A-B distances and one B-B distance, statistical considerations require the ratio of the products, A_2 : AB : B_2, to be 1:4:1. For a cluster of n pairs the same arguments show that the product ratio is $n(n-1)/2 : n^2 : n(n-1)/2$. In fact, this is the correct statistical weighting for any n pair cluster, even when the particles are not identical. The ratio prescribed by a deterministic treatment is always 1:2:1, irrespective of the number of pairs, which is the large n limit of the correct result. It is apparent, even from this simple analysis, that stochastic effects in nonhomogeneous kinetics can be substantial.

Stochastic techniques have been used to investigate a large number of problems of interest in radiation chemical kinetics.

5.1 Prototype systems: single species spurs

If the reaction scheme is simple, the stochastic master equation treatment is much more convenient for modeling kinetics than any other stochastic method, since it only requires a standard differential equation solver, and is far less computer-intensive than either simulation method. The most basic scheme possible is that for the prototype single species spur

$$R + R \rightarrow R_2.$$

The state of the spur is labeled by the number of particles that remain. If the particles are identically distributed then the probability that a spur has exactly n particles is described by the set of equations[57],

$$\frac{dP_n}{dt} = \tfrac{1}{2}(n+2)(n+1)\ell_{R+R}P_{n+2} - \tfrac{1}{2}n(n-1)\ell_{R+R}P_n \qquad (30)$$

where, under the independent pairs approximation, the time-dependent rate coefficient ℓ_{R+R} does not depend on n. Analytic solution is straightforward and the expectation value of the number of particles surviving is given by[57]

$$<n> = \sum_{n=n_m}^{n_o} \frac{2n-1}{2^n} \frac{\Gamma(n_o+1)}{\Gamma(n_o-n+1)} \frac{\Gamma((n_o-n+1)/2)}{\Gamma((n_o+n+1)/2)} \Omega^{n(n-1)/2} \qquad (31)$$

where $n_m = 2$ if n_0 is even and $n_m = 1$ if n_0 is odd. The predictions of the master equation model are found to be in good agreement with the results given by random flights (and IRT) simulation, which shows that the independent pairs approximation is acceptable, at least for this simple system.[57]

The master equation technique and the IRT simulation method both employ the independent pairs approximation. In the simulation model, the calculation considers a set of coordinates for the reacting particles at the start of each realization. The resulting spatial configuration maintains all the appropriate triangle inequalities. In contrast, the master equation approach extends the independent pairs approximation to the initial configuration, thus while the distribution of pair distances is correct, the triangle inequalities do not hold. Consequently, the predictions of the master equation and the IRT model differ slightly for random spatial distributions, e.g. Gaussian spurs, and are only equivalent for fixed configurations, e.g. four particles in a tetrahedral configuration.

5.2 Prototype systems: two species spurs

The study of the single species system serves only as a rudimentary test of the independent pairs approximation; it is of very limited chemical value. The simplest system in which the nature of reaction is significant is where the spurs are made up of pairs of two species (A/B). The chemistry of idealized multi-pair clusters has been investigated in depth.[52,54,64] These studies focused primarily on the chemistry of (A/B) pairs in spurs where the A and B concentration profiles are considered to be identical and independent.[54,64] Two different reaction schemes have been considered, (i) where the reaction distances, a_{A+A}, a_{A+B} and a_{B+B}, are the same, i.e.

$$
\begin{aligned}
A + A &\rightarrow A_2 \\
A + B &\rightarrow AB \qquad\qquad\qquad\qquad \text{Scheme (I)}\\
B + B &\rightarrow B_2
\end{aligned}
$$

and (ii) where 'like plus like' reaction does not occur,

$$
A + B \rightarrow AB \qquad\qquad\qquad\qquad \text{Scheme (II)}
$$

i.e. $a_{A+A} = a_{B+B} = 0$.

Studies for both schemes have been made for neutral pairs using the two Monte Carlo simulation models as well as the master equation method. The stochastic master equation for the probability that a spur has m A particles and n B particles is [54,64]

$$
\begin{aligned}
\frac{dP_{m,n}}{dt} = {}&\tfrac{1}{2}\ell_{A+A}((m+2)(m+1)P_{m+2,n} - m(m-1)P_{m,n})\\
&+\ell_{A+B}((m+1)(n+1)P_{m+1,n+1} - mnP_{m,n})\\
&+\tfrac{1}{2}\ell_{B+B}((n+2)(n+1)P_{m,n+2} - n(n-1)P_{m,n}).
\end{aligned} \tag{32}
$$

For scheme (I) when $\ell_{A+A} = \ell_{A+B} = \ell_{B+B}$, it is straightforward to obtain the yields of A, B, A_2, AB and B_2 from the solution for the single species spur, equation (31). For spurs initially with n_0 pairs, i.e. $2n_0$ particles

$$
<A>=<B>=\tfrac{1}{2}<N>=\sum_{i=2}^{2n_o}\frac{2i-1}{2^{i+1}}\frac{\Gamma(2n_o+1)}{\Gamma(2n_o-i+1)}\frac{\Gamma((2n_o-i+1)/2)}{\Gamma((2n_o+i+1)/2)}\Omega^{i(i-1)/2}. \tag{33}
$$

where N is the total number of reacting particles. Since the ratio of the products $A_2 : AB : B_2$ is $n_0(n_0-1)/2 : n_0^2 : n_0(n_0-1)/2$,

$$
<A_2>=<B_2>=\frac{(n_0-1)}{2(2n_0-1)}(n_0-\tfrac{1}{2}<N>) \tag{34}
$$

158

and

$$< AB >= \frac{n_0}{2n_0 - 1}(n_0 - \tfrac{1}{2} < N >)$$ (35)

A comparison of the predictions of random flights simulation and those of the IRT model and the master equation method for scheme (I) shows no significant differences.[54,64]

Under scheme (II) only the recombination reaction takes place and $\ell_{A+A} = \ell_{B+B} = 0$. The master equation for the probability that a spur has n pairs is reduced to[54,58]

$$\frac{dP_n}{dt} = (n+1)^2 \ell P_{n+1} - n^2 \ell P_n$$ (36)

which is easily solved to give the expectation value of the number of surviving pairs,[105]

$$< n >= \sum_{n=1}^{n_o} \frac{2n(\Gamma(n_o + 1))^2 \Omega^{n^2}}{\Gamma(n_o + n + 1)\Gamma(n_o - n + 1)}$$ (37)

As for scheme (I), agreement between the IRT model, the master equation formulation, and random flights simulation is good.[54]

5.3 Molecular product formation

In a real system, different chemical species do not normally have the same diffusion coefficients and reaction radii. However, it is possible to find systems in which the reaction scheme is essentially equivalent to schemes (I) or (II). One such scheme is the formation of molecular hydrogen and hydrogen peroxide in spurs initially comprised of hydrogen atom - hydroxyl radical pairs.[28] In this simple model for the radiolysis of aqueous acid, the scavenging of the hydrated electron by acid is assumed to be instantaneous and unscavengeable sources of molecular hydrogen are neglected.[30] The scheme incorporates three reactions

$$H + H \rightarrow H_2$$ (R7)
$$H + OH \rightarrow H_2O$$ (R8)
$$OH + OH \rightarrow H_2O_2$$ (R10)

and the resulting master equation describing the probability that a spur has m hydrogen atoms and n hydroxyl radicals is given by equation (32) (with $m_0 = n_0$). The time dependent rate coefficients, ℓ_{H+H}, ℓ_{H+OH}, and ℓ_{OH+OH}, are not equal, since the reaction distances for the three reactions and the relative diffusion coefficients are not equal, and the distributions of the particles, governed by the parameters σ_H and σ_{OH}, are not the same. In consequence, it is necessary to solve the master equation by numerical methods. Comparison of master equation calculations, IRT simulations and random flights simulations shows good agreement.[28]

Because stochastic calculations are sensitive to the number of particles in the spur, it is necessary to average the master equation results over a spur size distribution that is appropriate for a radiation track. Calculations averaged in this way have been optimized with respect to the unknown spur widths, σ_H and σ_{OH}, to match experimental escape yields of H, OH, H_2 and H_2O_2.[28] In these calculations, a single width parameter describes the nonhomogeneous spatial profile of a reactant in all spurs, irrespective of the number of pairs. For a spur of n_0 pairs, the width of the profile is related to that in a single pair spur by the prescription[4]

$$\sigma_i(n_o) = \sigma_i(1)n_0^{1/3}. \tag{38}$$

Calculations for several different distributions suggest unit radii in the ranges 0.4 nm $< \sigma_H(1) < 0.8$ nm and 0.2 nm $< \sigma_{OH}(1) < 0.3$ nm.[52,64] These radii are considerably smaller than those expected in reality, but it must be realized that the model system is a gross simplification of the true chemistry, and consequently it is not surprising that the parameterization is not realistic.[30] These preliminary calculations do, however, emphasize the potential significance of statistical effects in radiation chemical kinetics. A master equation treatment of the true short-time radiation chemistry of water incorporating a full reaction scheme, c.f. Table 2, is too complex to allow straightforward modeling.[49] This is an area that needs further development.

5.4 Reaction in multi ion pair spurs

Approximate stochastic theories might be expected to break down in systems where some of the particles are charged, because they are based on the independent pairs approximation. This means that although the reaction time of a pair is influenced properly by the electrostatic interaction of that pair, the Coulomb forces arising from the presence of other ions are neglected.

The accuracy of the independent pairs approximation in systems of multiple ion pairs has also been investigated in depth using stochastic models.[52,64] As in the studies of neutral systems, these calculations use two-species (A^+/B^-) pairs in spurs. Studies for both schemes

160

(I) and (II) have been made for cation - anion pairs in a high permittivity solvent, such as water. Calculations for two (A^+/B^-) ion pairs reacting according to scheme (II) are shown in Figure 27. The results of the IRT model are in good agreement with those of random flights simulation. Also included are the results for two neutral pairs with the same encounter distance. The Coulomb force has a considerable effect on the modeled kinetics.

For scheme (I), the statistical weighting of the reactions is $n_0(n_0-1)/2 : n_0^2 : n_0(n_0-1)/2$, however, even if a_{A+A}, a_{A+B} and a_{B+B}, are the same, the ratios of the product yields are different because the time-dependent rate coefficients ℓ are affected by the Coulomb forces. It is therefore not easy to estimate the correct product ratio analytically. The agreement between the approximate IRT model and random flights simulation shown in Figure 27 and found in reference [51] demonstrates that the independent pairs approximation is acceptable for describing the reaction of ions in high permittivity systems.[52,64]

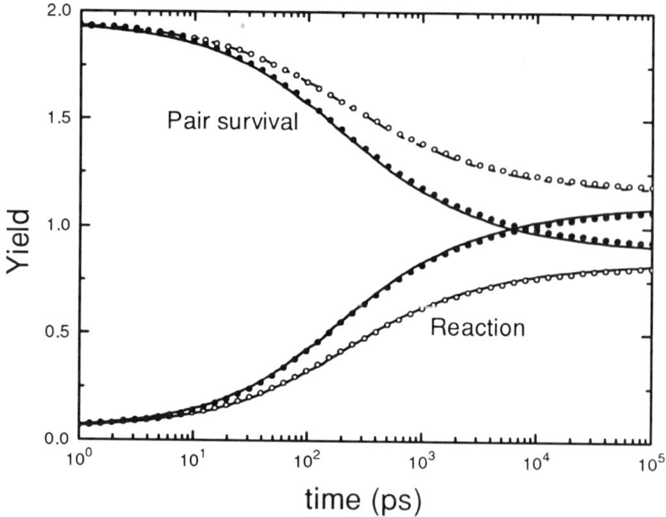

Figure 27. Comparison of random flights and IRT simulation for the kinetics of a two (A^+/B^-) ion pair spur (filled points and solid lines) and for the kinetics of a two neutral (A/B) pair spur (open points and dashed lines). The points refer to the predictions of random flights simulation and the lines to IRT modeling. The same reaction distance was used for the $A^+ + B^-$ reaction and for the A + B reaction.

In addition to the simulations testing the accuracy of the independent pairs approximation, investigations have also been reported to examine whether it is necessary to include charge effects explicitly in the model, or whether the effects can simply be parameterized by using effective reaction distances.[52] The reason for this is to check the long-standing claim that charge effects are not important in spur kinetics, since they are already accounted for in the experimental rate constants (through the effective reaction distance).[5] Calculations using the same effective reaction distance, defined by equation (18), but for neutral and for ionic pairs, show that the inclusion of charge has two observable consequences: the product ratios are altered and the time scale of the reaction is speeded up.[52]

5.5 Ion recombination in low permittivity media

In low permittivity solvents only scheme (II) is feasible, because the strong Coulomb forces prevent cation-cation and anion-anion encounters. In this instance, the agreement between the predictions of the two simulation models depends upon the initial spatial distribution of the ions.[64] The IRT model accurately predicts the free ion yield from multiple ion pair spurs, but noticeable errors are sometimes found in the time dependence of the kinetics.[51,64] Figure 28 shows the predicted ion recombination kinetics for a four ion pair spur.

The errors observed in the predictions of the IRT model are a reflection of the way in which the Coulomb forces reinforce one another. The reaction time generated for a pair in the IRT method only incorporates the force between the pair and there is a breakdown of the independent pairs approximation when the strong inter-ion forces act in cooperation. This breakdown appears under special circumstances, where the initial distribution is selected so that the Coulomb forces interfere with one another; for spurs where the spatial distribution of the reactants is spherically symmetrical the independent pairs approximation is found to hold reasonably well.[51,64]

The conventional picture of multiple ion pair spurs in low permittivity solvents considers the relatively slow-moving cations to be centrally located when compared to the highly mobile anions, which are in a spherically symmetric distribution about the cations.[106,107] The ion recombination kinetics in these spatially simple spurs are amenable to study using the master equation method.[58,108] The kinetics of a multiple ion pair spur are described by equation (37). This solution, although explicit, is not entirely analytic as it depends upon Ω, the time-dependent survival probability of an ion pair with the same distribution of separations, which is not known analytically.[109]

Comparison of the results of the master equation model with the simulation techniques reveals good agreement.[58] The predictions for a four ion pair spur are included in Figure 28. For two, three and four ion pair spurs the master equation model slightly overestimates the amount of reaction (about 2%) and gives marginally faster reaction kinetics than both simulation models [58]. While the two simulation models are in good agreement and predict the same escape yields, the IRT treatment gives slightly slower reaction than random flights simulation. The difference between the IRT simulation and the master equation results from the extension of the approximation to the initial configuration in the master equation analysis.

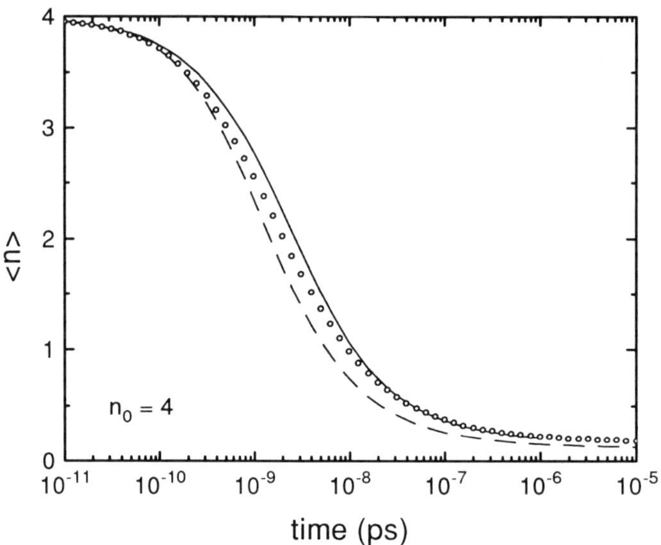

Figure 28. Ion recombination kinetics of a four ion pair spur in which the initial positions of the cations and anions are generated from Gaussians of standard deviation 2.5 nm and 6.0 nm, respectively. Onsager distance = 29 nm. The open points are the results of random flights simulations, the solid line of IRT simulation and the dashed line of the master equation.

5.6 Effects of electric field on ion recombination kinetics

There is a large experimental literature probing the effect of an applied electric field on the free ion yield following high energy electron radiolysis of hydrocarbons.[110-113] Recent energy loss distribution calculations have suggested that a significant fraction of the spurs contain multiple ion pairs.[13] The conventional model for ion recombination assumes that every spur rapidly degrades to a single ion pair and the observed kinetics are those of the final pair. The reaction probability of an ion pair in the presence of an electric field depends upon the angle between the electric field and the inter-ion vector and upon the inter-ion separation,[114,115]

$$W^E(\mathbf{r},a;\infty) = 1 - exp(-\alpha Er/r_c)\sum_{k=0}^{\infty}\Gamma(k+1,r_c/r)(\alpha Er/r_c)^k/(k!)^2 \tag{39}$$

where $\Gamma(i,x)$ is the incomplete Gamma function[95] and $\alpha = \beta r_c(1+cos\theta)$ with $\beta = e/(2k_BT)$. This equation reduces to[115]

$$W^E(\mathbf{r},a;\infty) = 1 - exp(-r_c/c)(1+\alpha E)-... \tag{40}$$

for weak fields. If the inter-ion vector has a spherically symmetric distribution, $g(\mathbf{r})$, the field dependence of the free ion yield is given by

$$\Omega^E(\infty) = \int_r g(\mathbf{r})(1-W^E(\mathbf{r},t))d\mathbf{r} \approx \Omega^{E=0}(\infty)(1+r_c\beta E) \tag{41}$$

and the angular dependence vanishes. Equation (41) predicts a linear increase in the free ion yield with field strength and a slope-to-intercept ratio of $r_c\beta$.

Random flights simulation and the master equation method have been used to probe the influence of applied field strength on the ion recombination kinetics in multiple ion pair spurs in low permittivity solvents.[59] In the master equation treatment the applied field is incorporated in the time-dependent rate coefficient ℓ, which is defined by

$$\ell^E = -\frac{1}{\Omega^E}\frac{d\Omega^E}{dt}. \tag{42}$$

An analytic approximation for the field dependence of the free ion yield from multiple ion pair spurs is obtained by substituting Ω^E for Ω in equation (37). The predictions of the master equation for the free ion yield from multiple ion pair spurs with realistic initial distributions are compared with the results of random flights simulations in Figure 29.

The figure shows a linear increase in the predicted free ion yield with applied field strength. Furthermore, the slope-to-intercept ratio for the effect of the field is numerically the same as the analytic result for a single ion pair, that is $r_c\beta$. The similarity between the multiple and single ion pair cases is easily explained. Consider the application of a weak applied field. In this limit, Ω^E is given by equation (41) and upon substitution into equation (37) it is apparent that[59]

$$<n^E> = <n^0> + r_c\beta E<(n^0)^2>. \tag{43}$$

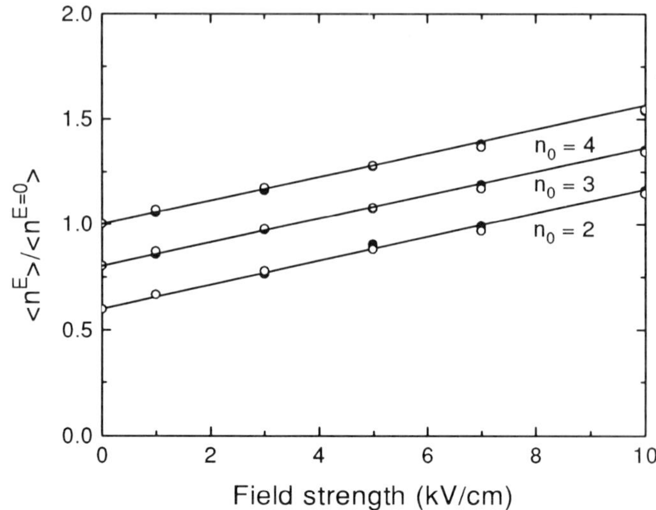

Figure 29. Effect of applied field on free ion yield from multiple ion pair spurs. The lines are the predictions of the Onsager analysis for a single pair. The solid and open points are the results of random flights simulation and the master equation model, respectively, for spurs with n_0 ion pairs. The data for $n_0 = 2$ and 3 are offset by 0.4 and 0.2 units, respectively.

In theory, equation (43) suggests that the field dependence of the free ion yield should provide information about the spur size distribution. Unfortunately this is not the case: for low permittivity solvents the Fano factor,

$$F = (<n^2> - <n>^2)/<n> \tag{44}$$

is close to 1 and $<n^2> \approx <n>$. In consequence, the slope-to-intercept ratio

$$<n^E>/<n^o> = \Omega^E/\Omega^o \tag{45}$$

and there can be no clear (experimental) distinction between the field dependences of single and multiple ion pair spurs.

5.7 Reactions which are not diffusion-controlled

In all of the systems discussed so far the reactions have been treated as diffusion-controlled. It has become clear recently from measurements of the temperature dependence of the rate constants that many reactions in the radiation chemistry of water are not diffusion-controlled.[88] At room temperature the reactions are all close to diffusion control, but at higher temperatures, such as pertain in a water-cooled nuclear reactor, rates deviate considerably from diffusion control.[88]

Elliot et al.[88] have identified two different types of temperature dependence: one in which the rate coefficient follows the Noyes expression describing the transition between diffusion and activation control,[56] and one in which the temperature dependence seems to follow the diffusion coefficient, but the rate is reduced by a constant factor. Most reactions fall in the former category, where the reaction can be described using the radiation boundary condition. The H+H reaction, however, falls into the latter category, because the rate is reduced by a spin statistical factor.[116]

5.7.1 Partially diffusion-controlled reactions

In a deterministic theory there is no distinction between a fully and a partially diffusion controlled reaction, since the only parameter affecting the reaction rate is the rate coefficient. A recent paper has been published testing whether this is also the case for stochastic theories.[61] Spurs were simulated with the same effective reaction distance, but with different true reaction distances. Once corrected for the obvious differences due to zero-time overlap between particles, this study concluded that the importance of using the correct description depends on the scale of the nonhomogeneous particle distribution. If it is narrow then the differences are significant, but for broader distributions it is sufficient to use the effective reaction distance and assume diffusion control, which is much faster to simulate. Investigations of partially diffusion-controlled reactions between ions are in progress. Of course, if the temperature dependences of the nonhomogeneous kinetics are of interest, it is necessary to take into account the effect of the temperature-dependent "activation" step on the effective reaction distance.

5.7.2 Spin-controlled reaction : the re-encounter approximation

The second reason for a reaction to be slower than diffusion control is that it may be spin-selective. Many of the reactive particles in a spur are free radicals, and it is well known from photochemistry that such reactions may be spin-controlled.[116] This not only affects the kinetics, but it may also give rise to coherent spin effects such as CIDEP or CIDNP.[117]

If reaction is spin-controlled, then the occurrence (or not) of reaction upon encounter is determined by the spin state of the pair at the encounter time.[118] The spur kinetics and the fractionation between products then depends not only on the initial spatial distribution of the particles, but also on the initial spin state of the spur, the mechanisms by which this spin state can evolve and the difference in reactivity between singlet and triplet. A preliminary investigation of these effects using the master equation method has been reported.[84,85] In this simplified model, the encounter rate is assumed to be given by the diffusion-controlled encounter rate, and to be unaffected by previous encounters. This approximation, applied after an unreactive encounter, is referred to as the *re-encounter approximation*. Only singlet encounters are assumed to be reactive. Four different limiting cases are identified, depending on the relative time scales the spin relaxation and the spur kinetics. In the slow relaxation limit spin correlation effects are found to have very large effects on the modeled chemistry.[84,85]

Recently, this master equation analysis has been tested by comparison with random flights simulations. These proceed in the normal manner except that the spin wavefunction is followed from encounter to encounter. The spin function is assumed not to evolve in time except when an encounter occurs. When an encounter takes place, the singlet probability is calculated from the spin wavefunction, and a random number is used to discover whether the pair has reacted. If the pair reacts, then the singlet projection operator for the encounter pair is applied to the wavefunction, which is then renormalized. This accounts for the fact that a reaction of one pair can affect the singlet probabilities of other pairs, because the overall spin state of the spur must be conserved. In the event of a triplet encounter, the singlet component of the wavefunction is removed, and the wavefunction is renormalized. Good agreement is found between the master equation results and the random flights simulations.

More unpublished studies using the IRT model have focused on the re-encounter approximation. Here, the first encounter time of each pair is generated using the equation appropriate for diffusion-limited reaction. The spin state of the encountering pair is then used to determine whether reaction occurs or a new re-encounter time is required. If the spin state of the spur cannot evolve between encounters, the many rapid re-encounters that necessarily follow the first encounter of a pair are unreactive, and need not be considered. The pair remains unreactive until another pair encounters, at which point its spin state may change.

According to the re-encounter approximation, the rate of re-encounter following an unreactive encounter is unaffected by the first encounter [119,120]. This approximation allows the generation of a re-encounter time from the original encounter time distribution function for the pair, conditioned to be greater than the first encounter time. However, the re-encounter approximation can be relaxed, as it is possible to generalize the IRT model to handle re-encounters exactly within the independent pairs approximation. As the spin state of the pair that first encountered at t_1 remains unreactive until the encounter of a second (different) pair at t_2, the original pair diffuses with a reflecting boundary for the time period $\Delta t = t_2 - t_1$. The cumulative probability distribution function for the separation, r, between a diffusing geminate pair is then given by

$$F(r, \Delta t) = erf(\xi) - 2\xi\eta \, exp(2\xi\eta + \eta^2) erfc(\xi + \eta) \tag{46}$$

where $\xi = (r-a)/(4D't)^{1/2}$ and $\eta = (D't)^{1/2}/a$. The function $F(r,\Delta t)$ can be inverted numerically to generate a random separation for the first encountering pair at the encounter time of the second pair and this distance is then used to generate a re-encounter time for the first pair.

It is found that the re-encounter approximation gives a small underestimate (of the order of 5%) in the overall amount of reaction in the spur. This error helps to counteract a similar error in the opposite direction incurred by the independent pairs approximation. Consequently the preliminary comparison with random flights simulations suggest that the master equation with the re-encounter approximation is slightly more accurate even than the full IRT method, as a result of these self-canceling errors.

5.8 Spatial correlations and clustering

Conventional wisdom suggests that although stochastic methods may be necessary to understand properly the kinetics of isolated spurs, if spurs are generated close to one another, as in a short track or at the track end, deterministic methods should be accurate. This supposition arises from the large number of possible combinations of particles, as the statistical weights are close to the deterministic limit. A recent study has indicated that this may not be the case,[121] because, although there are many possible combinations of particles, they are not equally likely to lead to reaction. A track-end is laid down event by event with spatial correlations resulting from the underlying physical processes. All the tests which imply that the deterministic theory is acceptable for medium to large numbers of particles assume that the spatial distributions of the particles of a given species are independent and identical, and independent of all other species.[122] This is clearly not true in a radiation track; the spatial distribution of a particle depends on the event from which it originated, and a particle of one species will be correlated with the other particles originating in the same event.

The effect of spatial correlations was investigated using a crude model track of overlapping spurs made up of H/OH pairs, equally spaced along a line. The resulting track is characterized by a spatial density for each species, which can be interpreted in two different ways. The first way takes account of the correlations caused by particles being born in correlated pairs. The second interpretation assumes that the track is characterized by a concentration for each species. This classical interpretation of a concentration is clearly unphysical in the case of a track. It results in random numbers of particles of each type being associated with each spur, and in particular the numbers of H atoms and OH radicals in a spur are not necessarily the same. Deterministic calculations should agree well with the classical interpretation of concentration. The spatial distribution of the track has considerable overlap between adjacent spurs; however, the kinetics predicted by random flights simulation show that there are important differences between the two interpretations, particularly in the product yields. The ratio of H_2 or H_2O_2 yields to the water yield is much lower in the track with the correlated distribution than in the purely random (classical) distribution. Similar results are also found with more realistic track-ends.

From these results, it seems likely that deterministic models will lead to significant errors in predicting the fractionation between the possible products of radical recombination, even under conditions where spurs overlap significantly. Furthermore, the problem lies in the very nature of a deterministic model, which applies classical kinetic rate laws to the spatial concentration profiles. The fundamental assumption is that the track can be characterized by the concentrations of each species. The macroscopic concept of concentration is inadequate to define the kinetics of track-ends: two spatial distributions with the same concentrations give different chemistry, because they have different pair correlations.

5.9 Critique of stochastic methods

Stochastic methods for modeling the diffusion-limited kinetics of spatially nonhomogeneous systems are inherently probabilistic. These techniques require more detailed information than conventional deterministic models about the initial spatial distribution of the particles, about their diffusive motion and about the inter-particle reactions. However, stochastic calculations provide a more acceptable approach for probing particular factors influencing the nonhomogeneous kinetics; they have the correct statistical weighting for the products of reaction and they can cope with spatial correlation effects.

The stochastic analyses are particularly good for investigating qualitative effects that are difficult (or impossible) to study using a deterministic approach. Studies discussed in the preceding sections considered the effects of charge, of partially diffusion-controlled and spin-controlled reactions and the influence of clustering, i.e. the "pairwise" formation of reactants, on the predicted chemistry. Calculations examining the dependence of spur chemistry on the number of particles comprising the spur demonstrate that the radical decay kinetics show only a small dependence on the size (content) of the spur, however, the formation of the (molecular) products depends critically on the spur size. Stochastic methods are also useful for testing the validity of the approximations used to model chemical kinetics. Specific examples discussed include the use of an effective distance scaling to account for the kinetic effects of Coulomb inter-ion forces and tests of the accuracy of the re-encounter approximation.

All the calculations presented here involve idealized or simple reaction schemes and the approximate stochastic methods are easily implemented for these schemes. However, this is not true for more complex reaction mechanisms, as it is difficult to include secondary reactions. Consequently, the methods, especially the master equation treatment, are not very flexible and require further development to make the application to realistic schemes routine.

The comparison of the predictions of stochastic calculations with experimental measurements is of limited value as the problem is over-parameterized. Stochastic calculations require detailed spatial distributions for the reactants and suitable distributions are not available. Furthermore, the random flights and IRT simulation techniques are computer-intensive. Thus far calculations have been limited to idealized track segments of high energy electron tracks, to low energy track-ends and to spur size distributions ignoring the effects of track-ends.

6 COMMENTS AND DISCUSSION

Stochastic and deterministic techniques for probing the short-time chemistry of spatially nonhomogeneous systems found in high energy electron radiolysis have been described in detail, and the use of these techniques to examine a variety of interesting problems has been discussed. The two different approaches are based on very different assumptions. The deterministic models consider a "typical, average" spur and describe the chemistry of this spur using a set of coupled diffusion-reaction equations and spatially nonhomogeneous concentration profiles for the reactants. The stochastic treatments recognize the individual identities of the reacting particles and employ statistically correct methods to model the kinetics of clusters with realizable initial configurations.

The results presented demonstrate that the deterministic treatment of the radiolysis of aqueous solutions provides a suitable method for analyzing experimental data. The deterministic approach is simple to understand and to use, and is sufficiently flexible that it can be applied to complex reaction mechanisms in a facile manner. A single set of consistent parameters gives wide agreement with an extensive set of experiments using scavengers. In addition to being easily used, the deterministic models provide a useful method for testing the some of the approximations employed in the analysis of experimental data, e.g. the Laplace transform relationship between the yields in scavenger experiments and the underlying decay and formation kinetics of the radiation-induced species. While the deterministic approach provides a simple and useful model for predicting the radiation chemical kinetics of aqueous systems, the track of a fast electron is condensed into an average spur with a few characteristic parameters. Consequently, the modeled kinetics are insensitive to the detailed structure of the track and to the spur size distribution.

Deterministic analyses of the nonhomogeneous kinetics of small clusters of reactants are subject to a number of limitations. The most important of these is the incorrect fractionation of the products. The predictions of deterministic calculations and stochastic simulations for identical spurs have been compared. Studies of spurs made up of hydrogen atom - hydroxyl radical pairs and of hydrated electron - hydrated proton - hydroxyl radical groups show that for small spurs (1, 2 or 3 ionization / excitation events) a deterministic analysis gives very different product yields from those suggested by the statistically correct stochastic methods.[54,122] For larger spurs there is very little difference between the predictions of the two different approaches unless spatial correlations are important.

In contrast to the conventional deterministic analyses of short time radiation chemical kinetics, stochastic analyses are sensitive to the distribution of spurs, to their contents and to the spatial distribution of the reactants. This sensitivity reflects the fact that the stochastic models recognize the individual particles. Stochastic calculations highlight the chemical consequences of track structure and of the detailed spatial correlations of the reactants. In particular, they emphasize the effect of clustering on the ratios of the products of nonhomogeneous reaction. As stochastic techniques consider the individual spur particles and their specific nature they can be used to study phenomena averaged by the macroscopic rate used in deterministic calculations. For instance they have been used to examine the effects of spin correlations within a spur and the influence of an applied field upon the escape of ions from multiple ion pair spurs.

The random flights and IRT simulation methods are (computationally) complex and the methodologies require careful testing before application to and consideration of chemically realistic systems. Furthermore, the formulation of a stochastic master equation for complex reaction schemes is difficult as each possible state (set of contents) of the spur has to be considered. Clearly while stochastic models have a number of limitations which must be overcome in future studies, these methods provide many potentially important insights which cannot be obtained from deterministic methods.

Acknowledgment

The authors wish to thank Dr. P. Clifford, Dr. A. Mozumder, Dr. J. A. LaVerne and Prof. M. J. Pilling for numerous interesting and informative discussions. The research described was supported by the Office of Basic Energy Sciences of the U.S. Department of Energy. This is contribution SR-165 of the Notre Dame Radiation Laboratory.

REFERENCES

(1)   Wilson, C. T. R. *Proc. Roy. Soc.* **1923**, *104A*, 1.

(2)   Wilson, C. T. R. *Proc. Roy. Soc.* **1923**, *104A*, 192.

(3)   Kara-Michailova, E.; Lea, D. E. *Proc. Camb. Phil. Soc.* **1940**, *36*, 101.

(4)   Samuel, A. H.; Magee, J. L. *J. Chem. Phys.* **1953**, *25*, 129.

(5)   Schwarz, H. A. *J. Phys. Chem.* **1969**, *73*, 1928.

(6)   Mozumder, A.; Magee, J. L. *Radiat. Phys. Chem.* **1975**, *7*, 83.

(7)   Bethe, H. *Z. Phys.* **1932**, *76*, 293.

(8)   Allen, A. O. *The Radiation Chemistry of Water and Aqueous Solutions*; Van Nostrand: New York, 1961.

(9)   Ashley, J. C. *J. Electron Spectrosc. Relat. Phenom.* **1988**, *46*, 199.

(10)   LaVerne, J. A.; Mozumder, A. *J. Phys. Chem.* **1986**, *90*, 3242.

(11)   Ashley, J. C.; Williams, M. W. Rome Air Development Center, Griffiss Air Force Base, NY, 1983.

(12)   Green, N. J. B.; LaVerne, J. A.; Mozumder, A. *Radiat. Phys. Chem.* **1988**, *32*, 99.

(13)   Pimblott, S. M.; LaVerne, J. A. *J. Phys. Chem.* **1991**, *95*, 3907.

(14)   Mozumder, A.; Magee, J. L. *J. Chem. Phys.* **1966**, *45*, 3332.

(15)   Pimblott, S. M.; LaVerne, J. A.; Mozumder, A.; Green, N. J. B. *J. Phys. Chem.* **1990**, *94*, 488.

(16)   Zaider, M.; Brenner, D. J.; Wilson, W. E. *Radiat. Res.* **1983**, *95*, 231.

(17)   Paretzke, H. G.; Turner, J. E.; Hamm, R. N.; Wright, H. A.; Ritchie, R. H. *J. Chem. Phys.* **1986**, *84*, 3182.

(18)   Paretzke, H. G.; Turner, J. E.; Hamm, R. N.; Ritchie, R. H.; Wright, H. A. *Radiat. Res.* **1991**, *127*, 121.

(19)   Hill, M. A.; Smith, F. A. *Radiat. Phys. Chem.* **1994**, *43*, 265.

(20)   LaVerne, J. A.; Pimblott, S. M.; Mozumder, A. *Radiat. Phys. Chem.* **1991**, *38*, 75.

(21)   Sanche, L. In *Excess Electrons in Dielectric Media*; C. Ferradini and J.-P. Jay-Gerin, Ed.; CRC Press: Boca Raton, 1991.

(22)   LaVerne, J. A.; Pimblott, S. M. *Radiat. Res* **1993**, *135*, 16.

(23)   Nikjoo, H.; Goodhead, D. T.; Charlton, D. E.; Paretzke, H. G. *Physics in Medicine and Biology* **1989**, *34*, 691.

(24)   Goodhead, D. T.; Nikjoo, H. *Int. J. Radiat. Biol.* **1989**, *55*, 513.

(25)   Turner, J. E.; Magee, J. L.; Wright, H. A.; Chatterjee, A.; Hamm, R. N.; Ritchie, R. H. *Radiat. Res.* **1983**, *96*, 437.

(26)   Zaider, M.; Brenner, D. J. *Radiat. Res.* **1984**, *100*, 245.

(27)   Brenner, D. J.; Zaider, M. *Radiat. Prot. Dosim.* **1985**, *13*, 127.

(28)   Clifford, P.; Green, N. J. B.; Pilling, M. J.; Pimblott, S. M.; Burns, W. G. *Radiat. Phys. Chem.* **1987**, *30*, 125.

(29)   Turner, J. E.; Hamm, R. N.; Wright, H. A.; Ritchie, R. H.; Magee, J. L.; Chatterjee, A.; Bolch, W. E. *Radiat. Phys. Chem.* **1988**, *32*, 503.

(30)   Pimblott, S. M.; Pilling, M. J.; Green, N. J. B. *Radiat. Phys. Chem.* **1991**, *37*, 377.

(31)   LaVerne, J. A.; Mozumder, A. *Radiat. Res.* **1993**, *133*, 282.

(32)   Pimblott, S. M.; Mozumder, A. *J. Phys. Chem.* **1991**, *95*, 7291.

(33)   Baker, G. J.; Brocklehurst, B.; Hayes, M.; Hopkirk, A.; Holland, D. M. P.; Munro, I. H.; Shaw, D. A. *Chem. Phys. Lett.* **1989**, *161*, 327.

(34) Long, F. H.; Lu, H.; Eisenthal, K. B. *Phys. Rev. Lett.* **1990**, *64*, 1469.

(35) Long, F. H.; Lu, H.; Shi, X.; Eisenthal, K. B. *Chem. Phys. Lett.* **1991**, *185*, 47.

(36) Pimblott, S. M. *J. Phys. Chem.* **1991**, *95*, 6946.

(37) Sander, M. U.; Luther, K.; Troe, J. *J. Phys. Chem.* **1993**, *97*, 11489.

(38) Gauduel, Y.; Pommeret, S.; Migus, A.; Antonetti, A. *Chem. Phys.* **1990**, *149*, 1.

(39) Burns, W. G.; Curtis, A. R. *J. Phys. Chem.* **1972**, *76*, 3008.

(40) Mozumder, A.; Magee, J. L. *Radiat. Res.* **1966**, *28*, 215.

(41) Burns, W. G.; Sims, H. E.; Goodall, J. A. B. *Radiat. Phys. Chem.* **1984**, *23*, 143.

(42) LaVerne, J. A.; Pimblott, S. M. *J. Phys. Chem.* **1991**, *95*, 3196.

(43) Trumbore, C. N.; Short, D. R.; Fanning Jr., J. E.; Olson, H. *J. Phys. Chem.* **1978**, *82*, 2762.

(44) Chance, E. M.; Curtis, A. R.; Jones, I. P.; Kirby, C. P. "FACSIMILE : A computer program for flow and chemistry simulation, and general initial value problems," AERE, Harwell, 1977.

(45) Lea, D. E. *Actions of Radiation on Living Cells*; C.U.P.: Cambridge, 1955.

(46) Lea, D. E. *Proc. Camb. Phil. Soc.* **1934**, *30*, 80.

(47) Green, N. J. B.; Pimblott, S. M. *Molec. Phys.* **1991**, *74*, 811.

(48) Green, N. J. B.; Pimblott, S. M. *Molec. Phys.* **1991**, *74*, 795.

(49) Green, N. J. B.; Pilling, M. J.; Pimblott, S. M. *Radiat. Phys. Chem.* **1989**, *34*, 105.

(50) Bartczak, W. M.; Hummel, A. *J. Chem. Phys.* **1987**, *87*, 5222.

(51) Green, N. J. B.; Pilling, M. J.; Pimblott, S. M.; Clifford, P. *J. Phys. Chem.* **1989**, *93*, 8025.

(52) Clifford, P.; Green, N. J. B.; Pilling, M. J.; Pimblott, S. M. *J. Phys. Chem.* **1987**, *91*, 4417.

(53) Clifford, P.; Green, N. J. B.; Pilling, M. J. *J. Phys. Chem.* **1982**, *86*, 1322.

(54) Clifford, P.; Green, N. J. B.; Oldfield, M. J.; Pilling, M. J.; Pimblott, S. M. *J. Chem. Soc., Faraday Trans.* **1986**, *82*, 2673.

(55) Pimblott, S. M. *J. Phys. Chem.* **1992**, *96*, 4485.

(56) Noyes, R. M. *Prog. React. Kinet.* **1961**, *1*, 129.

(57) Clifford, P.; Green, N. J. B.; Pilling, M. J. *J. Phys. Chem.* **1982**, *86*, 1318.

(58) Green, N. J. B.; Pimblott, S. M. *J. Phys. Chem.* **1990**, *94*, 2922.

(59) Pimblott, S. M. *J. Chem. Soc., Faraday Trans.* **1993**, *89*, 3533.

(60) Green, N. J. B. *Molec. Phys.* **1987**, *65*, 1399.

(61) Pimblott, S. M.; Green, N. J. B. *J. Phys. Chem.* **1992**, *96*, 9338.

(62) Milshtein, G. N. *Theory Prob. Appl.* **1978**, *23*, 396.

(63) Rumelin, W. *SIAM J. Numer. Anal.* **1982**, *19*, 604.

(64) Pimblott, S. M. D. Phil. Thesis, Oxford University, 1988.

(65) Rice, S. A. *Diffusion-limited Reactions*; Elsevier: Amsterdam, 1985; Vol. 25.

(66) Chandrasekhar, S. *Rev. Mod. Phys.* **1943**, *15*, 59.

(67) Green, N. J. B.; Pimblott, S. M. *J. Phys. Chem.* **1989**, *93*, 5462.

(68) Collins, F. C.; Kimball, G. E. *J. Colloid Sci.* **1949**, *4*, 425.

(69) Flannery, M. R.; Mansky, E. *J. Chem. Phys.* **1989**, *132*, 115.

(70) Magee, J. L.; Tayler, A. B. *J. Chem. Phys.* **1972**, *56*, 3061.

(71) Mozumder, A. *J. Chem. Phys.* **1974**, *61*, 780.

(72) Pedersen, J. B.; Sibani, P. *J. Chem. Phys.* **1981**, *75*, 5368.

(73) Rice, S. A.; Baird, J. K. *J. Phys. Chem.* **1978**, *69*, 1989.

(74)   Pedersen, J. B.; Larsen, J. E. *J. Chem. Phys.* **1982**, *77*, 1061.

(75)   Green, N. J. B. *Chem. Phys. Lett.* **1984**, *107*, 485.

(76)   Clifford, P.; Green, N. J. B.; Pilling, M. J. *J. Phys. Chem.* **1984**, *88*, 4171.

(77)   Ito, K.; McKean, H. P. *Diffusion Processes and their Sample Paths;* 2nd ed.; Springer: Berlin, 1974.

(78)   Devroye, L. *Non-uniform Random Variate Generation*; Springer-Verlag: New York, 1986.

(79)   Hong, K. M.; Noolandi, J. *J. Chem. Phys.* **1978**, *68*, 5163.

(80)   Noolandi, J. In *Kinetics of Nonhomogeneous Processes. A Practical Guide for Chemists, Biologists, Physicists and Material Scientists*; G. R. Freeman, Ed.; Wiley-Interscience: New York, 1987.

(81)   Umberger, J.; LaMer, V. *J. Am. Chem. Soc.* **1945**, *67*, 1099.

(82)   Green, N. J. B.; Pilling, M. J.; Clifford, P. *Molec. Phys.* **1989**, *67*, 1085.

(83)   Green, N. J. B.; Pilling, M. J.; Pimblott, S. M.; Clifford, P. *J. Phys. Chem.* **1990**, *94*, 251.

(84)   Green, N. J. B.; Pimblott, S. M.; Brocklehurst, B. *J. Chem. Soc., Faraday Trans.* **1991**, *87*, 2427.

(85)   Pimblott, S. M.; Green, N. J. B.; Brocklehurst, B. *J. Chem. Soc., Faraday Trans.* **1991**, *87*, 3601.

(86)   Chernovitz, A. C.; Jonah, C. D. *J. Phys. Chem.* **1988**, *92*, 5946.

(87)   Jonah, C. D.; Miller, J. R. *J. Phys. Chem.* **1977**, *81*, 1974.

(88)   Elliot, A. J.; McCracken, D. R.; Buxton, G. V.; Wood, N. D. *J. Chem. Soc., Faraday Trans.* **1990**, *86*, 1539.

(89)   *Pulse Radiolysis*; Tabata, Y., Ed.; CRC Press: Boca Raton, 1990.

(90)   Buxton, G. V. In *Radiation Chemistry. Principles and Applications*; Farhataziz and M. A. J. Rodgers, Ed.; VCH Publishers: New York, 1987.

(91)   Buxton, G. V.; Greenstock, C. L.; Helman, W. P.; Ross, A. B. *J. Phys. Chem. Ref. Data* **1988**, *17*, 513.

(92)   Pimblott, S. M.; LaVerne, J. A. *Radiat. Prot. Dosim.* **1994**, *52*, 183.

(93)   Balkas, T. I.; Fendler, J. H.; Schuler, R. H. *J. Phys. Chem.* **1970**, *74*, 4497.

(94)   Pimblott, S. M.; LaVerne, J. A. *Radiat. Res.* **1992**, *129*, 265.

(95)   Abramowitz, M.; Stegun, I. A. *Handbook of Mathematical Functions*; Dover: New York, 1970.

(96)   Pimblott, S. M.; LaVerne, J. A. *J. Phys. Chem.* **1992**, *96*, 8904.

(97)   LaVerne, J. A.; Pimblott, S. M. *J. Chem. Soc., Faraday Trans.* **1993**, *89*, 3527.

(98)   Draganic, I. G.; Draganic, Z. D. *J. Phys. Chem.* **1973**, *77*, 765.

(99)   Pimblott, S. M.; Schuler, R. H.; LaVerne, J. A. *J. Phys. Chem.* **1992**, *96*, 7839.

(100)   Schuler, R. H.; Hartzell, A. L.; Behar, B. *J. Phys. Chem.* **1981**, *85*, 192.

(101)   Elliot, A. J.; Chenier, M. P.; Ouellette, D. C. *J. Chem. Soc. Faraday Trans.* **1993**, *89*, 1193.

(102)   Christensen, H.; Sehested, K. *J. Phys. Chem.* **1986**, *90*, 186.

(103)   Sehested, K.; Christensen, H. *Radiat. Phys. Chem.* **1990**, *36*, 499.

(104)   LaVerne, J. A.; Pimblott, S. M. *J. Phys. Chem.* **1993**, *97*, 3291.

(105)   McQuarrie, D. A. *J. Appl. Prob.* **1967**, *4*, 143.

(106)   Mozumder, A. *J. Chem. Phys.* **1971**, *55*, 3020.

(107)   Dodelet, J.-P.; Freeman, G. R. *Radiat. Phys. Chem.* **1975**, *7*, 183.

174

(108) Green, N. J. B.; Pimblott, S. M. *Radiat. Phys. Chem.* **1991**, *37*, 161.
(109) Delaire, J. A.; Croc, E.; Cordier, P. *J. Phys. Chem.* **1981**, *85*, 1549.
(110) Gee, N.; Freeman, G. R. *J. Chem. Phys.* **1992**, *96*, 586.
(111) Gee, N.; Freeman, G. R. *J. Chem. Phys.* **1987**, *86*, 5716.
(112) Gee, N.; Senanayake, P. C.; Freeman, G. R. *J. Chem. Phys.* **1988**, *89*, 3710.
(113) Schmidt, W. F.; Allen, A. O. *J. Chem. Phys.* **1970**, *52*, 2345.
(114) Pimblott, S. M.; Mozumder, A.; Green, N. J. B. *J. Chem. Phys.* **1989**, *90*, 6595.
(115) Onsager, L. *Phys. Rev.* **1938**, *54*, 554.
(116) Saltiel, J.; Atwater, B. W. *Adv. Photochem.* **1988**, *14*, 1.
(117) Pedersen, J. B.; Freed, J. H. *J. Chem. Phys.* **1973**, *58*, 2746.
(118) Brocklehurst, B. *Int. Rev. Phys. Chem.* **1985**, *4*, 279.
(119) Brocklehurst, B. *Radiat. Phys. Chem.* **1989**, *34*, 513.
(120) Brocklehurst, B. *Faraday Discuss. Chem. Soc.* **1984**, *78*, 303.
(121) Green, N. J. B.; Harris, R. *Chem. Phys. Lett.* **1992**, *198*, 81.
(122) Pimblott, S. M.; LaVerne, J. A. *Radiat. Res.* **1990**, *122*, 12.

Research in Chemical Kinetics, Volume 3
R.G. Compton and G. Hancock (editors)
© 1995 Elsevier Science B.V. All rights reserved.

Effects of low frequency electromagnetic fields on crystal growth from solutions

Dario T. Beruto and Marino Giordani

Interdepartmental Centre for Materials Engineering and Institute of Chemistry,
Faculty of Engineering, University of Genoa
Piazzale J. F. Kennedy, Fiera del Mare Pad. D, 16129 Genova, Italy

1. INTRODUCTION

The effects of electromagnetic fields on the behaviour and the properties of matter have long been investigated since the corner-stone works by Faraday and Maxwell. When the intensities and the frequencies of such fields are low (ELF-fields), they often induce unique and puzzling effects. Galvani was the first to notice that a weak electrical current causes changes in the functional properties of muscles and in the precipitation of tartrate salts from a hydroalcoholic solution. The evidence that weak fields can affect the behaviours of both living [1] and non-living materials [2] is fascinating, and it is believed that this effect occurs because ELF fields can modify mass-transport phenomena and the related reactions of particles and/or ions, which play an important role in the overall process.

It is surprising that sometimes ELF fields have noticeable effects in comparison with the small amount of energy supplied by the applied fields. This aspect has not been adequately explored yet. To investigate these problems, one should try to know the nature and the role of the weak forces exerted by ELF fields on matter, as well as the kind of cooperative steps that take part in the general process. This is a difficult task and recently became the focus of interdisciplinary research activities on biological systems [3-10]. By contrast, the effects of these weak fields on the behaviour of inorganic materials have not been extensively studied, although this research may provide very useful bases for interpreting and predicting the properties of more complex systems, like biological ones.

This paper tries to compensate for this back off of information by dealing with the effects of low-intensity and electromagnetic low-frequency (ELF) fields on the nucleation and growth processes of inorganic crystalline salts from solutions.

When ELF fields are applied to solutions and/or to suspensions and/or to colloid systems, such media may become a space where a non-uniform electric field is generated even if the associate magnetic field remains uniform. Under the action of the non-uniform electric field charged particles and neutral ones can move from lower-field regions to higher-field ones. This motion might give rise to local differences in the spatial distribution of the particles in their ambient medium. If aggregation phenomena and/or reactions between the components occur, the compositions of the particle ensembles and the particle size might change.

Applications of this theory to various branches of science are possible and of great interest. An excellent book on dielectrophoresis (i. e., the motion induced by non-uniform electric field in neutral matter), has been published [11]. The phenomena described in this paper cannot be depicted as a dielectrophoresis process; nevertheless some considerations about the relationships between the particle size and the effects of non-uniform fields on particle dynamics can be useful also for our systems.

According to the pioneering works by Wrede [12] and to those by the Princenton and Cornel groups [13-23], it is possible to assume that the effects of non-uniform fields are negligible at the molecular level and progressively increase with the particle size.

When this general assumption is applied to the interactions between ELF-fields and the ions, nuclei, clusters and fine particles suspended in the liquid phase during the precipitation process, it is difficult to understand and explain the effects of low-intensity exogenous fields, as their action is further reduced by the medium size of the particle.

Accordingly, it becomes important both to define the variables that affect the nucleation and the growth of the solid phase from the liquid one and to establish which parameter may be significantly affected by the low-energy exchanges between a system and ELF-fields. Furthermore, it is mandatory to design experiments that may allow one to observe particle microstructures due to the effect of an ELF-field and not to spurious contributions.

The purpose of the present article is to make a survey of these topics, with special emphasis on our recent results in this field and on new data that are interesting and useful in understanding the mechanism of mineral formation in bone.

2. THEORY

The sequence of steps involved in the nucleation, crystallization and precipitation processes of a solid product in a solution to which a precipitating agent has been added is shown in Figure 1.

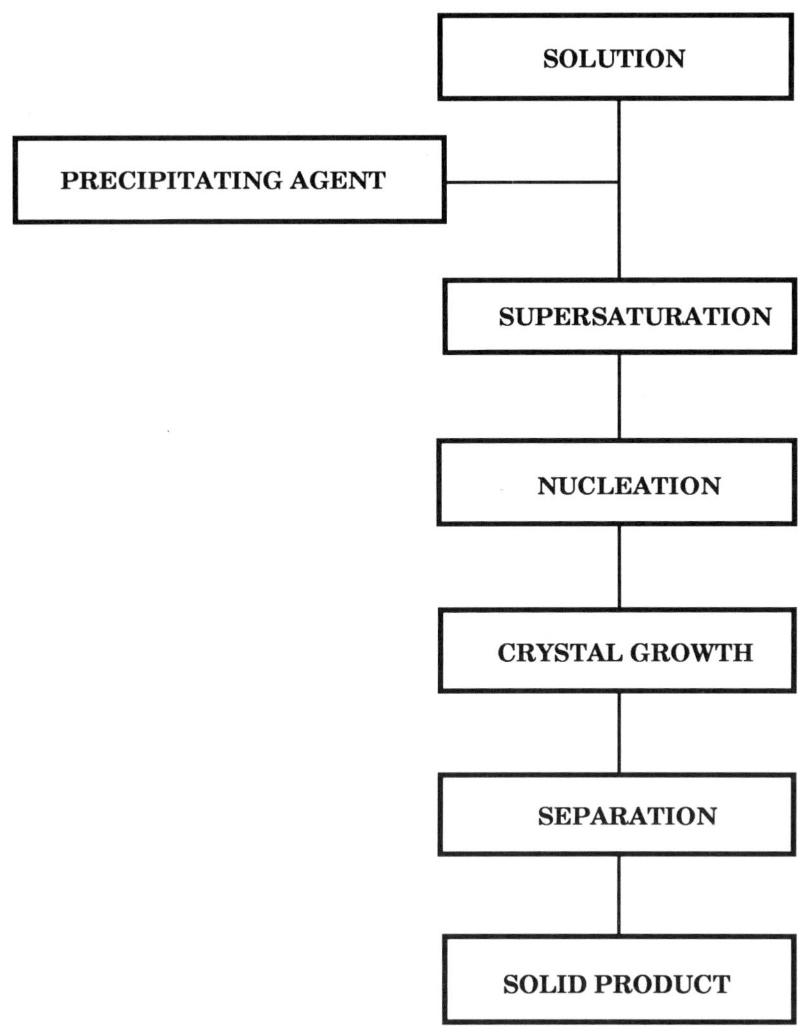

Figure 1. Sequence of the steps involved in the nucleation, crystallization and precipitation processes in a solution

In the medium formed by the initial solution and the added precipitating agent, different kinds of cations and anions may interact both with water molecules and with one another. If the solvent is water, the interactions between charged atoms and polar molecules are caused by strong electrostatic forces of attraction. These may lead to the formation of hydrated ions, whose nature and structure are rather stable and responsible for all static and kinetic properties of the solutions (viscosity, ionic mobility, diffusion, etc.) [24]. The nature of these species is certainly affected by the nature of the solvent, which generally leads to a solvation process. Once hydrated and/or solvated ions are formed, a set of dipoles appear in the medium. The interactions of these species with an applied electromagnetic field can be described, as a first approximation, by the electromagnetic Lorenzt force. The dipolar part of this force is proportional to the volume of the dipoles, and, according to previous findings [2], should not be significant, as the volume of the dipoles is determined by the coordination number of solvent molecules around the ions which should not to exceed the molecular size. Furthermore, when an ELF field is applied, the value of such a force is necessarily small, as the value of the applied field is small ($\mathbf{B} \approx 3$ mT or less). Accordingly, if one compares such a force with the strong forces responsible for the formation of hydrated and/or solvated ions, it seems reasonable to conclude that *ELF fields cannot affect the nature and the structure of hydrated/solvated ions.*

In a static solution, the movement of solvated ions is described only by a random walk process. For the sake of simplicity, let us assume that each solvated ion is a sphere of radius r. Manipulations of the Fick diffusion equation [25] lead to the well-known relationships:

$$D = kT/ 6\pi\eta\, r \tag{1}$$

and

$$\langle x^2 \rangle = 2Dt \tag{2}$$

where D is the species diffusion coefficient, k is the Boltzmann constant, T is the absolute temperature, η is the viscosity coefficient, $\langle x^2 \rangle$ is the average diffusion distance of the species, and t is the time.

When an electromagnetic field is applied, the average diffusion distance changes because of the Lorentz force, and, under some of assumptions, it has been demonstrated by Kinouchi et al. [25] that an approximate solution, for t >> 1, is equal to :

$$\langle x^2 \rangle = 2\,[D / (1 + b^2)]\, t \tag{3}$$

where b is a parameter can be expressed as:

$$b = q\,\mathbf{B} / (6\pi\eta\, r) \tag{4}$$

where q is the particle charge and $\mathbf{B}$ is the applied field.

From equations (3) and (4) it is possible to derive the value of the field for a known system that has suppressed the average diffusion distance. Estimated values of $\mathbf{B}$ that, for several substances, like calcium ions, lead to the suppression of the average diffusion distance indicate that the threshold field strength is of the order of 10^6 T. Hence low-intensity ELF fields should not induce any variation in the average diffusion path of the reactants. Accordingly, *the average probability of collision between cations and anions should not be affected by the action of ELF fields*

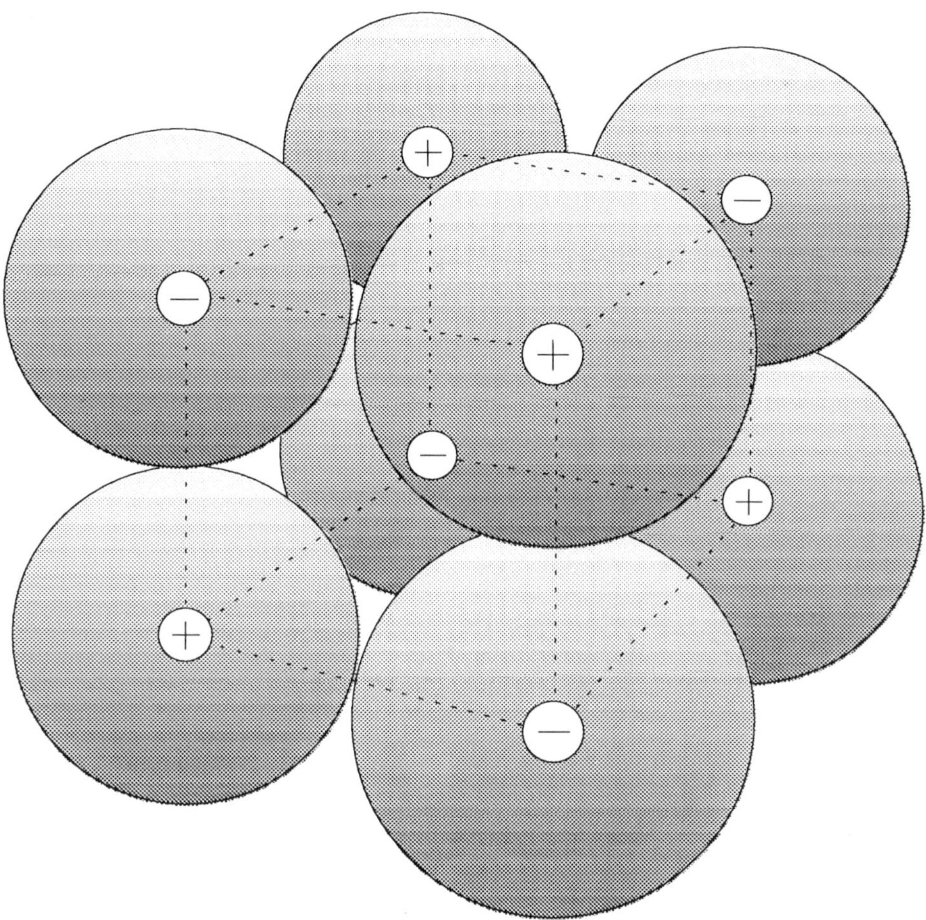

Figure 2. Ionic embryo model for ions in a solvent cavity.

The reaction mechanisms associated with the nucleation and the growth of the solid product formed by the reactants in solution need therefore to be

180

investigated to understand the effects of ELF fields on crystal growth from solutions.

These processes have been the subjects of many and important papers [26-28], and it seems appropriate outline here the most important points of those theories in order to decide on which reaction steps the ELF fields are acting.

Let S be a formed solid product and let us consider only homogeneous nucleation processes. Before reaching a well-defined crystalline shape, the solid product appears as a small nucleus. In order to grow, nuclei need to reach a critical size. An experimental and observable nucleus of the solid product formed from solution can be regarded as a droplet of critical size, made up of 50-250 atoms and/or ions. Let us concentrate on nuclei formed by ions, as they are closer to the nuclei of many inorganic salts.

The building-blocks of these nuclei (i. e., embryos) can be described at least by two models: ions in a cavity and diffused ions [29].

The formation of ions in a cavity requires strong interactions among ions so that the internal energy lattice solvation may be overcome and the configuration of the building block may become that of a small piece of crystal in a solvent cavity (Figure 2). In this configuration, the ions have a geometry that is very close to the final bulk one, but the molecular volume cannot be considered to be equal to that of the final crystallite.

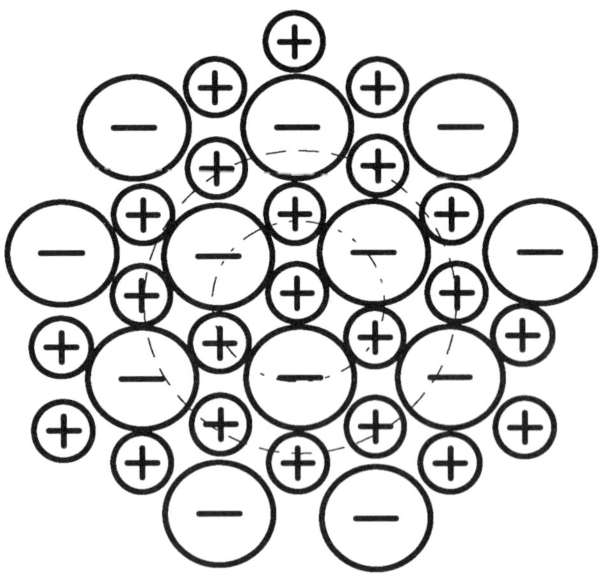

Figure 3. Model describing diffuse-ion embryos.

In the formation of diffuse-ion embryos, the ionic forces between cations and anions are not strong enough to overcome the solvation forces so that the structure of solvated ions remains virtually unchanged in the building-block. As a result, the embryo configuration can be depicted with a number of solvated ions equal to the number required by the unit lattice salts spread over a diffuse volume V (Figure 3).

These embryos have to be regarded as temporary groupings of ions analogous to those considered by the theory of associated solutions, which by statistical fluctuations get together according to the reaction :

$$n \, S \, (l) \; = \; S_n \, (s) \tag{5}$$

where $S(l)$ is a single embryo in the liquid phase (l) and $S_n(s)$ is the solid nucleus formed by n embryos.

How reaction (5) occurs is an intriguing problem. To comply with thermodynamic principles, the free-energy change in such a reaction has to be less than zero, but this requirement is not always met. The free-energy value is a function of the number n of embryos and of the shape of the nucleus, as follows:

$$\Delta G \; = \; n \, \Delta G_b \; + f \, \sigma \tag{6}$$

where ΔG_b is the bulk free-energy change per embryo, n is the number of embryos forming the droplet, f is a shape factor of the droplet, and σ is the interfacial energy per unit surface area of the droplet. The first term in eq. (6) is negative, as by convention, bond formation releases energy in the environment, whereas the second term is positive, as energy needs to be done on the system in order to form a solid/liquid interface. For a spherical nucleus of radius r formed by n embryos (each with a volume equal to v), if the ions-in-a-cavity model is applied, it is possible to write:

$$n \; = \; (4\pi/3) \; x \; (r^3/v) \tag{7}$$

It should be noted that equation (7) is dependent on the shape of the nucleus, and that it might become rather difficult to write it for a diffused-ion embryo model.

By combining equations (6) and (7), one gets:

$$\Delta G \; = \; n^{2/3} \; \Phi \; + \; n \, \Delta G_b \tag{8}$$

where Φ is proportional to the interfacial free energy per embryo. The first term of eq. (8) is always positive, whereas the second is always negative. For values of n very close to zero, the change in free energy is positive, whereas it becomes negative only for increasing values of n. Accordingly, reaction (5) is not thermodynamically possible for all values of n, smaller than a critical one, which can be derived from the condition:

$$(\partial \Delta G / \partial n)_{n = n^*} = 0 \tag{9}$$

By solving equation (9), one obtains :

$$n^* = h \left[(\sigma^3 \, v^2) / (\Delta G_b)^3 \right] \tag{10}$$

h being a numerical value equal to $32\pi/3$ for a spherical nucleus.

The obtained result defines a known phenomenon concerning the precipitation from a supersaturated solution. It is well known that a supersaturated solution completely free from dust particles does not yield any precipitate as long as the supersaturation remains below a critical limit. Above this limit value, a phase change occurs spontaneously. This behaviour is related to the increased solubility of very small crystals which is predicted by the Kelvin equation. In order to deal with a solid phase as distinguished by the liquid one, i. e., to take into account the solid-liquid interfacial energy contributions, one has to consider solid phases formed by a number of embryos larger than n^*.

A puzzling question is how embryos can reach this critical size, as the free energy change for reaction (5) in that region is positive. To solve this problem, one has to assume that, in this field, reactants are only in a liquid state, for a solid state cannot exist. This leads to the approach derived from the molecular association theory as applied to liquid states [30]. Embryos can form groups of polymers by means of a set of molecular steps described by the general equation:

$$n \, S(l) \iff S_n \, (l) \tag{11}$$

This reaction does not imply the formation of any interfaces and gives rise to an entropy of mixing, which is completely neglected in reaction (5). The mixing mechanism can provide the driving force that allows the reaction to occur. Reaction (11) can proceed up to the critical grouping cluster component defined by $n = n^*$, which is entitle to give the phase transformation :

$$S_{n^*} \, (l) \iff S_{n^*} \, (s) \tag{12}$$

A detailed description of embryo dynamics was given by I. Prigogine et al. [33]. From these studies, it is possible to deduce that nucleation kinetics can be expressed as:

$$J_N = \Omega \, (4 \, \sigma / kT)^{1/2} \, \exp \left[-(\Delta G^* / kT) \right] \tag{13}$$

where Ω is a variable including frequency factors and ΔG^* is the free-energy change in eq. (5) when n is equal to n^*. This equation shows that the free energy of formation of a critical nucleus plays a dominant role in determining the rate of nucleation. This value is easily derived, for n = n*, from equations (10) and (8) as:

$$\Delta G^* = h\,(\sigma^3\,v^2)\,/\,(\Delta G_b)^2 \tag{14}$$

where the contributions of surface-term are neglected, as the nuclei have now reached the critical size and, under this condition, the terms describing the bulk free-energy change are dominant. The bulk free-energy change can be derived from the well-known Gibbs-Kelvin equation as a function of the degree of saturation, as follows:

$$\Delta G_b = -\,(m\,k\,T\,/\,v^\circ)\ln s \tag{15}$$

where m is the number of ions in the neutral salt and v° is the molecular volume. If v° is assumed to be almost equal to the molecular volume of an embryo, then the nucleation rate can be written as:

$$J_N = \Omega\,(4\,\sigma\,/\,kT)^{1/2}\,\exp-[h\,(\sigma^3\,v^4)\,/\,(kT)^3\,(m\ln s^*)^2] \tag{16}$$

This nucleation rate is very slow for values of s less than s*, and rapidly increases when s approaches a critical value, which is obtained when nuclei of critical radii are formed.

Any possible discussion and understanding of the effects of ELF fields on the nucleation process should therefore start from an analysis of the different variables that appear in equation (16). This equation is the result of a particular and appropriate way of dealing with the complex mechanisms associated with the dynamics of embryos up to the formation of nuclei of critical size that allow the next crystal-growth steps. Furthermore, the same equation was tested experimentally [31-33] and a close agreement between theoretical and experimental results was obtained.

In principle, the variables that play an important role in eq. (16) and that might be affected by ELF fields at constant temperature are:
1)   the molecular volume of embryo and/or of the neutral salt;
2)   the degree of supersaturation s*;
3)   the interfacial energy of a nucleus of critical size σ.

Let us discuss these variables one by one.

As far as the influence of ELF fields on the molecular volume of an embryo is concerned, it should be noted that the value of v is dependent upon the force balance that holds for an ion-embryo configuration with opposite charges. These forces are very strong, as they are ionic in nature. It is hard to believe that low electromagnetic Lorentz forces, like the one exerted by ELF fields, may affect the structure of an ions-embryo configuration.

As far as the influence of ELF fields on the degree of supersaturation s is concerned, the low electromagnetic forces should be able to affect the actual concentration of ions in solution and/or the solubility product.

To change the actual concentration of the ions in solution the diffusion patterns of the correspondent ions should be affected by ELF fields.

As we reported previously this possibility is far away to be satisfied by low intensity ELF fields.

Consequently the only variable that in eq. (16) may be affected by an ELF field is the interfacial energy between the solid nucleus of critical radius and the liquid phase.

To explore how this can happen let us start with a discussion of the magnitude of surface energy in ionic materials. The specific surface-energy values of these salts are proportional to their Young modulus according to:

$$\sigma_{sv} = E\, d^\circ / \pi^2 \tag{17}$$

where E is the Young modulus, d° is the equilibrium separation distance among atoms, and the symbol sv concerns the solid/vapour interface.

The specific solid/liquid energy interfaces, σ_{sl} (sl = solid/liquid), can be related to the ones concerning σ_{sv} through the adsorption of the liquid phase on the solid nuclei by the equation:

$$\sigma_{sl} = \sigma_{sv} + \Delta g_{ads} \tag{18}$$

where Δg_{ads} is the free-energy change for the liquid-solid adsorption step expressed per cm^2 of nucleus surface.

For an exothermic adsorption step Δg_{ads} is negative, therefore, σ_{sl} will be less than the work σ_{sv}.

The heat of adsorption of liquid water on an ionic solid surface barely exceeds - 34.3 kJ mol^{-1} [34]. Neglecting the entropy terms and assuming that a water molecule has a cross-section equal to about 0.1 nm^2 [35], the change in free energy per cm^2 can be estimated to be of the order of 50 erg cm^{-2}. Accordingly, the values of σ_{sl} can be derived from the known values of σ_{sv} [36].

Table 1 presents the results. In the same table (last column), the corresponding value of the free energy per nucleus of critical size is given, assuming that 100 atoms form a critical nucleus according to a f.c.c. lattice. It is worth noting that these values are large enough to overcome the thermal disorder term kT which, at T = 300 K, is equal to about 26 meV, but that they do not exceed a few hundred meV.

This means that solid/liquid interfaces cannot be destroyed by thermal fluctuations, but that they might be affected by low ELF fields of the same magnitude.

To evaluate the magnitude of the effect produced by an ELF field let us to refer to a simple model like the one illustrated in Figure 4a, b.

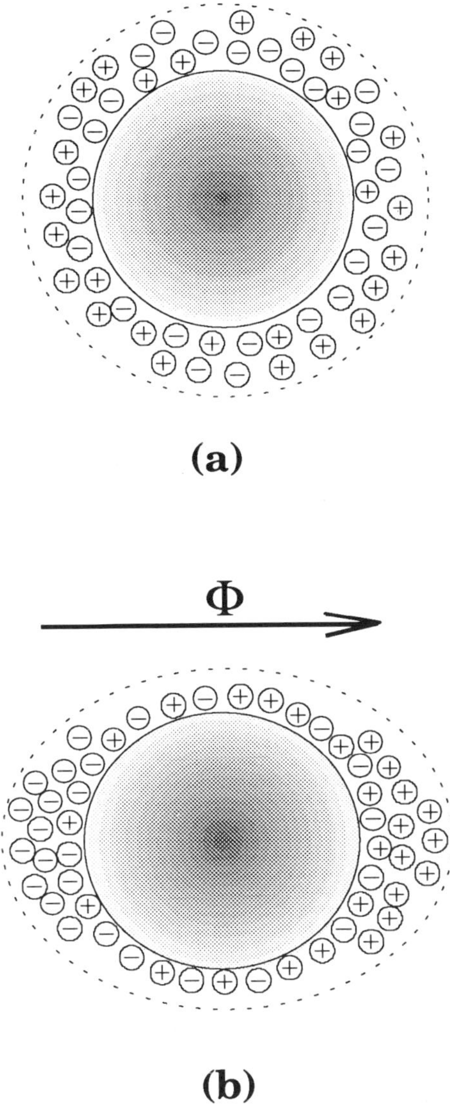

(a)

(b)

Figure 4. Model to illustrate how ELF field can strain the solid/liquid interface of a nucleus of critical size.

Generally speaking, the surface of the solid/liquid interface nucleus can be described as a surface characterized by a uniform charge density (Fig. 4a). When an external electric field ϕ is applied, the charge density is modified by a polarization effect (Fig. 4b). If the applied field has a low frequency, the new configuration should be stable over a certain period of time. Because of this charge redistribution, the surface is strained, so the final interfacial energy value will be equal to the initial interfacial energy σ_{sl} plus the strain energy due to the deformation.

Table 1
Solid/vapour and solid/liquid specific surface free energy.

Compound	Experimental σ_{sv} (erg cm^{-2})	Calculated σ_{sl} (erg cm^{-2})	Calculated σ_{sl} *for nucleus of critical size* (meV)
CaCO$_3$	230	157	125
NaCl	300	227	180
LiF	340	267	212
BaF$_2$	280	207	165
CaF$_2$	450	377	300

This last term can be calculated as a vectorial product of the induced strain ε and the corresponding stresses γ.

To calculate the stresses corresponding to the strains caused by the field, an approach equal to the one recently developed by one of the authors [37] can be used.

Let Δn_q be the charges per unit area that can be introduced or removed from the interface by the electric field ϕ. When the field does not affect the total amount of charges but it simply redistributes them, Δn_q might be regarded as the difference between the maximum and the minimum charge densities at the interface. In any case, the surface will be strained by a two-dimensional state of, ε_q, equal to :

$$\varepsilon_q = \Delta n_q \, \Omega_s \, L_s \tag{19}$$

where Ω_s is the area occupied by a unit charge and L_s is a two-dimensional shape tensor accounting for the symmetry of the charge with respect to the surface lattice. For now, the effect of orientating the charge distribution will be neglected and the state of strain ε_q will be considered isotropic. To keep the surface layer joined to the underlying bulk, an elastic strain ε_{el} will oppose the inelastic one and the resulting effective strain will be given by :

$$\varepsilon = \varepsilon_q + \varepsilon_{el} \tag{20}$$

Then, the strain energy interface, W_s, associated with the deformation will be equal to:

$$W_s = 4 \pi r^2 \gamma \varepsilon \tag{21}$$

where r is the radius of the nucleus of critical size and γ is the stress corresponding to the inelastic strain caused by the electric field ϕ. The two-dimensional state of stress γ is further associated with the state of elastic strain ε_{el} by [38]:

$$\gamma = [E_s \, d_s / 1 - \upsilon_s] \, \varepsilon_{el} \tag{22}$$

where E_s and υ_s are, respectively, the Young and Poisson moduli of the solid-liquid interface, and d_s is the thickness of the surface layer.
It can be shown [38] that, on account of the spherical symmetry and of the well-known Laplace equation, the actual strain is a function of the charge-induced inelastic strain according to:

$$\varepsilon = [1 / (1 + \alpha)] \, \varepsilon_q \tag{23}$$

where α is a constant that is dependent upon microstructural properties of the nuclei and that can be taken approximately equal to $\alpha \approx r/d_s$.
By combining equations (20)-(23), it is easy to obtain :

$$W_s = 4 \pi r^2 \theta \varepsilon_q^2 \tag{24}$$

where θ is a constant equal to :

$$\theta = [\alpha / (1 + \alpha)^2] [E_s \, d_s / (1 - \upsilon_s)] \tag{25}$$

The values of the Young and Poisson surface moduli in eq. (25), at least as orders of magnitude, can be taken equal to the ones tabulated for the bulk phases. Experimental evidence [39] concerning the rheological behaviour of dispersed aqueous systems allows us to establish that a reasonable value of d_s is equal to around 1-2 nanometers. Accordingly, for a critical nucleus of about 100 atoms, the value of the constant θ turns out to be of the order of 10^{-1}.

To be a plausible experimental variable, the value of the strain energy associated with the effect of the electrical field has to exceed 20 meV, i. e., the value of the term kT at 300 K. Application of equation (24) yields a value 50 meV when the inelastic strain ε_q is equal to 10^{-2}. This deformation should be produced by the local electrical field Φ applied on the solid particles. Actually, the evaluation of the local field in a solution with dispersed solid particles is a severe problem. Studies are currently in progress to determine this value [40].

As a first attempt, let us consider a relationship like:

$$\varepsilon_q = P\phi \tag{26}$$

where P is a constant that takes into account the amount of surface charges that can be polarized and ϕ is the applied local electrical field. If the value of ϕ is equal to 1 V m^{-1} it is possible to demonstrate [41] that, when the surface polarizable charges is equal to about same ten of electrons, then (eq. 26) the correspondent value of ε_q is equal to 10^{-2} and the value of the strain energy is about 50 meV.

Hence it seems very reasonable to produce a strain energy of the same order of magnitude as that of the solid/liquid interfacial energy under the effects of electrical field as low as 1 V m^{-1}.

Another suitable way of proving that a strain energy of the order of 50-100 meV can be produced by a low ELF field, lies in estimating the electromagnetic energy that is ready to be used by a nucleus of critical size for straining its surface/liquid interface.

By applying of the well known equation:

$$U = (1 / \mu^\circ) \, (\mathbf{B}^2) \, (V_{tot} / 2) \tag{27}$$

where μ° is the magnetic permeability in a vacuum, $\mathbf{B}$ is the applied magnetic field, and V_{tot} is the geometrical volume of the reactor to which the ELF field has been applied, it is possible to determine the total electromagnetic energy (U) that is given to the system formed by dispersed solid and liquid phase.

If the number of critical nuclei in solution is known, one can to derive the amount of electromagnetic energy per nucleus. When these calculations are made for a value of $\mathbf{B}$ equal to 3 mT and under our experimental conditions (i. e., 10^{15} nuclei per cm^3 of solution), an electromagnetic energy of the order of 100 meV per nucleus can be derived.

Accordingly, the exchange of energy between ELF field and each nucleus is greater enough to provide the energy requested to strain the nucleus/liquid interface by an amount larger than the noises produced by the thermal fluctuations.

This effect is not at all negligible, as it might ncrease the value of the interfacial energy of a nucleus by a factor of two. Such an increase would involve several important consequences for both the nucleation process and the subsequent growth step.

From eq. (14), it follows that doubling the interfacial energy value corresponds to multiplying by a factor of 8 the apparent activation energy of the nucleation process. This may have a considerable impact on the nucleation rate, as this rate depends exponentially on the apparent activation energy according to eq. (16). As a consequence the effect of ELF-field will be that of slowing down the nucleation rate process and that of producing nuclei with a larger critical size because the number of elementar embryos that need to be joined to reach the critical size will increase according to eq. (10).

The subsequent crystal-growth step may be influenced by these effects. To describe how this may occur, let us refer to a simple growth model, like the one described in Figure 5.

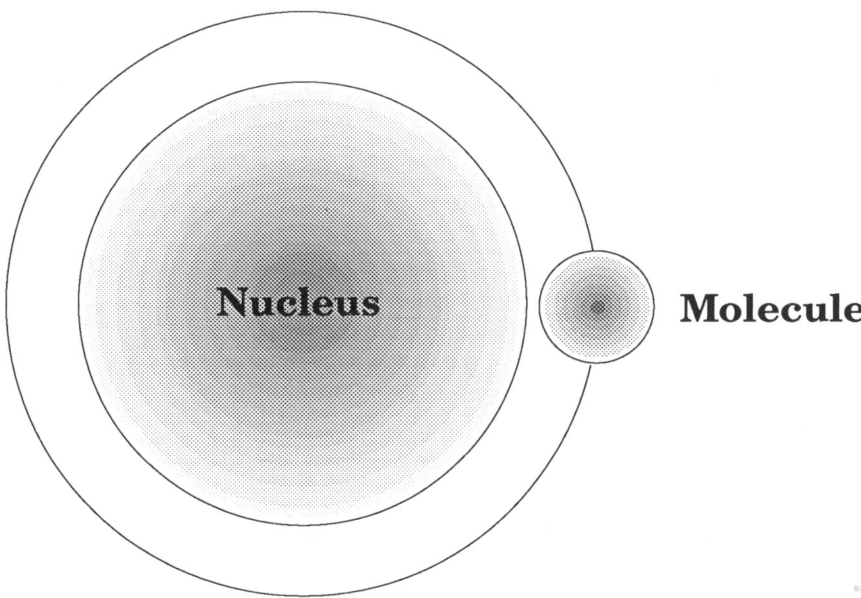

Figure 5. Model for crystal growth.

The molecules in solution will first adsorb on the critical nuclei and they will proceed into the nucleus bulk phase to produce the growth process. This sequence of steps can be described by the equations:

$$A \text{ (sol)} \quad \overset{k_1}{\underset{k_2}{\Leftrightarrow}} \quad A \text{ (ads on } S_{n+}) \tag{28}$$

$$A \text{ (ads on } S_{n+}) \quad \overset{k_3}{\Rightarrow} \quad A \text{ (into the lattice of } S_{n+}) \tag{29}$$

where k_i is the kinetic constant and A denotes the molecules in solution. Actually, the nature of these molecules may be the same as the one that formed S_{n+} or, more generally, it might be different. In the latter case, coprecipitation and/or other reactive processes can be described.

Under steady-state conditions one can write :

$$d[A(\text{ads on } S_{n+})]/dt = k_1[A(\text{sol})] - k_2[A(\text{ads on } S_{n+})] + k_3[A(\text{ads on } S_{n+})] = 0 \tag{30}$$

solving this equation allows us to express the growth rate as :

$$J_G = [A \text{ (sol)}] \times [k_1 k_3 / (k_2 + k_3)] \tag{31}$$

If $k_2 \ll k_3$, then :

$$J_G = [A(\text{sol})] \times k_1 \tag{32}$$

According to equation (32), the growth rate increases with the kinetic constant of the adsorption step. Since the nuclei produced under ELF fields have a higher interfacial energies, the adsorption step *must be facilitated*. Hence it may be predicted that the grain growth rate of the solid product will be enhanced by the effects of ELF fields. Actually, if the nature of the molecule A is different from the nature of the solid nuclei S, equation (32) leads to predict that also the kinetic rate between A and S will be increased. In the following, we shall test these theoretical predictions.

3. EXPERIMENTAL

3.1. Techniques

The electromagnetic forces acting between solid nuclei of critical size formed in a solution and an applied low-intensity ELF are the Lorentz forces and the Maxwell stresses. If the solution flows through a circuit, a static magnetic field can be used as a source of exogenous fields, whereas, if the solution is static (i. e., the barycentric rate is zero), the source of ELF fields can be an alternating electric current flowing in Helmoltz coils. In both cases, the resulting electromagnetic forces acting on the nuclei are the same; therefore, the observed effects on the behaviour of solid nuclei should be equivalent.

In Figure 6 and in Figure 7, two examples of laboratory apparatus are given. The apparatus shown in Figure 6 uses a Helmoltz-coils source, whereas, in the laboratory apparatus in Figure 7, a static magnetic field device is used as an ELF source.

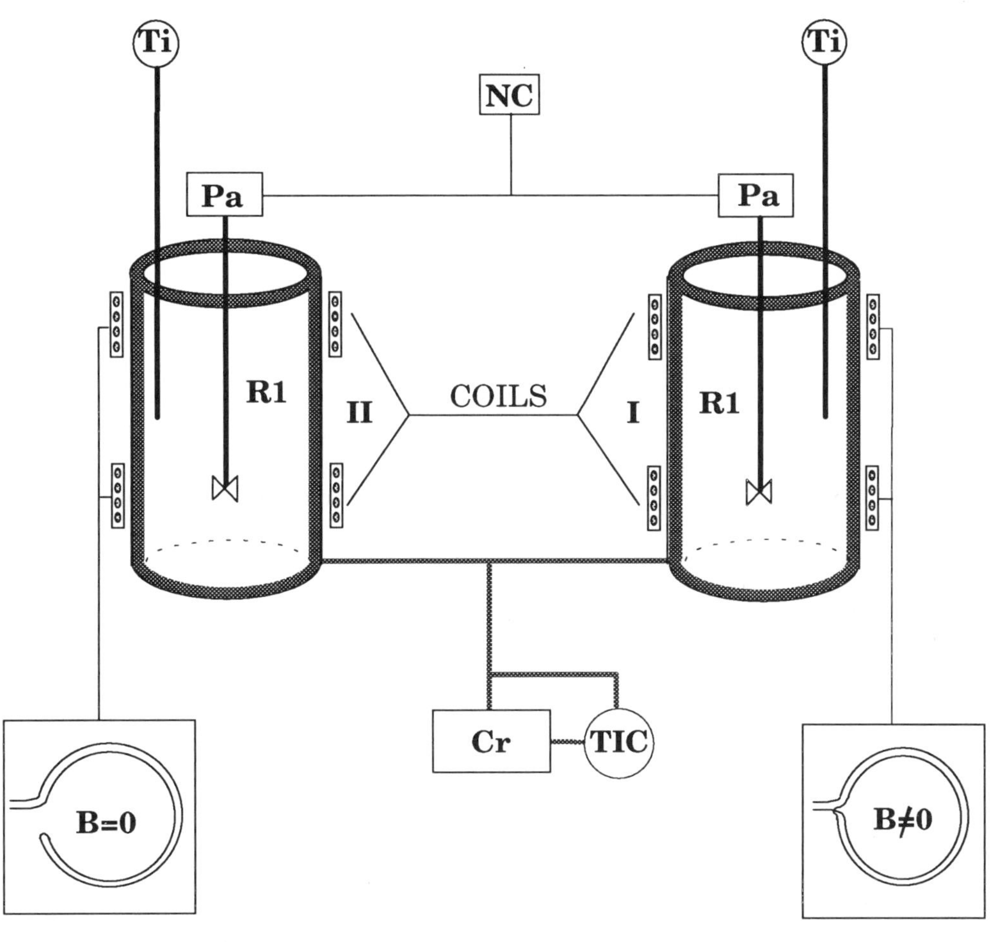

R1	**Jacketed beaker**	**Cr**	**Cryotermostat**
Pa	**Plastic or glass stirrer**	**TIC**	**Temperature control**
NC	**Motor control**	**Ti**	**Thermometer**

Figure 6. Experimental apparatus with Helmoltz-coils source.

The treatment of a solutions (with and without the applied ELF field) must follow a very precise protocol. It is very important that, during the treatment of the same solution (with and without the applied ELF field), all the variables should be the same so that a comparison of results may point out only the effect of the exogenous field. This is not an easy task because the application of the electromagnetic field might introduce some variables and/or change the values of other variables.

For instance, in the case of an ELF field produced by Helmoltz coils, extrathermal effects due to the electric current flowing in the coils should be avoided. To this end, the same liquid solution should be poured into two vessels (I and II in Fig. 6), placed midway between two dummy coils. On the active-coils side (I), the alternating current generated by a power amplifier (driven by a sinusoidal signal generator) flows to produce a preselected constant a. c. magnetic field superimposed upon the earth's field. On the passive coils side (II), wires of the same diameter and loop as utilized for the active coils must be used, but the loop windings must be such as to generate an internal field **B** equal to zero. Accordingly, if there is any extrathermal effect due to the flow of the electric current, this effect should be equal both for the experiments carried-out under the action of the ELF and for those carried-out without applying any ELF.

When the liquid and/or the suspension circulates through a static magnetic field device it should be stressed that the dynamics of the flowing system must be the same in the ELF-treated and non-treated experiments. To this end the same liquid solution must circulate through two devices of equal geometry (III and IV in Fig. 7), and the static magnetic field must act only in one of these devices.

In addition, the pH value, the temperature, the concentration, the degree of supersaturation, the digestion time, the stirring conditions etc, must be equal in the ELF-treated and non-treated solutions/dispersion.

If further experimental operations, like the drying process, need to be performed on the precipitate powders, before the final check on the microstructural and chemical grain properties, the sequences and the conditions to be used for the required steps must be equal for the ELF-treated and non-treated samples.

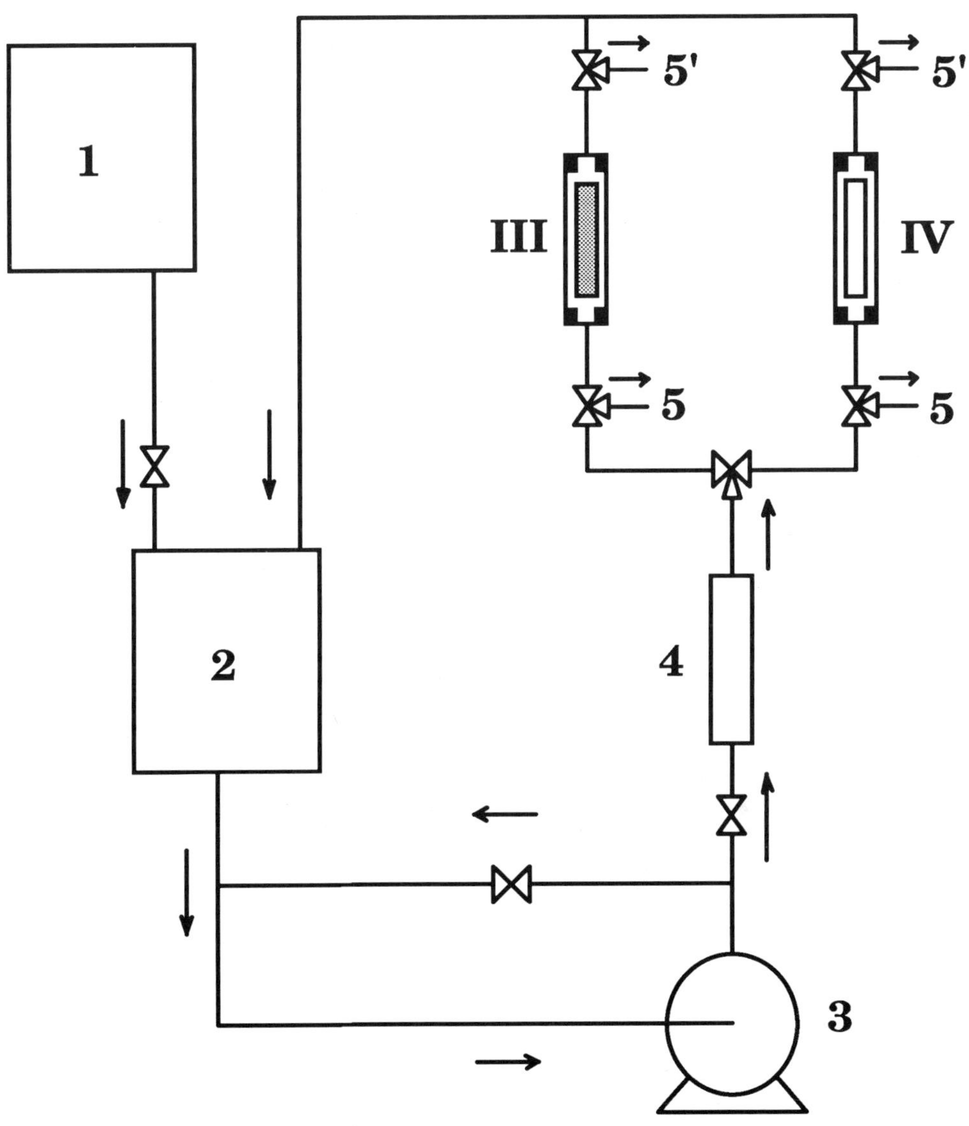

1-2 Solution reservoir   **III Device with permanent magnet**
3 Pump   **IV Device without permanent magnet**
4 Flowmeter   **5-5' Sample for analysis**

Figure 7. Experimental apparatus using static magnetic field.

3.2. Results

Direct experimental observations of the effects of ELF fields on the solid/liquid interface energies of solid nuclei of critical size (formed during precipitation processes from solutions) are very difficult because of the nuclei dimensions. Nevertheless, any modification to the nature and to the microstructure of the nuclei interfaces may strongly affect any subsequent reaction and/or process where the nuclei are involved.

To test the theory defined in the Section 2 of this work, we shall refer to experimental data that can be considered significant for three classes of precipitation processes. The main differences among the three classes of processes concern the steps that can be coupled with the growth mechanism.

In the first class of processes, the particle-growth step follows directly the nucleation one, without any interactions with other solid-state mechanisms. For the sake of clarity, let us call this process *nucleation step and non-reactive growth process*.

The second class of processes, includes the process where the growth of a solid product may proceed along two different lattice structures. Let us name this process *nucleation step and oriented growth process*.

Finally, the third class of processes is characterized by a reaction between a nucleus and other reactants; this reaction yields a solid product whose nature is different from that of the nucleus. Let us call this process *nucleation step and reactive growth process*.

In all three classes of processes, any modifications to the solid/liquid interfacial energies of the nuclei produce significant and experimentally observable variations in the nature and the microstructure of the final solid product. This evidence will be used to test the theory previously defined.

3.2.1. Nucleation step and non-reactive growth process

As an example of this class of processes, let us consider the precipitation of Barium Oxalate from aqueous and ethanol/water solutions [42, 43].

Aqueous solutions of Barium Nitrate containing 22 g l^{-1} were used, along with Ammonium Oxalate 0.35 M as precipitating agent. The experimental apparatus where the precipitation process was performed was the one described in the previous section (Fig. 6). The precipitation experiments were carried out under and without low-intensity ELF sinusoidal exposure, which yielded a magnetic induction **B** equal to 3 mT.

In Figure 8, the Barium concentration left in the aqueous solution vs. time is plotted for two sets of experiments carried out under and without ELF exposure.

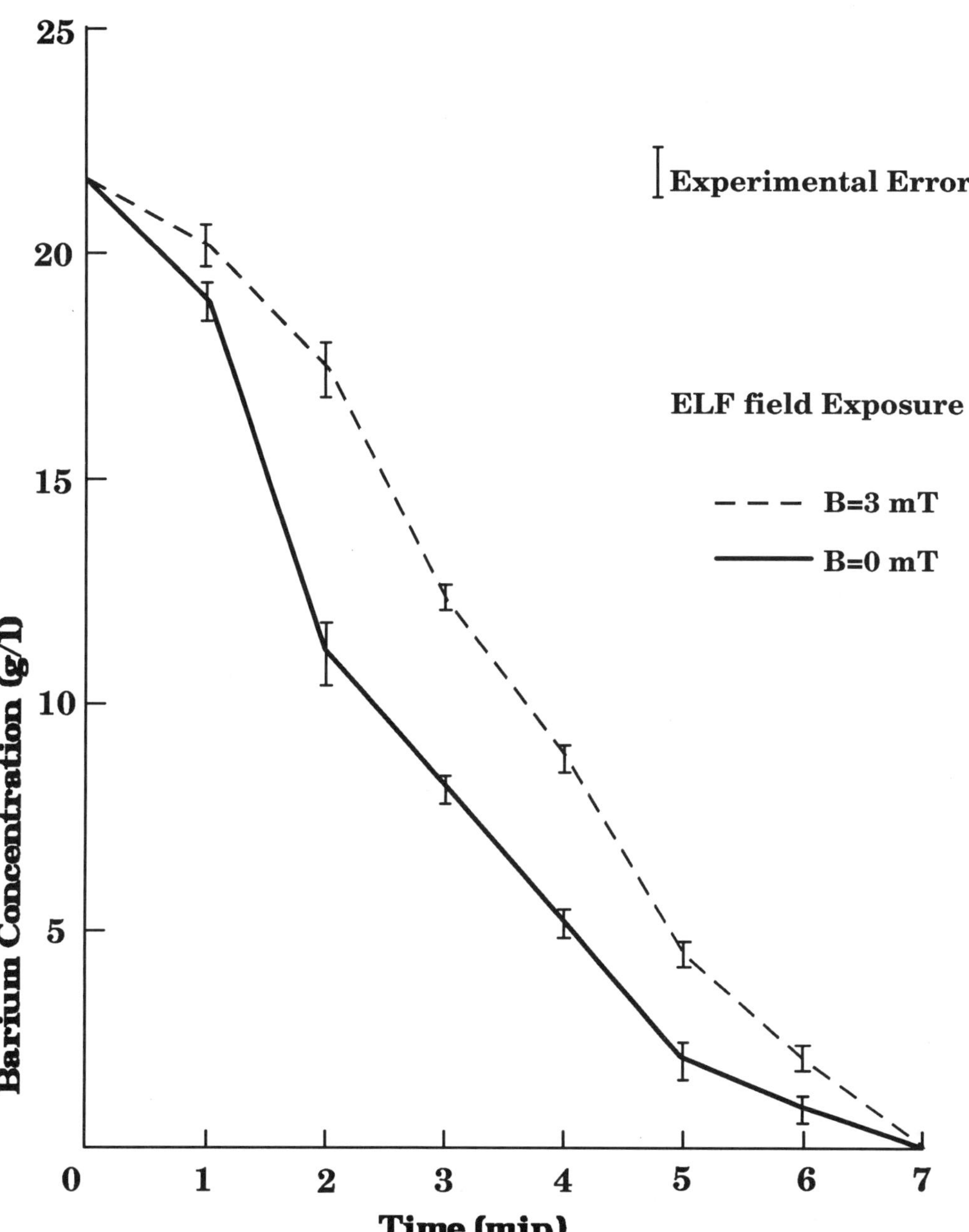

Figure 8. Barium concentration left in aqueous solution vs. time.

Taking into account all sources of experimental errors, the experimental results were such that we can state:

I. The effect of an applied ELF field does not change the total time required to get a complete precipitation.

II. The barium concentration left in an aqueous solution significantly increases when an ELF field is applied.

Both effects confirm the predictions of the theory on the effects of ELF fields on the solid/liquid interfacial energies of the formed nuclei. Indeed, if this parameter is increased by the exposure of the nuclei of critical size to an ELF field, according to eq. (16) the number of nuclei of barium oxalate present in solution at a certain time should decrease. Therefore, the corresponding barium-ion concentration in solution should increase, hence, in accordance with eq. (32), the rate of crystal growth should increase too. If so, as the total precipitation time turned out to be equal for all experiments, the final microstructure of the precipitate particles obtained under the ELF exposure *should be larger*. Figure 9a, b proves these considerations.

The average size of the precipitate particles obtained when the precipitation process was performed under the ELF exposure is five times larger than the size of the particles obtained when no ELF field was applied.

We regard these experimental results as reliable data that support our theoretical predictions.

Ethanol/water mixture was used in proportion 1:3 (v/v) as solvent for the preparation of solutions of Barium Nitrate containing 22 g l^{-1}, analogous concentration to which used for the test carried out in aquesous solutions. As precipitating agent Ammonium Oxalate 0.35 M was used. The precipitation experiments were carried out under and without low-intensity ELF sinusoidal exposure, which yielded a magnetic induction **B** equal to 3 mT.

Figure 10 shows the precipitation data obtained without the ELF field vs time, and those obtained by similar runs but in presence of a low-frequency, low-intensity sinusoidal field.

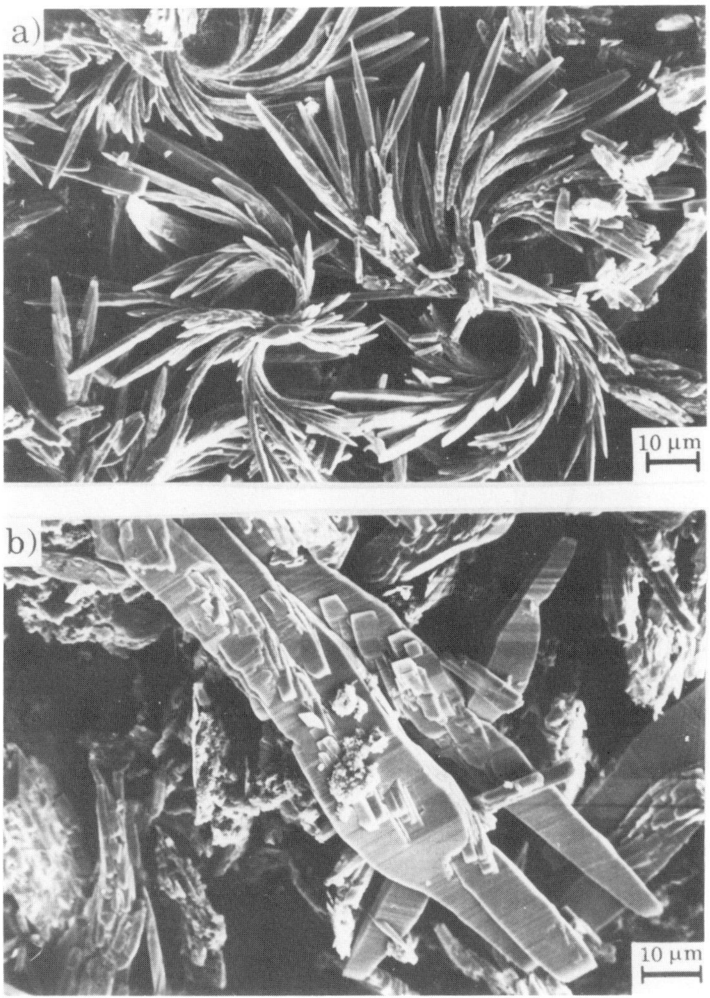

Figure 9. SEM micrographs of precipitate particles obtained in aqueous solutions, with and without ELF field applied: a) $\mathbf{B} = 0$ mT, b) $\mathbf{B} = 3$ mT.

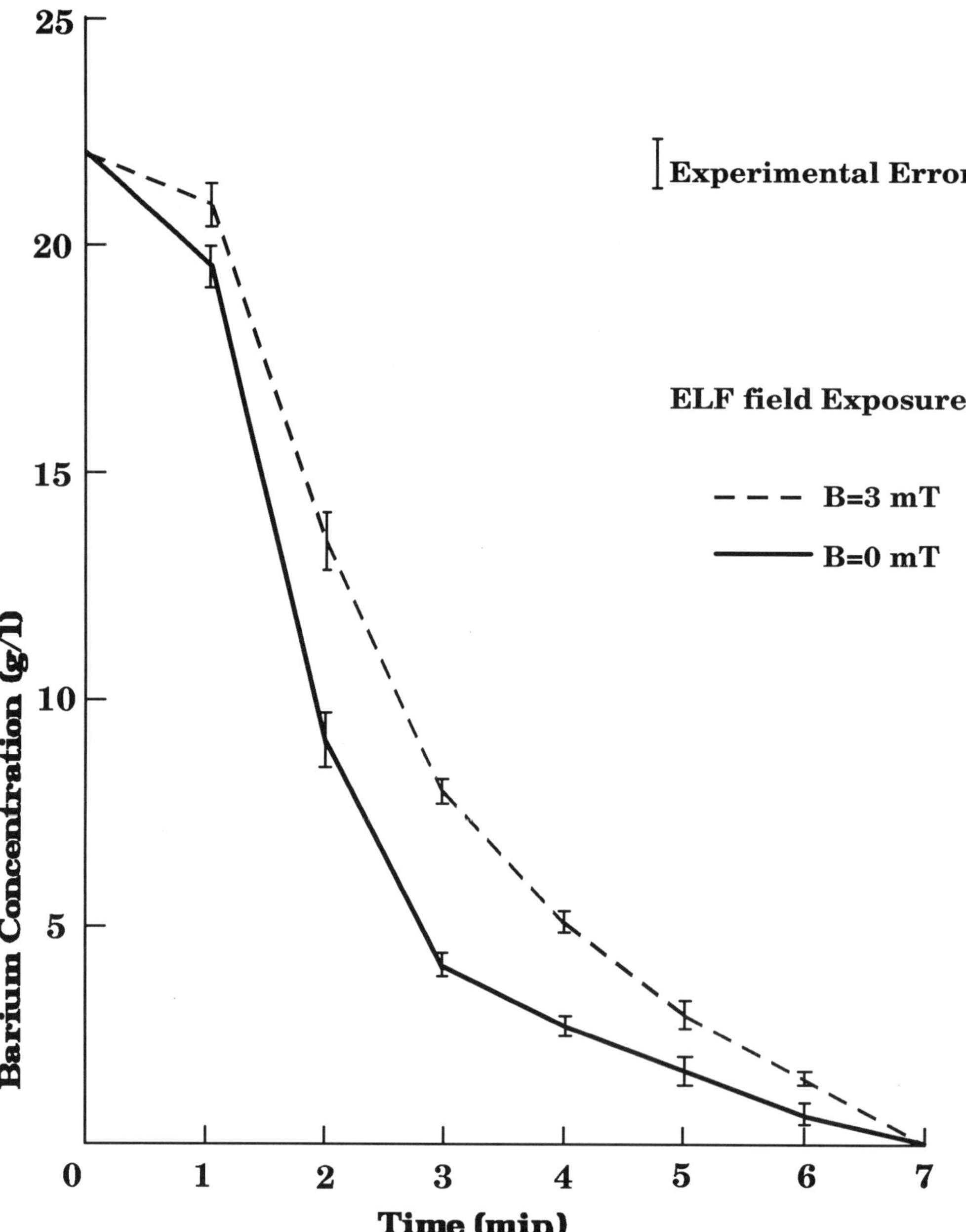

Figure 10. Barium concentration left in ethanol/water solution vs. time.

The results obtained with these solutions are in agreement with the previous one and reflect only the fact that the solubility product is less than the one in pure water [43]. Accordingly the precipitation rate will be faster.

3.2.2. Nucleation step and oriented growth process.

Calcium hydrogencarbonate solutions were chosen to test this class of reactions, as their behaviour is important for both inorganic and biological systems. In these solutions, small nuclei of calcium carbonate can be present at room temperature, for the standard Gibbs free energy associated with the reaction:

$$Ca^{2+}(ao) + 2(HCO_3)^-(ao) = CaCO_3(s) + H_2O(l) + CO_2(ao) \qquad (33)$$

(ao = aqueous solution, s = solid phase, l = liquid phase) is negative and equal to -24 kJ mol^{-1}. Furthermore, it has been proved that the growth of nuclei can be favoured, by thermal heating in the temperature range 283-373 K and at a total pressure of 0.1 MPa, in the directions of the solid phases corresponding to aragonite and calcite. Evolution in the direction of the stabler solid phase (calcite) becomes easier step if the nuclei growth reaction is a fast process [44].

As a consequence, it might be inferred that if the solid-liquid interfaces of the nuclei are increased by exposing the calcium hydrogencarbonate solutions to a low-frequency ELF field, their growth should proceed at a faster rate and the final precipitate should be formed mainly by the calcitic phase. In a previous research work, we explored this possibility and results turned out to be close agreement with predictions [45].

Solutions with an equal chemical composition were made to circulate through an apparatus like the one illustrated in Fig. 7. Treatments up to point 5 (Fig. 7) were equal for all solutions; the passage through a magnetic device, which induced on the charged surface of calcium carbonate nuclei an ELF field at a frequency of 10-30 Hz, was utilized only for the solutions to be tested. Subsequent thermal treatments to produce calcium carbonate precipitates were all equal for treated and non-treated solutions. The X-ray diffractogram patterns and the SEM micrographs of the precipitates concerning treated and non-treated solutions are shown in Figure 11a, b and in Figure 12a, b. As it can be noticed, the effects of the ELF fields lie in increasing the dimensions of the precipitate grains and in orienting the growth process in the direction of the calcitic phase formation. As has previously been observed [45], the effects of an ELF field are dependent on the magnitude of the applied field in a non-linear way.

These experimental results further confirm the assumed kinetic mechanism according to which ELF fields can be regarded as an external source that can affect the surface properties of solid nuclei and all other processes that are dependent on the existence of active critical nuclei.

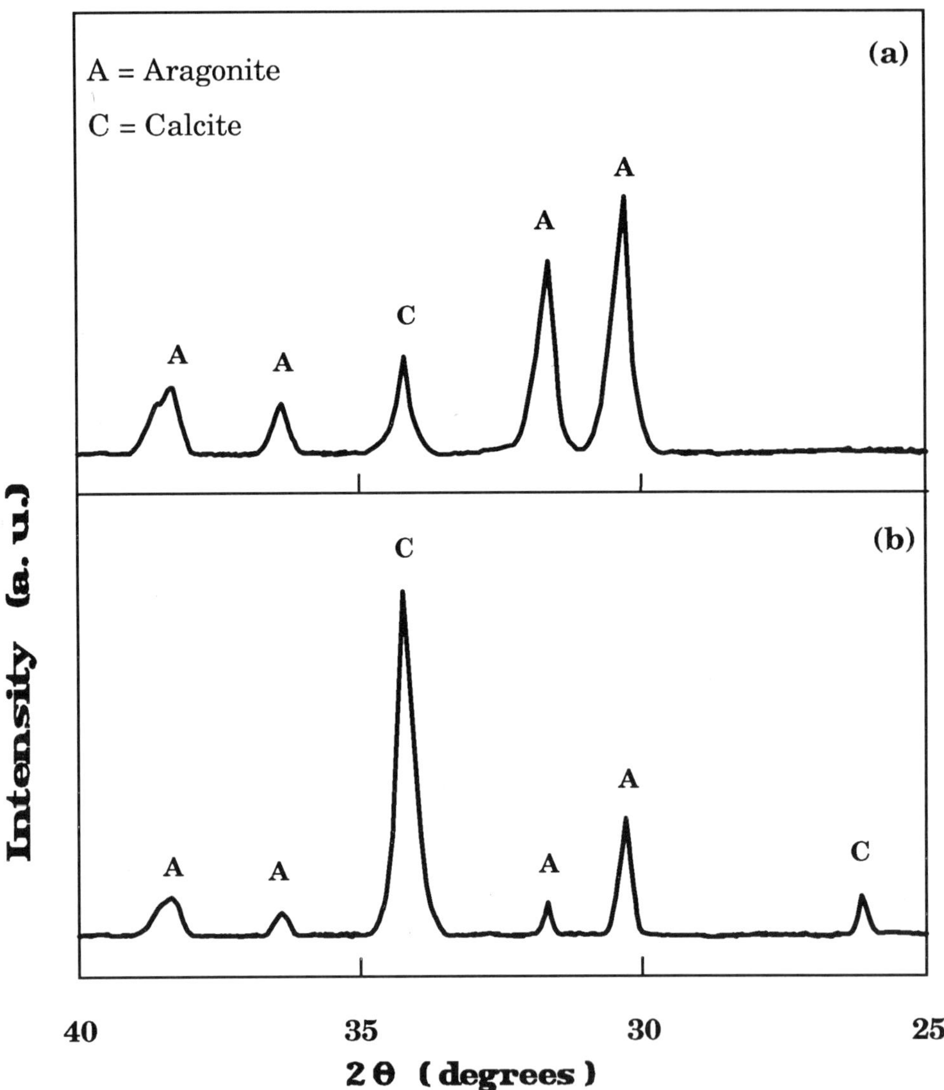

Figure 11. X-ray diffractogram (λ = 1.7902 Å) of the precipitate obtained by evaporation, at a pressure of 0.1 Mpa and a temperature of 353 K, of aqueous calcium hydrogencarbonate solutions, at a rate of 0.03 g s^{-1}. a) aqueous solution before treatment with static magnetic field; b) aqueous solution after treatment with static magnetic field.

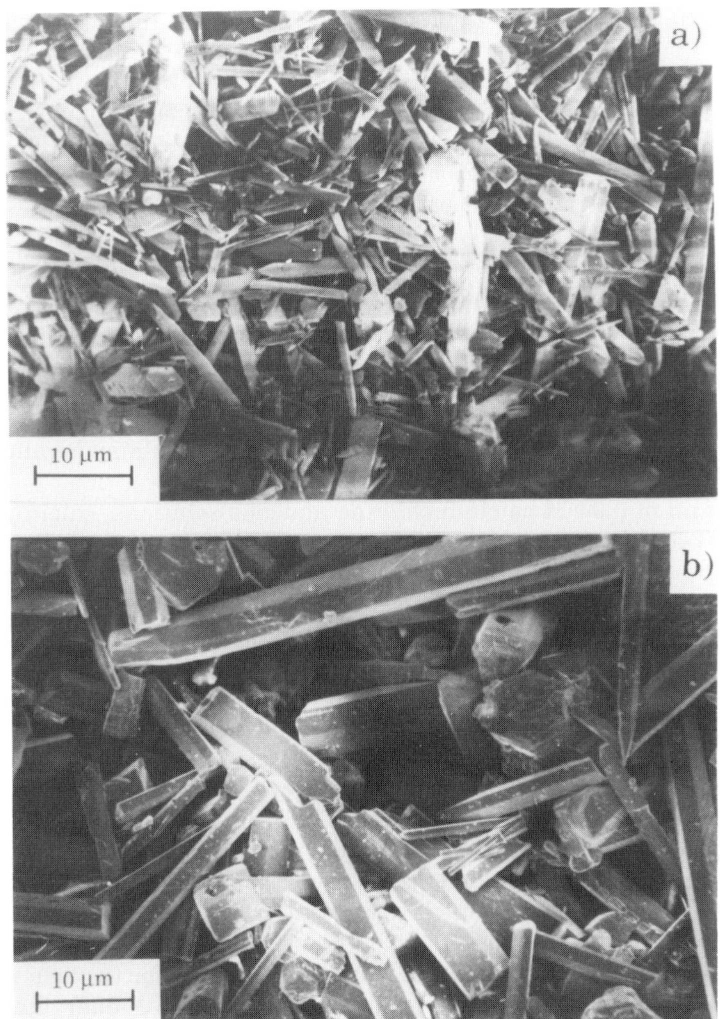

Figure 12. SEM micrographs of precipitate particles obtained under the same conditions as in Fig. 11.

3.2.3. Nucleation step and reactive growth process

This is rather an attractive class of reactions, which include some important phenomena that occur in the biomedical field as well as in the production of ceramic powders to be used as raw materials for ceramic sensors [46]. In the following we shall report new data on the formation of hydroxyapatite through the precipitation process of calcium phosphate from aqueous solutions.

The effect of low-frequency ELF fields on the mechanisms of bone repair has been focused in many important papers [7-9].

Experimental results strongly support the assumption that bone repair occurs at a faster rate when ELF fields are applied; however, no definitive explanations have been provide to understand the mechanisms of the complex interactions between ELF fields and biological systems. A deeper insight into these phenomena can be gained by studying the process step by step. Therefore, the mineral formation of hydroxyapatite has been recognised to be of major importance for the overall process.

It is very interesting to analyze 'in vitro' how such step may be affected by the exposure of the precursory solutions to ELF fields.

From 225 ml of 0.04 M calcium chloride aqueous solutions, calcium phosphate were precipitated by using Na_2HPO_4 as precipitating agent, at a concentration 0.036 M. The temperature and the pH value were kept constant, during the process at 37° C and 7.40, respectively. As soon as nuclei and crystallites of calcium phosphate were formed, a hydrolysis process occurred which transformed the calcium phosphate into hydroxyapatite. More than likely, this step occurs also 'in vivo'; hence the conclusions that can be drawn from 'in vitro' experiments may provide useful information to understand the behaviour of a biological system.

The experiments were carried out with and without application of a 3 mT ELF field, for different precipitation times. The obtained precipitates were first filtered, then dried, and finally examined with TG, XRD and N_2 adsorption techniques to derive information on the evolution of the solid-product microstructure and of the chemical composition. A detailed description of these experiments is given elsewhere [47]; here we report only the most important results in order to discuss the implications of our theory for this set of reactions.

If ELF fields can modify the solid-liquid interfacial energies of calcium phosphate nuclei, then not only the crystal growth step should proceed at a faster rate but also the kinetics of the subsequent and/or the parallel calcium phosphate hydrolysis step should be enhanced. Accordingly, experimental results should point out the formation of larger precipitate crystallites with a chemical composition richer in hydroxyapatite.

Nitrogen adsorption-desorption isotherms at 78 K were taken on precipitated powders obtained at different digestion time with and without ELF fields applied. From these data, it is possible to deduce information on the value of the powder specific surface areas, on the total porosity, and on the pore size distribution (see, for instance, [48-50]).

Figures 13a, 14a and 15a illustrate the evolution of the adsorption-desorption isotherms as a function of the digestion time.

Figures 13b, 14b and 15b give the asme informations at correspondent digestion time for samples obtained under the influence of 3 mT electromagnetic field.

Comparison among these data allows to reach a number of informations. First the all will should be noted that all isotherms since to be long to a 4th type class [35]. The applied ELF field does not seem to change this kind of behavior. Newertheless, the values of the specific surface area, total porosity and hysteresis shape, because the effect of the ELF field, seemto be significantly different.

For our purpose we focus the attention on the specific surface area values and not on the porosity value because this last parameter might be also be due to intraparticles porosity, 1. e., to powder packing factor.

Accordingly in Table 2, the specific surface area data are summarized.

Table 2
Specific surface data.

	ELF field exposure	
	$B = 0$ mT	$B = 3$ mT
Digestion time (min)	S_{BET} (m^2 g^{-1})	S_{BET} (m^2 g^{-1})
15	131.07	107.80
60	110.62	71.99
120	116.43	74.18

Experimental error less than 10%

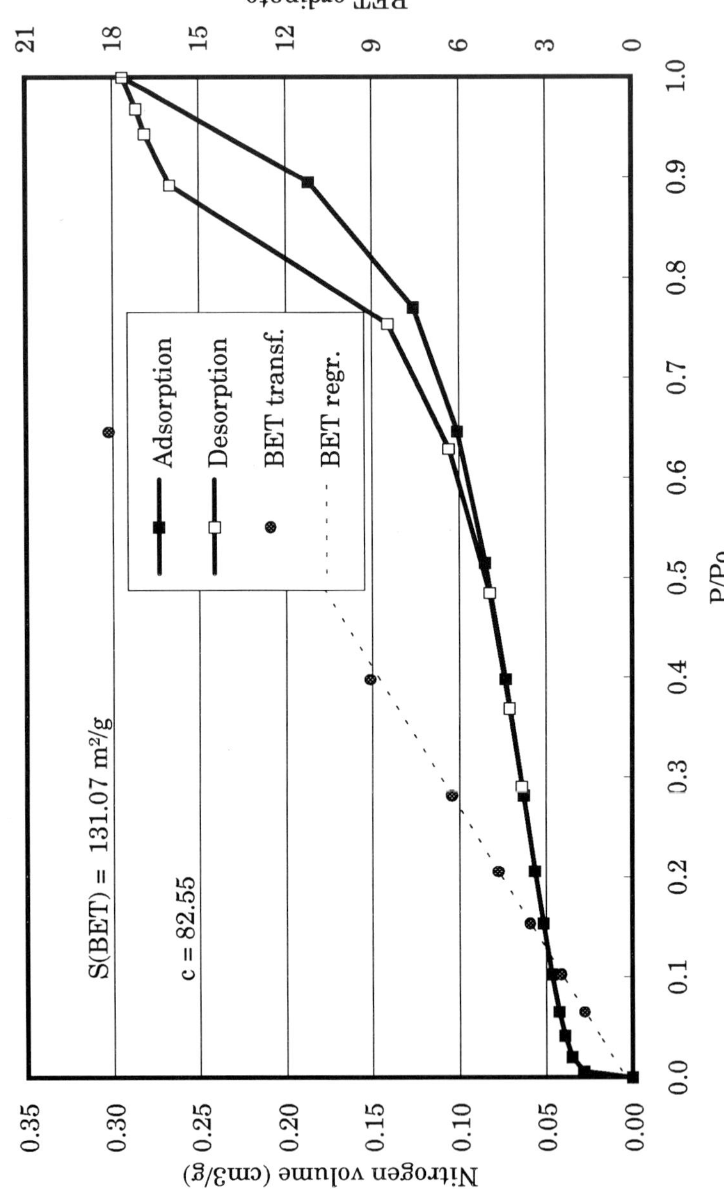

Figure 13a.   Nitrogen adsorption isotherms at 78 K for precipitates obtained without ELF field for a digestion time of 15 min.

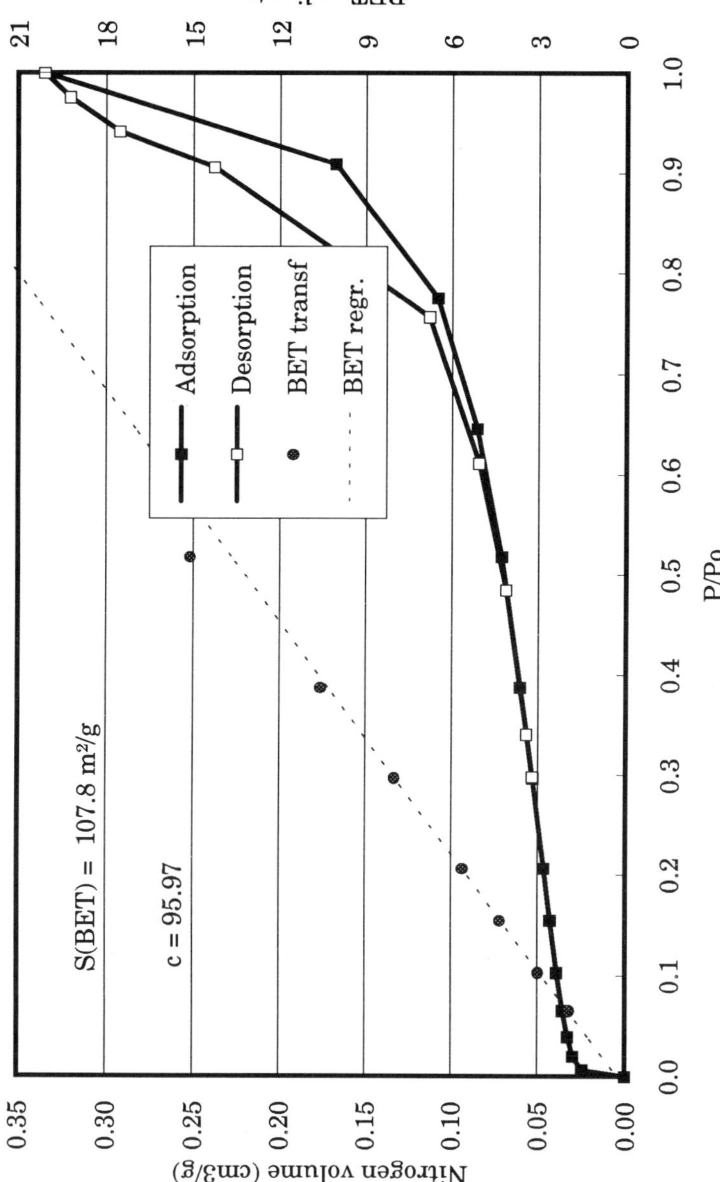

Figure 13b.   Nitrogen adsorption isotherms at 78 K for precipitates obtained with ELF field for a digestion time of 15 min.

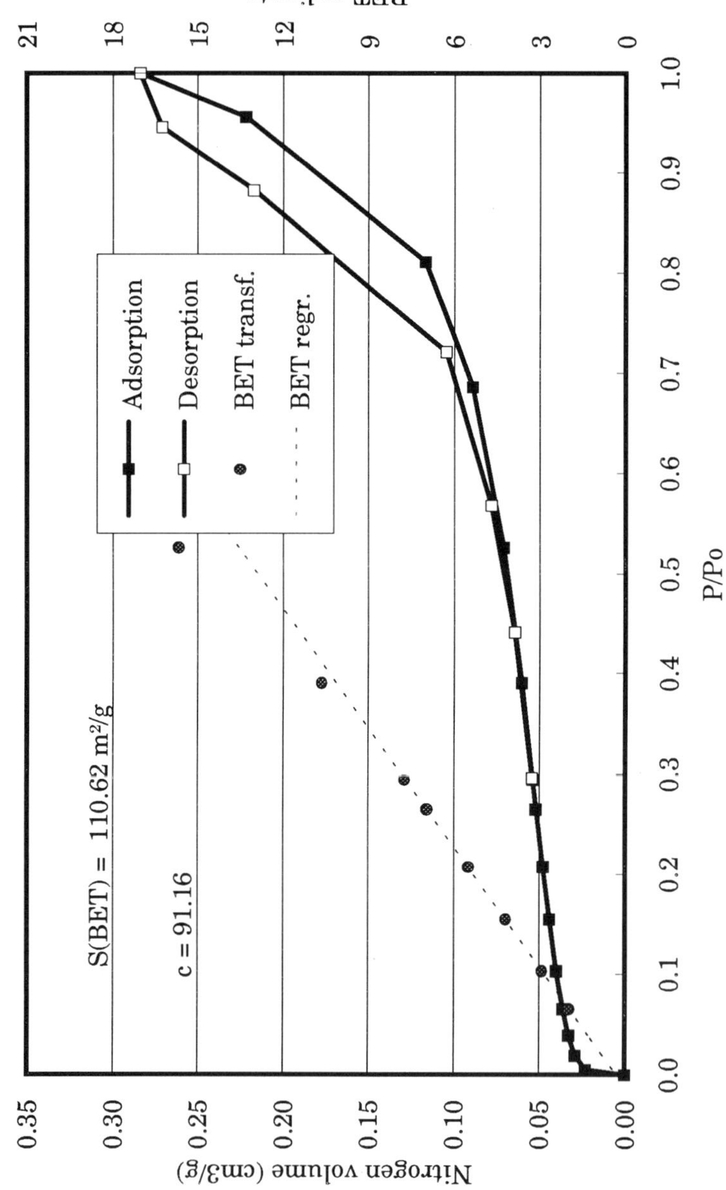

Figure 14.a. Nitrogen adsorption isotherms at 78 K for precipitates obtained without ELF field for a digestion time of 60 min.

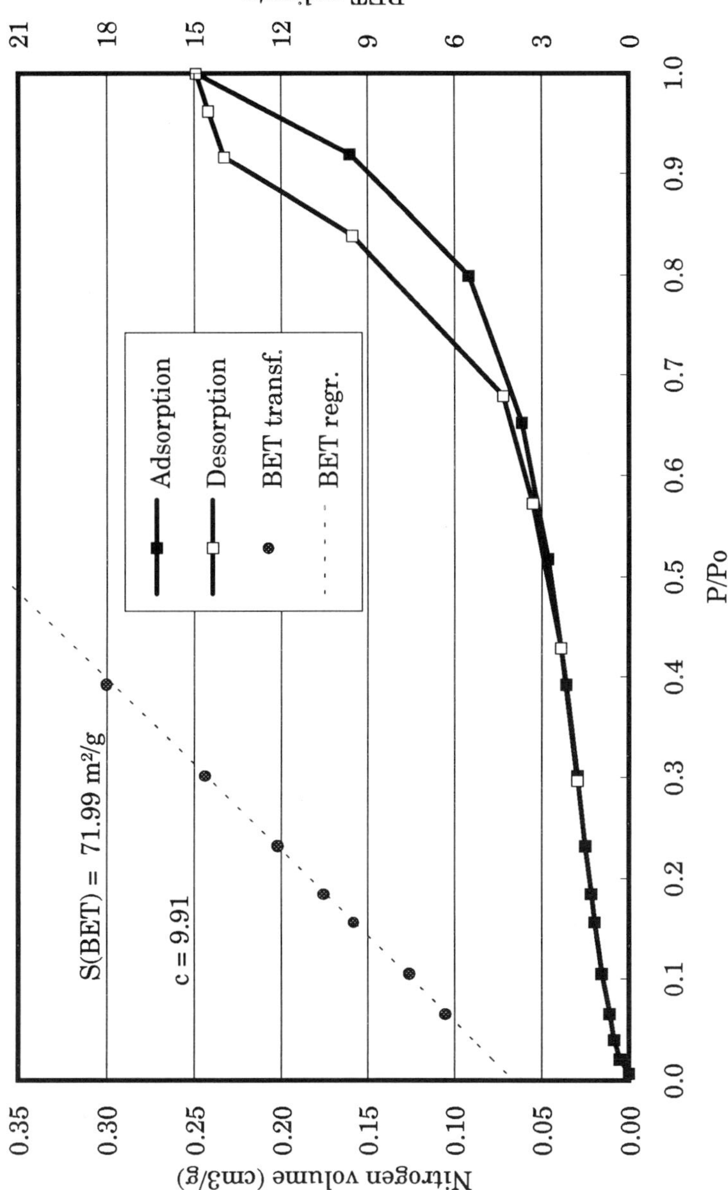

Figure 14.b. Nitrogen adsorption isotherms at 78 K for precipitates obtained with ELF field for a digestion time of 60 min.

208

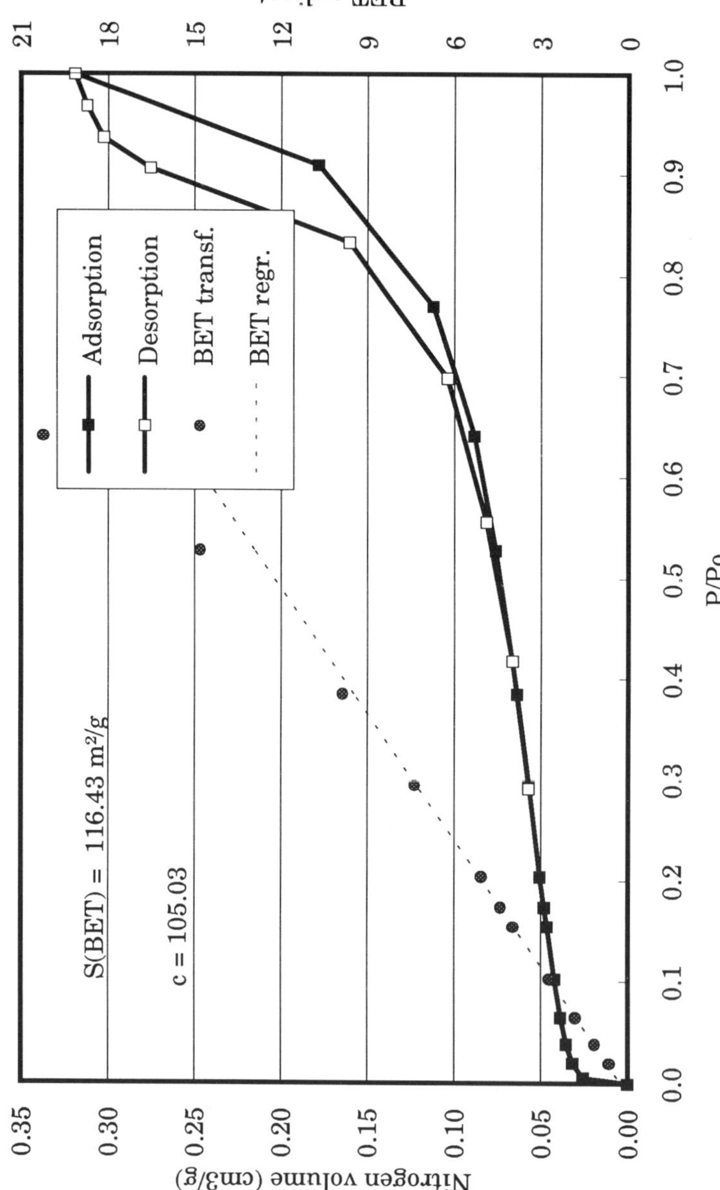

Figure 15.a. Nitrogen adsorption isotherms at 78 K for precipitates obtained without ELF field for a digestion time of 120 min.

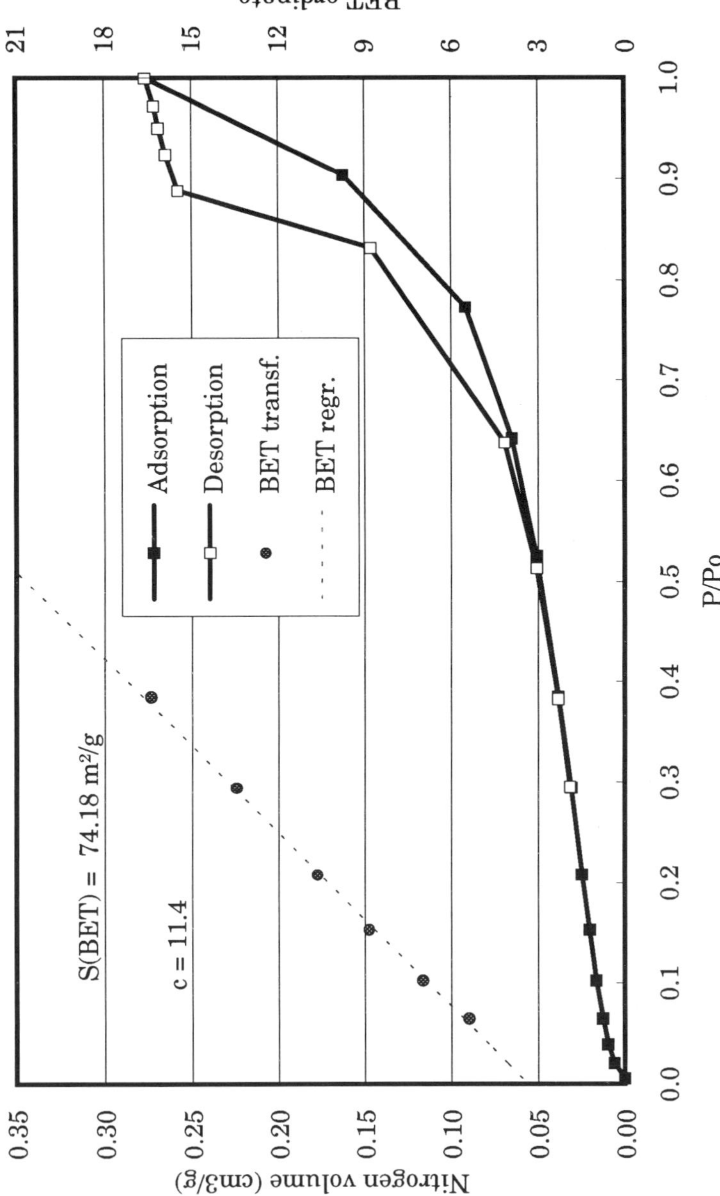

Figure 15.b. Nitrogen adsorption isotherms at 78 K for precipitates obtained with ELF field for a digestion time of 120 min.

The reported data clearly show that the value of the specific surface areas of the products obtained under the action of a 3 mT ELF field are definitely smaller than the corresponding values obtained for samples precipitated without applying an ELF field. These results strongly suggest that the particles produced under the action of an ELF field are larger than those obtained without applying any ELF field.

In Table 3 the hydroxyapatite powder contents resulting from the TG and XRD analysis are summarized.

The experimental error concerning XRD analysis may exceed the 5 %, accordingly no definitive comparison can be done between the amount of hydroxyapatite content in samples obtained with and without applied ELF field.

The experimental error related to the TG analysis is included within 1%, accordingly from data of Table 3 (column first and second) it seems that the amount of hydroxyapatite content obtained with applied ELF field is higher than the one concerning the sample obtained without applied ELF field.

Table 3
Hydroxyapatite content.

| | Hydroxyapatite content | | | |
| | Calculated from TG data* | | Calculated from XRD data | |
Digestion time (min)	$\mathbf{B} = 0$ mT	$\mathbf{B} = 3$ mT	$\mathbf{B} = 0$ mT	$\mathbf{B} = 3$mT
15	45 %	50 %	40 %	43 %
60	75 %	78 %	71 %	75 %
120	76 %	80 %	80 %	86 %

*Experimental error less than 1%

Hence the data on the microstructure of the solid hydroxyapatite/calcium phosphate crystallites and on their initial rate of conversion in to hydroxyapatite due to the hydrolysis of the calcium phosphate, are all in agreement with the theoretical predictions previously reported.

Extension of these results to the bone-repair process under the influence of a low ELF field is still to be managed with care, as different interactions between living cells and ELF field may be involved. Nevertheless, it worth noting that in agreement with the results above reported 'in vitro', also the grain concerning thin film section of bone treated with an ELF field seems to have a greater size than that concerning bone non-treated with electromagnetic field [51].

From a practical point of view the evidence that repaired bone can be characterized by a larger size of hydroxyapatite grains not necessarily can correspond to an increase in bone mechanical resistance. If so the biomedical features concerning the application of an ELF field to the bone repair operations needs to be evaluated with a global approach.

4. CONCLUSIONS

The accumulated experimental evidence for three different classes of homogeneous nucleation and subsequent steps phenomena, is in close agreement with theoretical predictions according to which application of low-intensity ELF fields *can modify the solid/liquid interfaces of initial solid nuclei of critical size.* Even though such a modification cannot be directly verified by experiments, nevertheless, results seem to confirm this phenomenon.

The possibility that a modification to the initial surface properties of solid nuclei may affect all other subsequent steps can be considered plausible *only if* these steps are irreversible in nature (eqs. 28, 29, 31 and 32). This is consistent with the evolution of the irreversible phenomena [52-54], in which boundary conditions play a central role in determining the final states.

From this point of view, low-intensity ELF fields can be regarded as an exogenous source of energy that, through modifications to the solid-liquid interfacial energies of solid nuclei is able to produce non-equilibrium conditions during *the growth steps* that lead to the formation of final solid products.

ACKNOWLEDGEMENT

Discussions with Dr. T. Aste and S. Mezzasalma and with Proff. B. Bianco, M. Capurro and A. Chiabrera were very helpful in the development of this research work.

This paper was supported by the Italian MURST 40% - (Project on Advanced Materials), by the National Council of Research (CNR) (Technological Committee) and by the P.F. MSTA (CNR).

REFERENCES

1. A. Chiabrera, C. Nicolini and H. P. Schwan Eds., Interactions between Electromagnetic Fields and Cells, NATO ASI Series, Series A, Life Sciences, Vol. 97, Plenum Press, New York, 1985.
2. S. Takashima, Electrical Properties of Biopolymers and Membranes, Adam Hilger, Philadelphia, 1989.
3. B. Bianco and A. Chiabrera, Bioelectrochemistry and Bioenergetics, 28 (1992) 355.
4. A. Chiabrera, B. Bianco, J. J. Kaufmann and A. A. Pilla, in B Norden and C. Ramel Eds, Interaction Mechanisms of Low Level Electromagnetic Fields in Living Systems, Oxford Press, Oxford, 1992, 164.
5. A. Chiabrera, B. Bianco, J. J. Kaufmann and A. A. Pilla, in C. T. Brighton and S. R. Pollack Eds., Electromagnetics in Medicine and Biology, San Francisco Press, San Francisco, 1991, 21.
6. A. Chiabrera, B. Bianco, J. J. Kaufmann and A. A. Pilla, in C. T. Brighton and S. R. Pollack Eds., Electromagnetics in Medicine and Biology, San Francisco Press, San Francisco, 1991, 27.
7. C. A. Basset, Critical Reviews in Biomedical Engineering, 17(5) (1989) 451.
8. C. A. L. Bassett, M. Schink-Ascani and S. M. Lewis, Clinical Orthopaedics, 246 (1989) 172.
9. W. J. W. Sharrard, J. of Bone and Joint Surgery, 72-B(3) (1990) 347.
10. M. Hinsenkamp and M. Rooze Eds., Current Concept in Bioelectromagnetics, Bioelectrochemistry and Bioenergetics, 30 (1993).
11. H. A. Pohl, Dielectrophoresis, Cambridge University Press, Cambridge, 1978.
12. E. Wrede, Z. Physik, 44 (1927) 261.
13. H. A. Pohl, J.Appl. Phys., 29 (1958) 1182.
14. H. A. Pohl and J. P. Schwar, J.Appl. Phys., 30 (1959) 69.
15. H. A. Pohl, J.Electrochem. Soc., 107 (1960) 386.
16. H. A. Pohl and C. E. Plymale, J.Electrochem. Soc., 107 (1960) 390.
17. H. A. Pohl and J. P. Schwar, J.Electrochem. Soc., 107 (1960) 383.
18. H. A. Pohl, J.Appl. Phys., 32 (1961) 1784.
19. H. A. Pohl, J.Electrochem. Soc., 115 (1968) 115c.
20. P. J. W. Debye, Phys. Rev., 91 (1953) 210.
21. P. Debye, P. P. Debye, B. H. Eckstein, W. A. Barber and G. J. Arquette, J. Chem. Phys., 22 (1954) 152.
22. P. Debye, P. P. Debye and B. H. Eckstein, Phys. Rev., 94 (1954) 1412.
23. W. A. Barber, P. Debye and B. H. Eckstein, Phys. Rev., 94 (1954) 1412.
24. O. Ya. Samoilov, Structure of Aqueous Electrolite Solutions and the Hydration of Ions, Consultant Bureau, New York, 1965.
25. Y. Kinouchi, S. Tanimoto, T. Ushita, K. Sato, H. Yamaguchi and H. Miyamoto, Bioelectromagnetics, 9 (1988) 159.

26. W. K. Burton, N. Cabrera and F. C. Frank, Royal Soc. Phil. Trans., 243A (1951) 299.
27. W. F. Dunning, in W. E. Garner Ed. Chemistry of the Solid State, Butterworths Scientific Publications, London, 1955, Cap. 6.
28. T. N. Rhodin and D. Walton, in Metals Surfaces: Structure, Energetics and Kinetics, American Society for Metals, Metals Park OH, 1962, Cap. 8.
29. A. G. Walton, The Formation and Properties of Precipitates, Interscience Publishers, New York, 1967.
30. R. Defay, I. Prigogine, A. Bellemans and D. H. Everett, Surface Tension and Adsorption, Longmans, London, 1966, Cap. XVIII.
31. D. Turnbull, J. Appl. Phys., 20 (1949) 817.
32. G. M. Pound and V. K. La Mer, J. Amer. Chem. Soc., 74 (1952) 2323.
33. W. Jacobi, Z. Naturforsch., 10 A (1955) 322.
34. P. F. Rossi and C. De Asmundis, Ann. Chim., 67 (1977) 65.
35. S. J. Gregg and K. S. W. Sing, Adsorption, Surface Area and Porosity, Academic Press, London, 1982.
36. J. J. Gilman, J. Appl. Phys., 31 (1960) 2208.
37. D. Beruto and M. Capurro, J. of Materials Science, 28 (1993) 4693.
38. M. Capurro, personal communication.
39. D. Baldovino, S. Mezzasalma and D. Beruto, send for publication.
40. B. Bianco and A. Chiabrera, work in progress.
41. S. Mezzasalma, personal communication.
42. R. Berton, D. Beruto, B. Bianco, A. Chiabrera and M. Giordani, Bioelectrochem. and Bioenerg., 30 1993 13.
43. R. Berton, D. Beruto, B. Bianco, A. Chiabrera and M. Giordani, in M. Blank Ed. Electricity and Magnetism in Biology and Medicine, San Francisco Press, S. Francisco, (1993) 902.
44. D. Beruto and M. Giordani, J. Cryst. Growth, 84, 679 (1987).
45. D. Beruto and M. Giordani, J. Chem. Soc. Faraday Trans., 89 (14), (1993), 2457.
46. D. Beruto, R. Botter, M. Giordani and G. Giannetti, in H. Hausner, G. L. Messing and S. Hirano Eds. Ceramic Powder Processing Science, Deutsche Keramische Gesellschaft, (1989) 77
47. D. Beruto and M. Giordani, send for publication.
48. D. Beruto, L. Barco, G. Belleri and A. W. Searcy, J. Amer. Cer. Soc., 64 (1981) 74 .
49. D. Beruto, L. Barco and A. W. Searcy, J. Amer. Cer. Soc., 67 (1984) 512.
50. D. Beruto, R. Botter and A. W. Searcy, J. Phys. Chem., 91 (1987) 3578 .
51. V. Canè, P. Botti, D. Farneti and S. Soana, J. of Orthopaedic Research, 9 (1991) 908.
52. I. Prigogine, From Being to Becoming, W. H. Freeman and Company, San Francisco, 1980.
53. P. Glansdorff and I. Prigogine, Structure, Stabilité et Fluctuations, Masson et Cie, Paris, 1971.
54. S. Carrà, La formazione delle strutture, Bollati Boringhieri, Torino, 1989.

Research in Chemical Kinetics, Volume 3
R.G. Compton and G. Hancock (editors)
© 1995 Elsevier Science B.V. All rights reserved.

The kinetics of the hydrogen evolution reaction at nickel electrodes

J. Tamm and L. Tamm

Institute of Physical Chemistry, University of Tartu,
Tartu, EE2400, Estonia

1. INTRODUCTION

The study of cathodic hydrogen evolution reaction has played a significant role in the formation of modern electrochemistry as a branch of science [1-15].
At the beginning of this century the discharge of hydrogen ions was considered to be a simple reaction understanding of which would enable to build up the general theoretical principles of electrochemical kinetics. By the middle of the century it was already clear that the hydrogen evolution reaction is anything but a simple process [14].
The cathodic hydrogen evolution is an electrocatalytic process influenced both by the properties of the surface of the electrode and of the solution. This reaction may be regarded as involving at least five consecutive steps as follows:
1. Transport of the reactants to the surface.
2. Adsorption of the reacting particles on the surface.
3. Reaction(s) on the surface.
4. Desorption of the products.
5. Transport of the products away from the surface.
Any of those steps could be rate-determining of the over-all process. If the concentration of reactants is high enough or specific measures are taken to accelerate the transport processes (for instance using a rotating disk electrode, flowing solution etc.[16]), the transport of reactants is fast enough not to be taken into consideration. In this case the mechanism of the chemical reactions proceeding directly on the surface becomes available for studying.
The cathodic hydrogen evolution reaction (HER) in acidic solutions

$$2\,H_3O^+ + 2\,e^- = H_2 + 2\,H_2O,$$

or in alkaline solutions

$$2\,H_2O + 2e^- = H_2 + 2\,OH^-$$

is a two-electron process proceeding most likely in two consecutive steps with adsorbed atomic hydrogen as an intermediate. The first step of the reaction in acidic solutions is the discharge of hydrogen ions (the Volmer's reaction) [4,9]:

$$H_3O^+_{ads} + e^- = H_{ads} + H_2O_{ads}, \tag{I}$$

followed by electrochemical desorption (the Heyrovsky's reaction) [3]:

$$H_3O^+_{ads} + H_{ads} + e^- = H_{2,ads} + H_2O_{ads},\tag{II}$$

or recombination (the Tafel's reaction) [1]:

$$H_{ads} + H_{ads} = H_{2,ads}\tag{III}$$

of the originated atomic hydrogen.

In alkaline solutions the first step is usually the discharge of protons from water molecules:

$$H_2O_{ads} + e^- = H_{ads} + OH^-_{ads},\tag{IV}$$

followed by the electrochemical desorption

$$H_2O_{ads} + H_{ads} + e^- = H_{2,ads} + OH^-_{ads},\tag{V}$$

or recombination (reaction III).

The index *ads* is added to the symbols of reactants in the reaction equations above to emphasise the greater or smaller influence of the surface on the energetic properties of the particles taking part in the reaction.

In the case of alkaline solutions another mechanism is possible with the discharge of alkali metal cations on the electrode as a first step followed by the reaction of originated active metal atoms with water molecules [17,18], for instance:

$$Na^+ + e^- = Na,\tag{VI}$$

$$2\,Na + 2\,H_2O = H_2 + 2\,Na^+ + 2\,OH^-.\tag{VII}$$

The study of cathodic hydrogen evolution is significant not only from the theoretical point of view. The reaction is widely used in the production of pure hydrogen by means of water electrolysis. Furthermore, the anodic process of the production of several substances (such as chlorine, manganese dioxide, peroxides etc.) is usually accompanied by the cathodic hydrogen evolution reaction.

The regularities of cathodic hydrogen evolution have been investigated nearly since the very beginning of this century starting with the historical work of Tafel [1]. Comparatively much success from the theoretical aspect has been achieved in the investigations carried out on mercury and other metals or alloys of this type [19,20]. The particular advances in this field are reasonable as the surface of liquid metals is well reproducible, energetically homogeneous and can be modelled more accurately than the surface of solid electrodes.

The utmost consequence has, certainly, the studying of electrochemical processes on catalytically active metals of great application, such as platinum-metals, iron, nickel etc. A thorough review by S. Trasatti "Electrolysis of Hydrogen Evolution: Progress in Cathode Activation." has been published recently [21]. The relationships between fundamental knowledge and practical achievements in this field with the technological aspects being under deeper investigation are discussed in this review.

Nickel belongs to electrocatalytical metals and is the most widely used cathode material for water electrolysis and other processes of practical interest. In this report emphasis is to be placed primarily on the problems, the settlement of which is needed for the deeper understanding of the regularities of the cathodic hydrogen evolution reaction from the theoretical point of view. Therefore the results of the investigations of the hydrogen evolution reaction obtained on pure nickel are of main interest of present work. The review does not claim to be an ultimate treatment of the problem as the interface nickel/solution is a very complicated system and not enough *in situ* experimental data are available about the structure of this interface. For that reason it is necessary to confine with some simplified model approaches which reflect the properties of nickel electrode surface quite approximately.

2. MAIN CHARACTERISTICS OF THE HYDROGEN EVOLUTION REACTION

A parameter of great importance in the description of the process of cathodic hydrogen evolution is hydrogen overpotential (η) showing the shift of the electrode potential under cathodic current in the respect of reversible potential in the same solution E°:

$$\eta = E^\circ - E. \tag{1}$$

In this review we use the traditional approach: hydrogen overpotential is defined to have positive value.

The first quantitative relationship between the rate of an electrochemical reaction i.e. current density and hydrogen overpotential is the Tafel equation [1]:

$$\eta = a + b \ln i . \tag{2}$$

It is an empirical expression with experimental constants *a* and *b* called Tafel constants. These constants are convenient to characterise the cathodic hydrogen evolution process on a given metal.

The metals are divided into three groups on the basis of the value of hydrogen overpotential:
a) the metals of high overpotential such as mercury, tin, lead, bismuth, etc. for which *a* is greater than 1 V;
b) the metals of low overpotential, primarily the metals of platinum group with *a* less than 0.5 V;

c) the metals of a medium value of hydrogen overpotential between these groups mentioned above (0.5 V < a < 1 V). The typical metals of medium overpotential are the ones of iron group.

Tafel slope b represents the dependence of potential on the rate of an electrochemical reaction (current) and can be written in the form

$$b = RT/\alpha F, \tag{3}$$

where α is charge transfer coefficient. In terms of decimal logarithm Tafel slope b will be 0,118 V at α = 0,5 and T = 293 K.

The net current density i can be defined as the difference of forward ($\vec{i}$) and backward ($\overleftarrow{i}$) components.

$$i = \vec{i} - \overleftarrow{i}. \tag{4}$$

At the reversible potential where i = 0 and η = 0 this leads to the

$$\vec{i} = \overleftarrow{i} = i_0, \tag{5}$$

where i_0 is the exchange current density, which is another parameter often used to characterise the hydrogen evolution process.

Both current density i and the exchange current density i_0 are determined, as a rule, in respect of the apparent geometric surface area, although the real and apparent surface areas may differ to great extent. The relation between the real and apparent areas is characterised by the factor of surface roughness f_r, which may vary from almost 1 for the ideally smooth surface of monocrystals to remarkably high values, for instance 10^4 in the case of Raney nickel [21]. Consequently the differences in the values of surface roughness must be taken into account while comparing the real activities of different surfaces. As a rule, both i_0 and a are determined by extrapolation and therefore the accurate estimation of their values will be complicated if the polarisation curve consists of two or more regions with different values of Tafel slope b.

A parameter involved in the kinetic relationships of hydrogen evolution is the surface coverage of the metal by adsorbed hydrogen θ_H.

The rate of the reactions of (I), (II) and (III) if the double layer effects regarded as constant (and Langmuir isotherm is apparently hold for hydrogen adsorption) may be described as [22,23]:

$$v_1 = k_1 C_{H_3O^+}(1-\theta_H)\exp(\alpha_1 F\eta/RT) - k_{-1}\theta_H \exp[-(1-\alpha_1)F\eta/RT], \tag{6}$$

$$v_2 = k_2 C_{H_3O^+}\theta_H \exp(\alpha_2 F\eta/RT) - k_{-2}(1-\theta_H)\exp[-(1-\alpha_2)F\eta/RT], \tag{7}$$

$$v_3 = k_3 \theta_H^2 - k_{-3}(1-\theta_H)^2, \tag{8}$$

where k is rate constant, α is the transfer coefficient.

It is clear that θ_H is strongly dependent on the catalytic activity of the metal surface. The metals of the platinum group, possessing maximum catalytic activity in the process of hydrogen evolution, have the surface coverage with adsorbed hydrogen $\theta_H \approx 1$ [20]. The catalytic activity of nickel is somewhat lower but still considerably high and it is reasonable to suppose that θ_H for nickel cannot be too low either.

The value of θ_H and its dependence on potential is one of the most indicative parameters for establishing the mechanism of the hydrogen evolution reaction and therefore particular attention must be paid to the reliability and accuracy of the methods for the determination of θ_H.

3. DETERMINATION OF SURFACE COVERAGE WITH ADSORBED HYDROGEN

The experimental methods, more or less appropriate for the determination of the degree of the coverage of metal surface with the adsorbed intermediate (hydrogen), are galvanostatic charging, potentiodynamic transients, overpotential decay and electrochemical impedance.

A thorough analysis of these methods is given in [22] thus only a brief account of the underlying principles will be given here needed for the unambiguous interpretation of experimental results obtained on nickel.

(I) Galvanostatic charging curves [20,22,24,25].

The procedure involves charging using a rather small constant anodic current and determination of the electric charge required to ionise the hydrogen adsorbed on the surface of the electrode:

$$H_{ads} - e^- + H_2O = H_3O^+. \tag{VIII}$$

If there is molecular hydrogen in the solution then a significant re-adsorption of hydrogen from the solution to the electrode surface may occur,

$$H_2 = 2\,H_{ads}, \tag{IX}$$

and as a result a high apparent surface coverage of hydrogen will be obtained. In order to eliminate this source of error the dissolved hydrogen has to be removed from the solution by bubbling nitrogen or noble gases.

This method, although, cannot be applied to electrodes polarised cathodically, as the solution is saturated with molecular hydrogen in the close vicinity of the electrode and small hydrogen bubbles may cling to the electrode surface so that re-adsorption of hydrogen cannot be avoided. The influence of re-adsorption can be diminished or even eliminated using the fast charging method developed in [26,27]. A very high current density, of the order of 1 A/cm^2, was used and the pulse duration was reduced to 10^{-4}s.

220

The galvanostatic charging method can be successfully applied only if there is a considerable difference between the potential corresponding to the completion of hydrogen ionisation and that corresponding to the commencement of the chemisorption of oxygen or oxide formation. Therefore this method is applicable successfully only for noble metals. If the succeeding process overlaps the hydrogen ionisation as in the case of nickel, the galvanostatic charging method becomes inapplicable.

(II) Double charging method [28,29].

In order to overcome the complication mentioned above a procedure has been developed, named by the authors as "double charging method". During the galvanostatic transient anodic current density i_a is expended on charging the double layer and on all anodic Faradaic processes proceeding at potential E, consequently

$$i_a = C_{d.l.}\left(\frac{dE}{dt}\right)_1 + i_F ,\tag{9}$$

where $C_{d.l.}$ is the capacity of the double layer at a potential E, i_F - the sum of currents of all anodic Faradaic processes proceeding at the potential E. The current of Faradaic processes i_F can be divided into two parts:

$$i_F = i_H + i_{an}(1-\theta_H) ,\tag{10}$$

where i_H is the current density due to the ionisation of adsorbed hydrogen, i_{an} is the current density due to other anodic processes (primarily surface oxide formation).

In addition to the conventional charging curve a compensating curve is taken from a potential sufficiently anodic in respect of the hydrogen reversible potential in the same solution assuming that the surface coverage by adsorbed hydrogen is negligible at this potential. Only the Faradaic process of surface oxidation and double layer charging are assumed to proceed during a compensating curve. Then

$$C_{d.l.}\left(\frac{dE}{dt}\right)_2 = i_a - i_{an} .\tag{11}$$

By combining the equations (6) and (7) the following equation can be derived:

$$C_{d.l.}\left[\left(\frac{dE}{dt}\right)_2 - \left(\frac{dE}{dt}\right)_1\right] = (i_H - i_{an})\theta_H .\tag{12}$$

Neglecting the effect of term $i_{an}\cdot\theta_H$ it is possible to calculate the amount of adsorbed hydrogen.

The double charging method is applicable to electrodes upon which significant oxide formation commences at the potential where the adsorbed hydrogen has been ionised to a great extent and anodic dissolution of the metal occurs to a negligible extent [28].

The problems concerning the applying of this method for the determination of adsorbed hydrogen on nickel will be discussed later in more detail.

(III) Potentiodynamic methods [22,30,31].

The measurements of potentiodynamic transients is a very widely used method for the determination of the amount of adsorbed hydrogen for the metals of the platinum group. Under the conditions where the process of hydrogen ionisation and the following oxidation of the surface are not separated considerably the same difficulties arise as while using galvanostatic methods.

In this case it is also necessary to measure two curves, one of them starting from the potential investigated, the other as a compensating curve from a more anodic potential. The amount of adsorbed hydrogen can then be calculated from the difference of these curves.

(IV) Overpotential decay [32-40].

This method is based on the assumption that the Faradaic process taking place during the steady-state polarisation continues after the interruption of the polarisation causing self-discharge of the double layer.

The initial part of overpotential decay curves can be very simply used to determine the capacitance of the electrode at the initial overpotential:

$$C = -\frac{i_i}{(d\eta / dt)_{t=0}} \ , \qquad (13)$$

where i_i is the current density just before the interruption.

More information can be obtained from overpotential decay measurements by calculating electrode capacitance along decay curves in a wide range of overpotentials. In this case the electrode capacitance is given by:

$$C = -\frac{i_t}{d\eta / dt} \ , \qquad (14)$$

where i_t is the self-discharge current decreasing in time. Different methods have been used to determine the value of i_t along the overpotential decay curve. The simplest way is to combine the equation (11) with the Tafel one, providing the relation between i_t and η is the same as between i and η under polarisation [48].

If the polarisation curve is not linear, i.e. the Tafel equation is not operative, this way of calculation is not applicable. The most general way is to determine the values of i_t at the corresponding overpotential on the basis of experimental η, $\log i$ -curves [22,41,42]. The complications may arise in the application of polarisation curves if the considerable hysteresis is evident between the curves measured in opposite directions. The possibilities to overcome this problem will be discussed later.

The capacitance to be determined consists of two components - the double layer capacitance $C_{d.l.}$ and pseudocapacitance c_{ps}:

$$C = C_{d.l.} + C_{ps} .$$ (15)

The appearance of pseudocapacitance is induced by the variation of θ_H with potential, thus

$$c_{ps} = q_H \frac{d\theta_H}{d\eta} ,$$ (16)

where q_H is the charge of a monolayer of adsorbed hydrogen the value of which in the case of nickel is approximately 240 mC/cm² in average [43].

It is evident that only the change of the surface coverage can be determined by the overpotential decay measurements whereas the absolute value of θ_H cannot be determined in such a way. It must be emphasised the pseudocapacitance is not observed if θ_H is independent of the overpotential in the range investigated, although the absolute value of θ_H may be remarkably high.

The method of overpotential decay is more preferable for the determination of pseudocapacitance than the other ones described above, since the rate of the change of potential and the self-discharge current density are determined by the kinetic parameters of the electrochemical reaction sequence itself i.e. the true adsorption pseudocapacitance is observed experimentally [22].

(V) Impedance measurements [19,20,23,44-46]

The a.c. impedance measurement is one of the most widely used method for the determination of the capacitance of the metal-solution interface. Using suitable equivalent-circuits it is possible to calculate the values of the $C_{d.l.}$ and the pseudo-capacitance caused by the change of θ_H. It must be noted that the values of pseudocapacitance measured by this method are remarkably dependent on the a.c. frequency. The most reliable values of adsorption pseudocapacitance could be obtained by carrying out the measurements at an extremely low frequency. If the low-frequency measurements are carried out at cathodic potentials the proceeding of Faradaic processes will introduce the additional complications in the interpretation of the results.

It is of interest to consider the nature of the degree of coverage determined by different methods. Hydrogen adsorbed on metal surface can be divided into two parts:

$$\theta_H = \theta_H{}^\circ + \theta_H{}^c ,$$ (17)

where θ_H is the degree of coverage with hydrogen adsorbed on the surface at the reversible hydrogen potential i.e. at $\eta = 0$, $\theta_H{}^c$ is the degree of coverage with cathodic hydrogen added to the $\theta_H^{\,o}$ during the cathodic process due to the retarded process of the removal of adsorbed hydrogen. The amount of hydrogen determined by both galvanostatic charging curves and potentiodynamic cycles corresponds to the total value of adsorbed hydrogen θ_H providing the range of potentials investigated is wide enough.

The overpotential decay method enables to calculate the part of adsorbed hydrogen ionised during the overpotential decay i.e. only the value of $\theta_H^{\,c}$, but often even less as the overpotential decay curve, as a rule, does not reach the value $\eta = 0$. By the impedance measurements it is possible to determine only the part of the adsorbed hydrogen which follows the slight variation of potential in the conditions of the applied alternating current. It is evident that the lower is the frequency, the larger is the amount of adsorbed hydrogen enable to respond to the change of potential, i.e. the higher is the pseudocapacitance. Therefore the values of θ_H determined by different methods need not to coincide entirely due to the specific features of the methods.

4. EXPERIMENTAL RESULTS

4.1. Influence of the surface state.

The experimental results of the hydrogen overpotential on nickel obtained in different laboratories by different researchers are considerably variable as well as on most of other metals. Such a variation of results may be connected with the low reproducibility of the experimental data. In fact the situation is not so hopeless if to consider the results of reliable investigations and not to take into account the ones obtained in the conditions of obviously impure solutions or electrode material.

The kinetic regularities of a reaction on a solution/electrode interface are influenced by a remarkably greater number of factors than in the case of homogeneous processes in the bulk of solution. Therefore comparing different results the methods of preparation of electrodes must be taken into particular account. It must be noted that the standardising of experimental conditions as well as controlling of occasional variables is considerably complicated in the case of heterogeneous processes.

The essential role of the surface state of nickel in the rate of the hydrogen evolution reaction both in acidic and alkaline solutions can be clearly seen on Fig. 1 and 2.

In the series of measurements presented below all the possible factors influencing the hydrogen evolution reaction (such as the concentration and composition of the solution, the conditions of polarisation, temperature, etc.) are kept unchangeable except the method for preparation of the electrode surface. Whereas the following methods for the preparation of nickel electrodes have been used [47-49]:

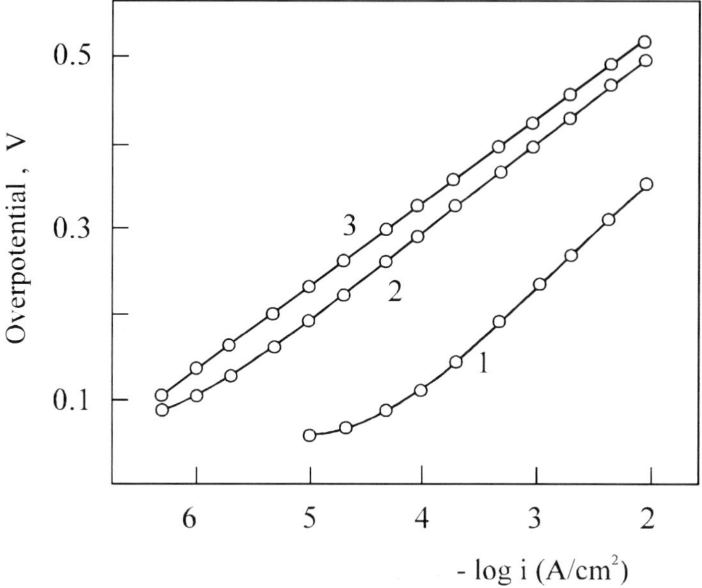

Fig. 1. Polarisation curves of differently prepared nickel electrodes in 0.25 M H_2SO_4 solution:
1 - mechanically polished, 2 - heated in hydrogen, 3 - chemically polished.

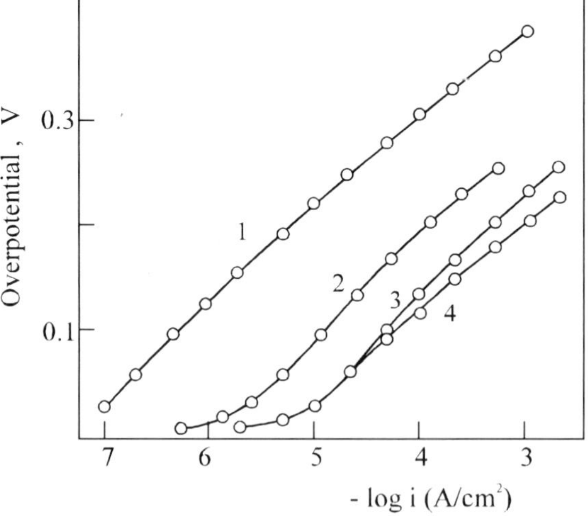

Fig. 2. Polarisation curves for different nickel electrodes in 0.5 M KOH solution:
1 - chemically polished, 2 - 4 - mechanically polished; 2 - corrected by the roughness factor, 4 - after anodic activation.

a) mechanical polishing of the surface with emery paper and the following finishing with fine quarz powder and rinsing thoroughly with tridistilled water before every experiment.
b) conventional mechanical polishing was followed by chemical polishing in a solution containing concentrated acetic acid, acetic anhydride and nitric acid, at temperature $60\,^{\circ}C$ during 15 s.
c) after mechanical polishing of electrodes heating in dry oxygen-free hydrogen at nearly 500°C for 2 hours.

The data characterising hydrogen overpotential on the described above types of nickel electrodes are presented in Table 1.

Table 1.
Characteristics of hydrogen evolution reaction on polycrystalline electrodes.

Electrode	a, V	b, V	$-\log i_0$ (A/cm^2)	C, $\mu F \times cm^{-2}$	a, V	$-\log i_0$ (A/cm^2)
	Apparent				Corrected	
Acidic solution						
Mech. polished	0.595	0.122	4.9	80	0.671	5.5
Chem. polished	0.685	0.092	7.5	20	0.685	7.5
Heated in hydrogen	0.705	0.100	7.1	40	0.725	7.4
Alkaline solution						
Mech. polished	0.585	0.115	5.1	80	0.655	5.7
Chem. polished	0.800	0.110	7.3	20	0.800	7.3

It is natural to assume that the significant dependence of the hydrogen overpotential on the preparation method of the electrode may be caused by the difference in the areas of real surface i.e. by the different surface roughness. The determination of the value of the real surface area of polycrystalline electrodes is quite complicated. On certain assumptions, the approximate determination can be done using conventional electrochemical measurements. For instance the determination of the real surface area may be based on the proportional relation between that with the capacitance of the double layer.

One of the most convenient methods for the determination of the capacitance of double layer under cathodic potentials is to apply the initial parts of overpotential decay curves provided the capacitance obtained corresponds to that of the double layer. This assumption is reasonable in the case of nickel electrodes at sufficiently high overpotentials, where the capacitance is approximately independent of overpotential (Fig. 3).

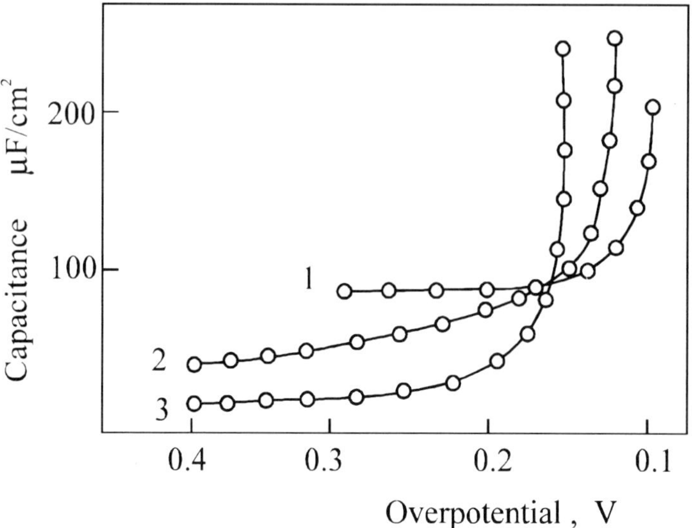

Fig. 3. The dependence of the capacitance on the overpotential for differently prepared nickel electrodes in 0.25 M H_2SO_4 solution:
1 - mechanically polished, 2 - heated in hydrogen , 3 - chemically polished.

The double layer capacitance data obtained from the overpotential decay measurements are presented in Table 1. It can be seen that the lowest value of the capacitance 18...20 μF per cm^2 of the apparent area of surface is exhibited by chemically polished nickel electrodes. If to compare this value with that obtained for liquid mercury at negative surface charge 16...18 $\mu F/cm^2$ [44,45], the chemically polished nickel seems to be ideally smooth, which is somewhat questionable. As the real double layer capacitance of ideally smooth nickel is not known exactly yet, it is not possible to determine the absolute value of surface roughness of the nickel electrodes used.

Fortunately, in this analysis we are interested in the relation of the real surface areas of differently prepared electrodes rather than in the absolute values of them. Therefore it is reasonable to take the double layer capacitance of chemically polished electrodes having the lowest value of that obtained for nickel in this work as a base of comparison, that means the factor of surface roughness of these electrodes is regarded conventionally to be equal to 1 unit. This value of capacitance can be used for the calculation of the relative factors of surface roughness of all other types of electrodes in respect of chemically polished nickel which yields us the opportunity to correct the values of current densities and the corresponding values of a and i_o on the basis of that (Table 1).

As can be seen from Table 1 the capacitance of the electrodes heated in hydrogen differs from that of chemically polished electrodes nearly twice and the relative factor of surface roughness of these electrodes can be considered equal to

2.   The relation of the values of capacitance of mechanically and chemically polished nickel is approximately 4 both for acidic and alkaline solutions and therefore it can be concluded that the factor of surface roughness of mechanically polished electrodes is nearly equal to 4 independently of the pH of the solution. This result implies that corrosion of nickel in acidic solutions under cathodic protection used in this work has negligible effect on the roughness of the surface. The polarisation curves corresponding to the corrected by the "real" current densities are shown in Fig. 2 and 4. It can be seen that the values of overpotential on different electrodes have come closer to each other if the relative surface roughness is taken into consideration. This approach is particularly obvious in the case of the polarisation curves of chemically polished electrodes and of electrodes heated in hydrogen, where the difference in overpotential has almost disappeared (Fig. 4), but the overpotential of mechanically polished electrodes has still remained remarkably lower of that on other electrodes in acidic solutions as well as in alkaline solutions (Fig. 2 and 4).

From these observations it can be concluded that the lower hydrogen overpotential, i.e., the greater activity in respect of the hydrogen evolution reaction on mechanically polished electrodes cannot be explained only by the greater real area of surface than that of other ones. The catalytic activity of this type of electrodes is obviously increased in the result of mechanical activating of the surface layer of metal by means of mechanical polishing, as this procedure introduces essential changes in the surface structure - the generating of a great number of defects and dislocations in it.   Chemical polishing of an electrode removes the surface layer deformed during the mechanical treatment by dissolving it in the electrolyte solution.   The renewed surface of metal has lower catalytic activity and consequently higher hydrogen overpotential.   Such a remarkable decrease of the metal activity becomes evident also in the case of heating the electrodes in hydrogen at temperatures high enough to recrystallise the surface layer.

In the case of alkaline solutions an additional method of the pre-treatment of nickel surface can be applied - the anodic activation of a mechanically polished or otherwise prepared electrode [48,51].   Such kind of anodic activation is common for the metals of platinum group [52-54].   During the anodic activation the potential of nickel electrode attains the value of approximately 1.5 V, i.e. passes through the regions of the formation of Ni(II) and Ni(III) hydroxides [55].   Under such highly oxidising conditions most of the impurities will be desorbed from the surface layer and after the following cathodic reduction the surface will be less contaminated.   The lower overpotential of anodically activated nickel electrodes, compared to the overpotential before activating (Fig 2), cannot be related to the higher factor of surface roughness as the capacitance of both types of electrodes is

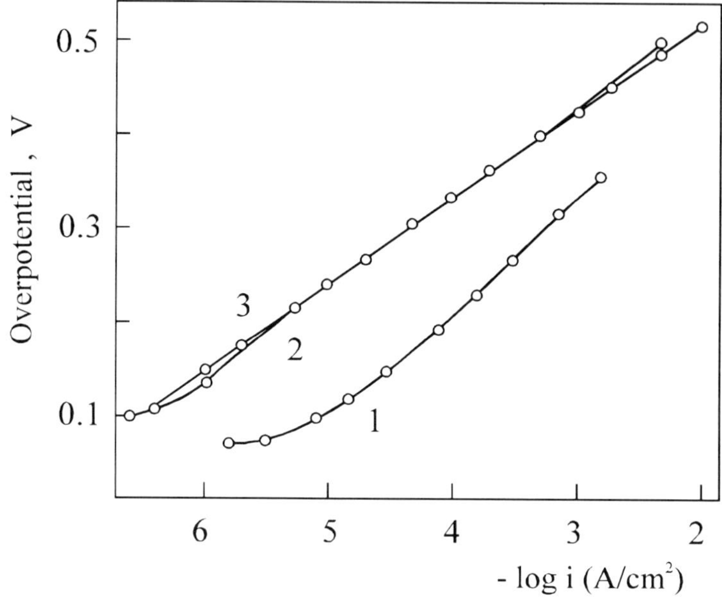

Fig. 4. Polarisation curves of nickel electrodes in 0.25 M H_2SO_4 solution
corrected by the surface roughness factor:
1 - mechanically polished, 2 - heated in hydrogen, 3 - chemically polished

almost the same. Questions may arise in respect of the completeness of the
following reduction during the following cathodic polarisation. Although the
increase of the surface roughness has not been observed as a result of anodic
activation, it cannot be stated firmly that all the oxygen-compounds originated at
high anodic potentials are entirely reduced having none catalytic effect on the
HER.

One of the factors which may influence the regularities of cathodic hydrogen
evolution is the ingress of hydrogen into metal [56]. The adsorbed hydrogen
atoms originated in the process of the discharge of protons can be transferred
across the metal surface into the metal: $H_{ads} \rightarrow H_{abs}$.

In this mechanism hydrogen entering the metal goes through the same adsorbed
intermediate as that leading to hydrogen evolution [57]. According to an
alternative mechanism hydrogen enters the metal in the same elementary act as
that in which it is discharged [20,58].

$$H^+ + e^- = H_{abs}$$

In this case the intermediate states for the entry and hydrogen evolution are
different. The first mechanism has been the more accepted one [56].

According to Dowden's theory the catalytic activity of d-metals is connected with the d-character of metals [59]. The 1s-electron of adsorbed hydrogen atom fills the 3d-holes of nickel and the catalytic activity of formed $Ni\text{-}H_{ads}$ is much lower. The similar decrease of the catalytic activity of nickel has been observed in the case of Ni-Cu alloy where 4s-electrons of copper fill the 3d-holes of Ni [60]. It is reasonable to consider that the ingress of hydrogen will increase the hydrogen overpotential.

The ingress of hydrogen into nickel investigated in the greater part of works, is induced by the use of various stimulators, having remarkable effect on the kinetics of hydrogen evolution and therefore these results cannot be considered directly. Nevertheless, in some reports published in recent years the hydrogenation of nickel has been realised without any stimulators [61,62]. For instance in [62] nickel has been hydrogenated galvanostatically in H_2SO_4 solutions with current density of 5×10^{-2} A/cm^2 during 2 hours. After interruption of the cathodic polarisation the sample was dissolved anodically. It was found that during the initial period of anodic dissolution hydrogen is ionised preferably while nickel dissolution is inhibited.

It is clear that hydrogen absorbed in nickel has remarkable influence on the energetic properties of nickel. Consequently, the possibility of hydrogenation of nickel during the measurements in pure conditions and the influence of it on the rate of hydrogen evolution must be taken into particular account. To elucidate the possible influence of hydrogenation on hydrogen overpotential the results of the HER measurements realised in different conditions must be compared. The results obtained on the electrodes heated in hydrogen [63-66] are valuable in this respect. In [66] the rapid measurements of overpotential were carried out in the direction of increasing current density only using a fresh electrode for every run. Contrary to this work the overpotential measurements in [63,64] were carried out after the potential had reached a steady state value. Therefore, it is reasonable to suppose the overpotential in the latter case may have a higher value as the result of the hydrogenation of the electrode. Unfortunately the data of [63,64] are not presented in detail enough to make reliable conclusions.

Special investigations have been carried out [47] to elucidate the dependence of overpotential on the polarisation conditions of electrodes previously heated in hydrogen. It has been established that it is possible to obtain a series of different polarisation curves on the same electrode depending on the magnitude of the current density to which the electrode has been polarised after the immersion of it into the solution. (Fig. 5). Increasing the polarising current density higher and higher values of hydrogen overpotential are achieved until the steady state is realised. These measurements show that different hydrogen overpotential values can be obtained by the variation of polarisation conditions of the electrode, being consistent with the supposition of the dependence of the degree of hydrogenation on the applied current density. As the hydrogen overpotential on chemically polished electrodes has approximately the same value as in the case of the electrodes heated in hydrogen (after correcting them on the factor of surface roughness) it is reasonable to suppose that chemically polished electrodes are also hydrogenated to similar extent.

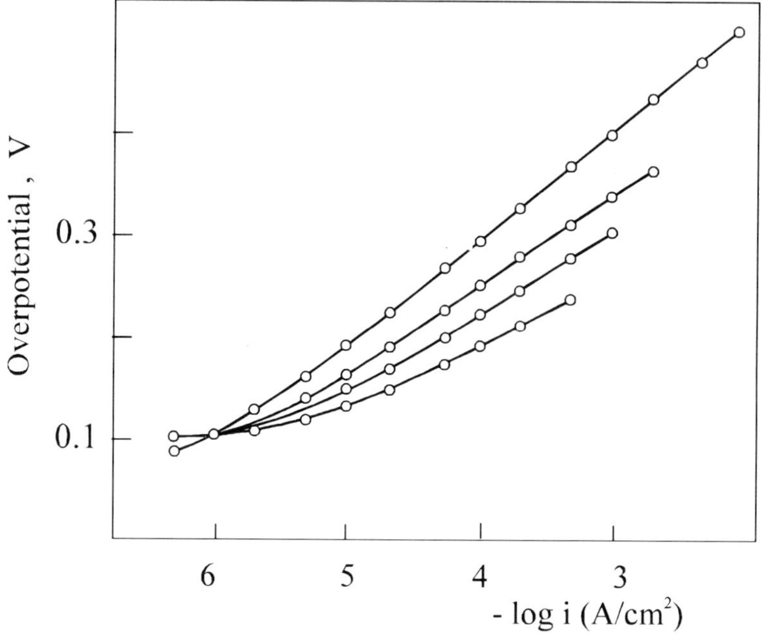

Fig. 5. Polarisation curves of nickel electrode heated in hydrogen in 0.25 M H_2SO_4 solution, polarised to different cathodic current densities.

4.2. Single Crystal Faces

The influence of various factors on the regularities of the process of hydrogen evolution on nickel can be more deeply considered by the investigations of this process on different faces of monocrystalline electrodes. The influence of nickel crystallographic structure on the HER has been studied only in some works [67-70] and the results obtained are considerably different. As a matter of fact the difference in quality of nickel single crystals may cause remarkable variation of the results obtained.

In our work the electrodes cut from highly pure nickel single crystal (about 99.999% Ni) produced by containerless vacuum zone melting were applied. The single crystals were aligned via X-ray diffraction. The misalignment angels of the studied surfaces of different electrodes were between 0.5 and 1.5°. Prior to each experiment the electrodes were polished chemically. The surface of prepared electrodes were perfectly smooth when viewed with a magnification of 450×. The polarisation curves measured in 0.25 M H_2SO_4 solution at the low-index faces are shown in Fig. 6 and the characteristics of the HER are presented in Table 2 [70]. One can see that except for face (111), hydrogen overpotential does not depend essentially on the crystallographic structure of nickel. This result is consistent with that obtained on some other metals [21].

Table 2.
Characteristics of hydrogen evolution reaction on single crystal nickel electrodes

Face	a, V	b, V	$-\log i_0$ (A/cm^2)
(100)	0.76	0.113	6.7
(110)	0.635	0.102	6.8
(111)	0.83	0.117	6.5

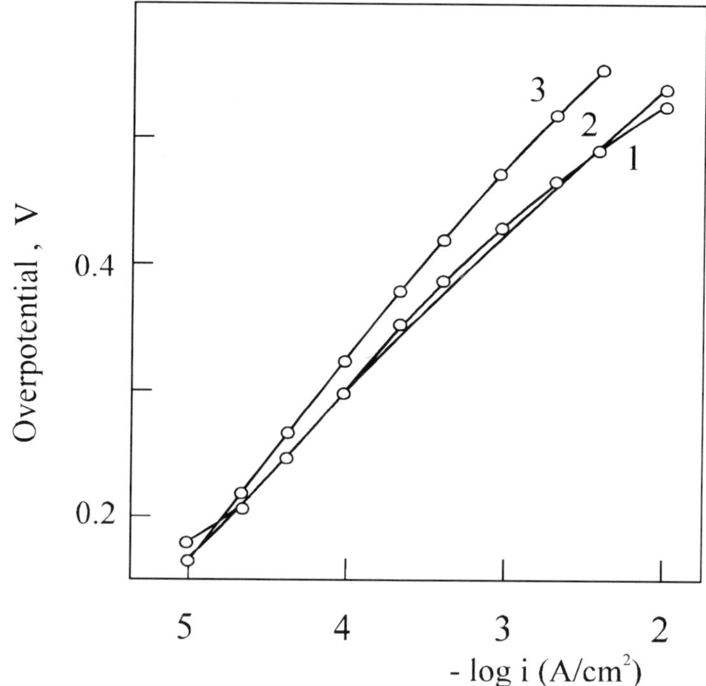

Fig. 6. Polarisation curves in the solution of 0.25 M H$_2$SO$_4$ solution for single crystal faces of nickel:
1 - face (110), 2 - face (100), 3 - (111).

Two possible explanations can be offered about the very slight effect of the crystallographic structure of the metal on hydrogen overpotential [21]. The high coverage with adsorbed hydrogen far from equilibrium may smooth down the differences in the hydrogen adsorption energy from face to face. Another possible explanation can be that the competition of solvent adsorption with hydrogen adsorption almost compensates the differences that would exist for the bare metal surface [71]. Electroreflectance spectra suggest that water molecules are chemisorbed on d-metals with partial injection of electrons into the d-band [71]. This circumstance may essentially effect the strength of Me-H bond, as the adsorbed hydrogen also injects its electron into the d-band. It is evident that the interaction of adsorbed hydrogen and water molecules can not be reduced to the competition of them for adsorption sites on the surface as the significant change of the energetic properties of the surface takes place.

It has been established [70] that the behaviour of nickel face (111) in respect of hydrogen evolution stands out if to compare with that of other faces. It was found that η on face (111) continues to increase after the immersion of the electrode in the solution for much longer time than on other single crystal faces.

According to [72] the amount of absorbed hydrogen in nickel is dependent on the crystallographic structure of metal and is more essential for the close-packed face (111). Therefore it is reasonable to conclude that the higher overvoltage of the HER on nickel face (111) observed in [70] is caused by the ingress of larger amounts of atomic hydrogen than for other faces.

4.3. Low and high current densities.

The investigation of the regularities of the cathodic hydrogen evolution under extreme conditions provides with additional information valuable in respect of the better understanding of the mechanism of overall process.

The measurements of the HER at extremely low current densities in acidic solutions are not available as in these conditions the value of stationary potential being set up is determined by the relation between the rate of cathodic hydrogen evolution and the rate of the dissolution of nickel, as both processes are proceeding under these conditions with considerable rate .

In the case of alkaline solutions the rate of dissolution is essentially lower and the potential of the reversible hydrogen electrode can be attained with great precise [74]. Thus the polarisation measurements reflect first of all the regularities of the processes of cathodic evolution and anodic ionisation of hydrogen. At sufficiently low overpotentials where $\eta \ll RT/F$ the relationship between η and i is linear (Fig. 7) and the slope of this straight line can be used for the determination of the stoichiometric number v [16,75,76]. The stoichiometric number shows the number of acts of the rate-determining step necessary for one act of the overall reaction and is given by

$$v = 2i_0 / (di / d\eta)_{\eta=0} \, . \tag{18}$$

According to this definition a value 2 for v shows that the discharge step is rate-determining of the process [76].

If the value of extrapolated exchange current density i_o', obtained by extrapolating i from remarkably larger overpotentials to $\eta = 0$, is used for calculation, the apparent stoichiometric number v' is to be obtained [50,76,77];

$$v' = 2i_0' / (di / d\eta)_{\eta=0} \, . \tag{18a}$$

The apparent stoichiometric number is arbitrary to a certain extent since it depends on the section of the polarisation curve chosen for the determination of

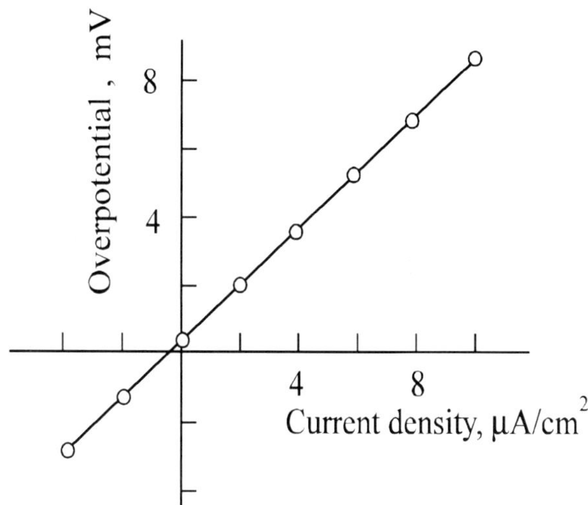

Fig. 7. Dependence of hydrogen overpotential on current density at low polarisation in 0.5 M KOH solution on mechanically polished nickel.

i_o'. The problems connected with interpretation of the values of v are discussed in detail in [50].

The experimental values of v for nickel determined in different works are very close to 2 [50,66,73], implying that the proton discharge is the rate-determining step of hydrogen evolution. Nevertheless, it must be pointed out that the determined parameter is actually the apparent stoichiometric number v', thus the unambiguous conclusions are not reasonable on the basis of that [50].

The results of measurements at high current densities are of especially great interest in terms of the elucidation of the mechanism of the hydrogen evolution reaction.

The equation (8) shows that the rate of recombination of hydrogen atoms is governed only by the degree of coverage with adsorbed atomic hydrogen. In the case of the retarded step of removal of adsorbed hydrogen θ_H increases with overpotential. If θ_H attains the value of 1, the saturation current i_s has to be observed and further increasing of the applied potential does not induce the increase of the rate of hydrogen evolution. Therefore the investigation of hydrogen evolution at sufficiently high current densities will enable to realise if the recombination step might be operative in the overall process.

According to the theoretical principles of recombination reaction worked out by Horiuti [78] the saturation current i_s for nickel has to be exhibited at 2×10^2 A/cm^2 at 25°C. Several attempts have been made for the determination of saturation current. In [79] the tendency of current to approach a constant value near $i=10$ A/cm^2 on nickel in 5 M HCl solution have been observed, although the reproducibility of that has not been too good. In [80,81] the rapid increase of η

with the increase of i beyond 1 A/cm^2 with the saturation current around 100 A/cm^2 has been observed.

It must be emphasised that the realisation of polarisation measurements at such high current densities is connected with great difficulties. The main factors complicating the measurements are the following: the shift of the potential due to the ohmic drop of potential between the electrode surface and the tip of Luggin's capillary ΔE_1, the concentration overpotential ΔE_2, the shift of potential ΔE_3, caused by the heating up of the electrode and the solution in the vicinity of the electrode. Thus, in addition to the desirable real value of hydrogen overpotential η, three more terms will be involved in the value of experimentally measured E:

$$E = \eta + \Delta E_1 + \Delta E_2 + \Delta E_3 \qquad (19)$$

The predominating of them is the ohmic drop of potential ΔE_1 as this term is varying linearly with current contrary to the overpotential of electrochemical process depending on the logarithm of current. On the other hand the linear dependence between ΔE_1 and i makes it possible to find the value of ΔE_1 at very high current densities by linear extrapolation. To diminish the influence of concentration polarisation comparatively concentrated solutions are used, which decreases also the ohmic drop of potential.

The heating of the solution can be diminished to a great extent either by shortening of the polarisation time, for example using the galvanostatic pulse method [80] or by carrying out the measurements in flowing solution [79.82,83]. When the terms ΔE_1 and ΔE_2 deviate the polarisation curves towards the higher overpotentials, ΔE_3 has an opposite effect as the increase of temperature accelerates the electrochemical reaction and increases the conductivity of the solution. Fig. 8 presents the polarisation curves measured current densities up to 100 A/cm^2 [83] showing the effect of various methods on diminishing the deviation of the polarisation curve due to the ohmic drop of potential and heating of the solution. The flowing solution of velocity 5 m/s has been used and the ohmic drop of potential has been determined by oscillographic switch-in and switch-off measurements for the correction of the polarisation curves. It has been established that the change of overpotential ΔE_1 is essentially greater if determined by switch-off measurements. A possible reason for it may be the additional drop of potential, caused by small bubbles of hydrogen appearing into the layer of solution between the electrode and the tip of Luggin's capillary. The rate of the generating bubbles is so high that even the stream of the flowing solution is not able to remove them entirely.

Corrections presented above will make it possible to linearise the polarisation curves approximately up to 10 A/cm^2. At higher current densities still essential deviations can be observed (Fig. 8). Remarkable fluctuations of overpotential values have been elucidated by oscillographic measurements, the amplitude of fluctuations being dependent on the rate of the flow of the solution. If the velocity of the flow was 5 m/s the fluctuations of E were observed at current

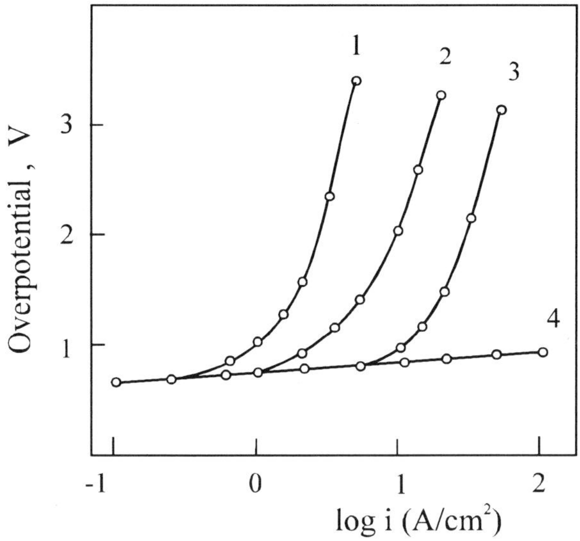

Fig. 8. Polarisation curves for mechanically polished nickel at high current densities in 0.5 M KOH solution:
1 - usually measured, 2 - 4 using flowing solution, 3,4 - corrected by the ohmic drop of potential, using switch-off curves, 4 - and additionally using linear extrapolation of the ohmic drop of potential at i > 10 A/cm^2.

densities above 10 A/cm^2. At such high current densities the formation of molecular hydrogen is so rapid, that a gaseous film of hydrogen appears on the surface of the electrode which can not be removed even by the stream of solution. As a result an additional ohmic resistance becomes apparent. The possible explanation of the fluctuations of overpotential observed in [80] at high current densities may be the formation of a such thin gaseous film on the electrode surface. Although the measurements in this work have been carried out by quite short galvanostatic impulses ($10^{-4}...10^{-3}$ s), this time is long enough to produce a hydrogen layer of thickness 20...100 μm at current densities 10...80 A/cm^2 inducing an additional ohmic drop of potential (moreover, flowing solution was not applied in this work).

It has been established in previous works [83] that the deviation of overpotential at current densities over 10 A/cm^2 is proportional with current density. This relation has been used for the correction of polarisation curves at very high current densities, as can be seen on Fig. 8. On the basis of the review presented above it may be concluded that the appearance of saturated current on nickel at high overpotentials has not been experimentally proved yet.

One possibility to overcome the complications discussed above is to diminish the surface area of electrode. Nowadays different micro and ultra micro electrodes are widely used [84], having surface areas of about 10^{-4} cm^2. In this case the experimental realisation of current densities of about 100 A/cm^2 is not a problem. However, the application of very small electrodes is connected with very strict requirement for the purity of the solution, as the relation between the volume of solution (i.e. the extent of impurities) and the surface area of electrode is larger in this case. Another complicating factor of measurements at very high current densities is the ingress of hydrogen, stimulated by the impurities adsorbed on the electrode.

4.4. Influence of the composition of solution.

(I). General principles.

Hydrogen evolution reaction is a typical heterogeneous reaction proceeding on the metal/solution interface the rate of which is dependent on the surface concentration of reactants and the existence of free sites on the surface.

The adsorption of reactants, including the intermediate of the reaction atomic hydrogen, is of great importance in the determination of the kinetic relationships of the reaction. In addition to the reactants the adsorption of other particles, especially anions, may have a significant effect on the process. The anions can be divided into two groups:

1) indifferent anions having only electrostatic interaction with the surface

2) specifically adsorbed anions with their adsorption being different from the one calculated on the basis of electrostatic interaction only [23] and is mainly of chemical nature.

Such a division is, of course, somewhat conventional, and is dependent on the sensitivity of the methods used for the determination of the specific character of adsorption. For the metals behaving in a certain range of potentials as "ideal polarised electrodes" (i.e. the charge is expended only on the charging of the double layer and no electrochemical reaction is occurring on the surface) the specific adsorption can be determined with satisfactory precision on the basis of the potential-dependence of the adsorption of ions [45]. Nickel does not behave as an ideal capacitor in at any range of potentials, therefore, such kind of data are not available. For that reason the establishment of the specific character of the adsorption of anions on nickel is far from being an easy problem. It is reasonable to assume that oxoanions, such as SO_4^{2-}, ClO_4^-, etc., are adsorbed on nickel surface mainly non-specifically, but the adsorption of halide ions especially of I$^-$- and Br$^-$-ions is of specific character [85-93].

The surface of nickel in acidic solutions is primarily covered by adsorbed water molecules and to a smaller extent by hydrogen ions and specifically adsorbed anions. The total potential difference between the metal and the solution E can be regarded as being formed of two parts: the difference between the bulk of the solution and the plane of reacting particles ψ_1 and the difference from that plane to the metal surface E-ψ_1.

The work required to transfer an ion from the bulk of solution to the metal surface can be described as consisting of two contributions, one determined by the potential of electrode, the other by the chemical interaction, similar to that of electrochemical potential. Taking into consideration that the surface concentration of hydrogen ions is comparatively small [94,95], the simplified relation can be used:

$$c^s_{H_3O^+} = c_{H_3O^+} \exp[(-g_{H_3O^+} + F\psi_1)/RT] \quad , \tag{20}$$

where $g_{H_3O^+}$ is the standard free energy of the specific adsorption of hydrogen ions, $c^s_{H_3O^+}$ is the surface concentration of hydrogen ions and $c_{H_3O^+}$ is the concentration of hydrogen ions in the bulk of solution.

In general protons are to be reduced from a proton-donor HB:

$$HB + e^- = H_{ads} + B^-_{ads} \quad , \tag{X}$$

and consequently

$$c^s_{HB} = c_{HB} \exp[-g_{HB} + zF\psi_1)/RT] \quad , \tag{21}$$

where z is the charge of the proton-donor.

It can be seen from the expression (21) that the surface concentration of an uncharged particle being a proton-donor is independent of the ψ_1-potential.

If the Langmuir isotherm is valid to describe the adsorption of atomic hydrogen, the overpotential governed by the slow discharge of protons can be given in general by the expression [96]:

$$\eta = const - \frac{1-\alpha}{\alpha} \frac{RT}{F} \ln c_{HB} - \frac{RT}{F} \ln c_B - - \frac{1-\alpha}{\alpha} \frac{g_{HB}}{F} - \frac{g_H}{F} - \frac{g_B}{F} - \left(1 - \frac{z_0}{\alpha}\right)\psi_1 - \frac{RT}{F} \ln \frac{\theta_H}{1-\theta_H} - \frac{RT}{\alpha F} \ln i \quad . \tag{22}$$

The expression may be simplified for further analysis. Assuming the adsorption of reactants (HB, B^-, and H) does not depend on the concentration of solution and potential and θ_H=const the equation (22) for acidic solutions can be presented as follows:

$$\eta = const + \frac{1-\alpha}{\alpha} \psi_1 - \frac{1-\alpha}{\alpha} \frac{RT}{F} \ln c_{H_3O^+} + \frac{RT}{\alpha F} \ln i \quad , \tag{23}$$

and for alkaline solutions

$$\eta = const - \psi_1 - \frac{RT}{F} \ln c_{OH^-} + \frac{RT}{\alpha F} \ln i \quad . \tag{24}$$

It can be seen that the two terms of the equation (23 and 24) depend on the composition of solution. Unfortunately, precise data characterising the ψ_1-potential are not available. It is commonly assumed that ψ_1-potential does not differ much from the ψ_0-potential (the potential of the outer plane of Helmholz layer) which may be determined from the double layer measurements.

ψ_0-potential can be expressed as follows [97]:

$$\psi_0 = \frac{2RT}{F} \text{arcsh} \left(\frac{\sigma}{2\sqrt{\varepsilon RT/2\pi}} \right) \quad , \tag{25}$$

where σ is the surface charge and D is dielectric permeability,

At low surface charge, i.e. near to the potential of zero charge, the ψ_0-potential is given by:

$$\psi_0 \approx \frac{RT\sigma}{F\sqrt{\varepsilon RTc2\pi}} \quad . \tag{26}$$

For sufficiently high surface charge, i.e. far from the potential of zero charge

$$\psi_0 = \text{const} + \frac{2RT}{F}\ln\sigma - \frac{RT}{F}\ln c \quad , \qquad \text{if } \sigma > 0 \quad , \tag{27}$$

and

$$\psi_0 = \text{const} - \frac{2RT}{F}\ln\sigma + \frac{RT}{F}\ln c \quad , \qquad \text{if } \sigma < 0 \quad . \tag{28}$$

Consequently, providing $\Psi_1 \approx \Psi_0$, Ψ_1-potential is proportional to $1/\sqrt{c}$ near to the potential of zero-charge and the Ψ_1-potential exhibits a logarithmic dependence on the concentration of electrolyte far from the potential of zero-charge.

According to the relationships presented above an analysis can be made about the influence of the composition of the solution on the hydrogen overpotential under the conditions of absence of specific adsorption. The concentration-dependence of Ψ_1-potential for three limiting conditions will be considered:
1) the surface charge is approximately 0 (or, in general, Ψ_1=const if the concentration of the solution is high enough),
2) the range of high negative surface charge where the expression (28) is operative,
3) the range of high positive surface charge, where the expression (27) can be used.

The measurements of hydrogen overpotential can be carried out either in pure acid or alkali solution or in the solutions of acid or alkali at the presence of the excess of salt. In the case of pure alkali or acid solution the concentration of the reactant and Ψ_1-potential are both determined by the concentration of the electrolyte. In the second case Ψ_1-potential is mainly determined by the excess of salt, consequently $\Psi_1 \approx$ const and only the concentration of the reactant depends on the concentration of acid or alkali. Possible values of $(\partial\eta/\partial\ln c)_{i=\text{const}}$ for all occasions discussed in case of $\alpha = 0.5$ are presented in Table 3.

Table 3

Values of $(\partial\eta/\partial\ln c)_{i=const}$ for various conditions

solution	$\sigma=0$	$\sigma<0$	$\sigma>0$
acid	-RT/F	0	-2RT/F
acid + salt	-RT/F	-RT/F	-RT/F
alkali	-RT/F	-2RT/F	0
alkali + salt	-RT/F	-RT/F	-RT/F

It can be seen that the overpotential may vary with tenfold dilution of the solution from 0 to 118 mV (at t = 25°C and $\alpha=0.5$) depending on the surface charge and the composition of the solution. The treatment presented above is consistent with the mechanism of slow discharge (reaction I). The same relations are valid if the rate-determining step is the electrochemical desorption (reaction II). If the recombination of hydrogen atoms (reaction III) is rate-determining of the process, the composition of the solution has no direct influence on the rate of the hydrogen evolution.

Consequently, the investigation of the dependence of hydrogen overpotential on the composition of solution (provided the specific adsorption of anions is not-pronounced) enables to distinguish only the non-electrochemical recombination reaction and an electrochemical step as rate-determining of the overall reaction, but not to distinguish proton discharge and electrochemical desorption as the rate-determining step.

(II) Experimental results.

The influence of the concentration of solution on hydrogen overpotential on nickel has been studied in several works in acidic as well as in alkaline solutions [63-66,70,98-103]. It must be noted, however, that the results do not coincide well. The remarkable variations of the results may be caused by some unnoticed variation of the conditions of the experiment when the concentration dependence of overpotential is studied. One parameter which easily may change during the measurements is the surface state of the electrode. In order to compare the results obtained in different works one must make sure the conditions of carrying out of the measurements are the same.

The investigation of the dependence of overpotential on the concentration and composition of the solution may be carried out differently: on one electrode in several solutions or on different electrodes in different solutions. In the latter case it is not excluded that the surface state may be remarkably different in different solutions, disturbing the dependence of hydrogen overpotential on the composition of solution. These errors can be diminished by measuring overpotential on the same electrode in several solutions, while the cathodic polarisation of the electrode is not interrupted and the inflow of air into the cell is excluded during the experiment [70,100,102]. The second technology described gives doubtless more reliable results if the polarisation in the solutions of different concentration introduces the variation of the surface state.

Most of the measurements in acidic medium have been carried out in the solutions of H_2SO_4 and HCl with pH in 0...3.5.

It has been established that for mechanically polished electrodes $(d\eta/d\log c)_{i=const}$ has a value in the range of 58...10 mV, whereas the more high values have been obtained for more concentrated solutions (Fig. 9) Such dependence has been obtained for H_2SO_4 solutions measured on different electrodes as well as on the same electrode, and for HCl solutions if measured on the same electrode. For the nickel electrodes heated in hydrogen and the chemically polished ones the value of $(d\eta/d\log c)_{i=const}$ in H_2SO_4 solutions is considerably smaller and in the latter case does not exceed 10 mV.

In [63-66] the dependence similar to that on mechanically polished electrodes in H_2SO_4 solutions has been exhibited also in solutions of HCl for the electrodes heated in hydrogen (measured on different electrodes in different solutions). The more essential concentration-dependence of overpotential obtained in [63,64] may be caused by the variation of the surface state of the electrode with the concentration of the HCl solution as pointed out in [104,105].

The value of $(d\eta/d\log c)_{i=const}$ in the solutions of H_2SO_4 for single crystal electrodes has been found to be dependent on the structure of the face [70]. In the case of the face (100) overpotential is almost independent of the concentration of solution while η on the face (110) and on face (111) increases by 15...25 mV with the tenfold dilution of acid solution without any change in the shape of η, log i-curve [70].

Only few data are available about the influence of the concentration of hydrogen ions on hydrogen overpotential in the solutions of constant total concentration of electrolyte (in the solutions of acid with the excess of salt) [63,64,70,101,103]. The Tafel slope b for the electrodes heated in hydrogen is found to be greater if the excess of salt is used and the corresponding polarisation curve intersects with the one measured in pure acid solution [63]. The similar effect has been obtained also on single crystal electrodes [70]. The results presented above imply that the measurements in acidic solutions are carried out in the range of potentials near to the potential of zero charge [63,70].

The polarisation measurements in alkaline medium exhibit slight dependence of overpotential on the concentration of alkali [63-66,98] or no dependence at all [73,99] (Fig. 10). The elucidation of concentration-dependence of overpotential in the intermediate region of pH is of great interest, however, the measurements in comparatively neutral solutions are connected with serious difficulties.

As hydrogen ions are used in the process of cathodic hydrogen evolution the concentration polarisation may become apparent in diluted acid solutions shifting the value of pH of the solution in the vicinity of the electrode surface (pH_S). The increase of pH_S is especially obvious in the case of diluted acid solutions with the excess of salt [103] essentially disturbing the shape of polarisation curves (Fig. 10).

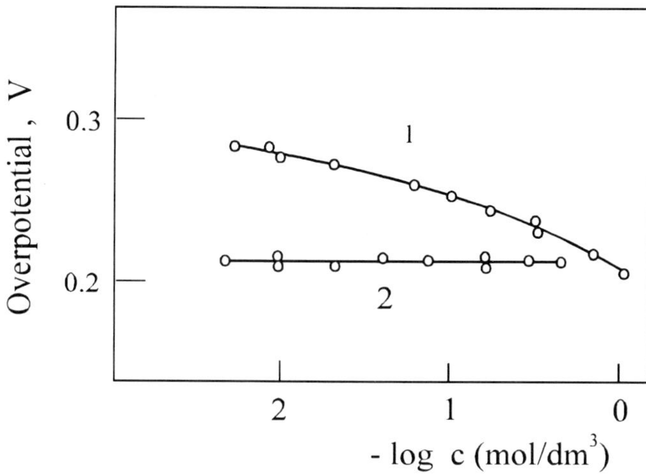

Fig. 9. Dependence of hydrogen overpotential on the concentration of solution on mechanically polished nickel electrodes:
1 - H_2SO_4 solution, 2 - KOH solution.

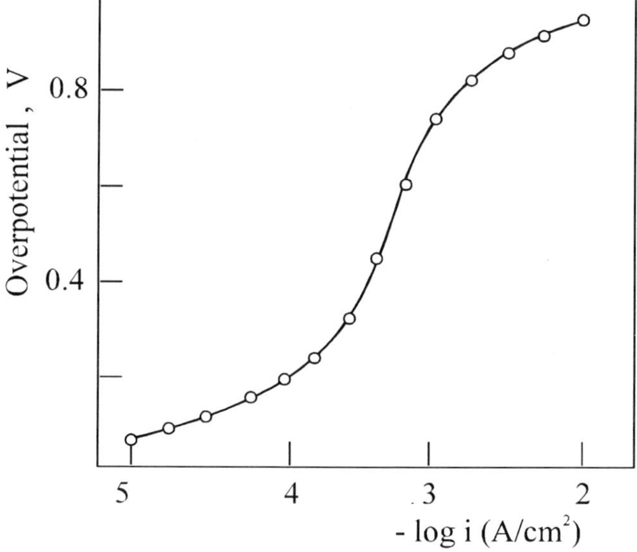

Fig. 10. Polarisation curve for mechanically polished nickel in the solution of 0.005 M H_2SO_4 + 0.5 M K_2SO_4

The value of potential ΔE measured in the conditions of concentration polarisation can be regarded as consisting of several terms:

$$\Delta E = \eta + \Delta E_{pH} + \Delta E_{IR} \quad , \tag{29}$$

where η is hydrogen overpotential, ΔE_{pH} - the shift of the potential due to the difference in pH between the solutions in the reference electrode compartment and in the close vicinity of the surface of the electrode studied. ΔE_{IR} is the ohmic drop of potential between the surface of the electrode and the tip of Luggin's capillary

At sufficiently high concentration of salt the difference of hydrogen overpotential from ΔE is determined mainly by ΔE_{pH}. Applying a pH probe it is possible to elucidate the value of a pH_S at different current densities. However, the application of pH probe is connected with some complications. The probe must be sufficiently small, otherwise the value of pH in a certain distance of the electrode is measured rather than the desirable value of pH_S in the close vicinity of the electrode. Besides, it must be taken into consideration that the comparatively thick tip of pH probe will cause some screening effect if transferred very close to the electrode. As a result essential deviations from the real value of pH_S will be obtained.

In diluted acid solutions with the excess of salt pH_S may attain even the value of 10-11, i.e. the HER is proceeding really in alkaline solution and the corresponding term ΔE_{pH} achieves the value of 400...450 mV. It can be concluded that the precise correction of polarisation curves on account of the change of pH_S is unachievable.

The influence of concentration polarisation can be essentially diminished carrying out the measurements after switching on the polarising current in a time short enough for the pH in the close vicinity of the electrode not to be changed remarkably yet. The polarisation curve built up on the basis of switch-over curves (from a sufficiently low cathodic current density to different higher current densities) is plotted in Fig. 11 [102,103]. It can be seen that the polarisation curve in the region of higher current densities is continued with approximately the same value of Tafel slope exhibited in the region of lower current densities.

In principle, the hydrogen overpotential measurements are possible to carry out also in buffer solutions, but the results obtained in these solutions cannot be easily interpreted. As a rule, the measurements in buffer solutions introduce additional ions and molecules in the solution which may adsorb on the surface of the electrode and thus to influence the rate of hydrogen evolution. Besides, it must be taken into consideration that not only water molecules or hydroxonium ions, but also the molecules of weak acid or other hydrogenated particles used in buffer solution may behave as proton-donors.

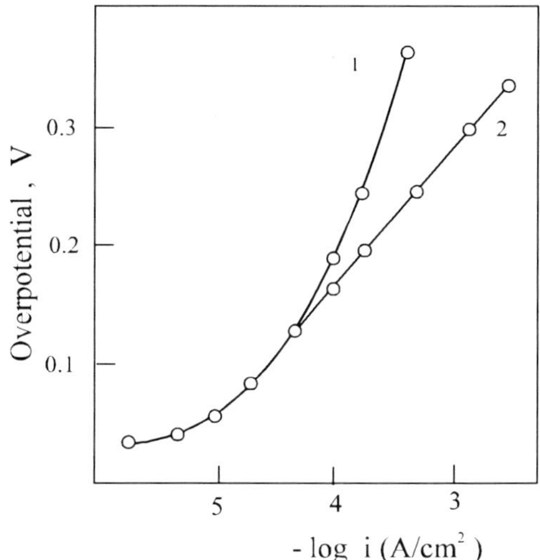

Fig. 11. Polarisation curves for mechanically polished nickel in 0.005 M H_2SO_4 + 0.5 M K_2SO_4:
1 - usually measured, 2- polarisation curve corrected on the basis of switch-over curves.

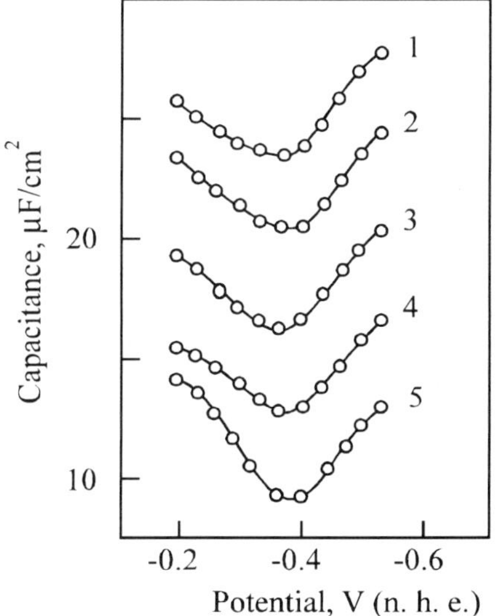

Fig. 12. C vs. E curves for nickel single crystal face (100) in $HClO_4$ solutions of different concentration:
1 - 0.25 M, 2 - 0.04 M, 3 - 0.014 M, 4 - 0.0065 M, 5 - 0.003 M.

244

Obviously the most reliable results of the influence of the concentration of solution on hydrogen overpotential have been obtained in pure acid or alkali solutions.

4.5. Potential of zero charge of nickel.

The potential of zero charge $E_{\sigma=0}$ is one of the most important characteristics of the metal-solution interface [106]. Adsorption ability of a large scale of compounds including such very important ones as corrosion inhibitors is strongly dependent on the value of $E_{\sigma=0}$ of the metal. This parameter is of great importance for the analysis of the influence of different factors, especially the composition of solution, on the kinetics of the hydrogen evolution reaction.

Different methods have been used for the determination of $E_{\sigma=0}$. The detail analysis of the subject has been carried out in [106]. The best experimental results have been achieved on liquid metals and on faces of single crystal electrodes. The precise establishment of $E_{\sigma=0}$ by electrochemical methods is available if the metal can be regarded as an ideal polarised electrode in the corresponding region of potentials. This condition is hardly satisfied for nickel as the region of potentials without any Faradaic processes cannot be found on this metal. Nevertheless several attempts have been made to determine the potential of zero charge for nickel [107-112].

The following methods have been used for the determination of $E_{\sigma=0}$ of the nickel/electrolyte interface: electrochemical impedance, galvanostatic pulses, adsorption of organic compounds, scraping at closed circuit, static friction between the surface and the solution. Some of these results are presented in Table 4.

Table 4
Potential of zero charge of polycrystalline nickel.

pH	$E_{\sigma=0}$(V)	Method	Reference
0.94	-0.224	impedance	[107]
1.4	-0.252	impedance	[107]
2.14	-0.237	impedance	[107]
2.97	-0.331	impedance	[107]
3.5	-0.11	galvanostatic pulse	[111]
5.8	-0.37	galvanostatic pulse	[111]
3.0	-0.24	scraping	[110]
5.6	-0.37	scraping	[110]
10.3	-0.24	impedance	[109]
13.0	-0.52	adsorption	[109]
12.5	-0.30	friction	[109]

One can see that, as a rule, there is essential dependence of the value of $E_{\sigma=0}$ on the pH of the solution. In the first approximation it is possible to resume that in acidic solutions $E_{\sigma=0}$ for polycrystalline nickel is within -0.22...-0,37V. The value for alkaline solutions within -0.24...-0.32V cannot be attributed to pure nickel surface as these potentials (0.3...0.4V in the respect of the potential of reversible hydrogen electrode in the same solution) correspond to the region where the surface of nickel is considered to be covered with $Ni(OH)_2$.

In the work [112] the determination of $E_{\sigma=0}$ on chemically polished nickel single crystals has been carried out. Electrochemical impedance has been used to determine the minimum of capacitance in 0.25...0.0025 M $HClO_4$ solutions. Fig. 12 shows the C vs. E curves for face (100) at different electrolyte concentrations. It can be seen that the minimum in the curves becomes more pronounced upon the dilution of the solution. Similar results have been obtained for face (110). The capacitance has been observed to decrease also with the dilution of solution if measured on face (111) or chemically polished polycrystalline electrodes but no marked minimum can be seen on C vs. E curves.

According to Grahame's model if the specific adsorption of ions is absent the measured capacitance can be described as being consistent of two capacitors connected in series: the capacitance of compact layer C_H and the diffuse-layer capacitance, thus:

$$1/C=1/C_H+1/C_D$$

At constant surface charge a plot of $1/C$ against $1/C_D$ (the Parsons-Zobel co-ordinates) [113] should be described by a straight line of 1 unit slope [113]. Fig. 13 shows such plots of corresponding capacitance values for the minimum of the C vs. E curves [112].

It has been obtained that for face (100) and face (110) these plots are described by straight lines with the slope close to 1 unit. It can be claimed therefore that the minimum in the C vs. E curves is caused by the diffuseness of the electric double layer and that the potential of the minimum is the potential of zero free charge with the values of $E_{\sigma=0}$=-0.39±0.02V for face (100) and -0.53±0.02V for face (110) [112].

The different behaviour of nickel face (111) can be related to more essential ingress of hydrogen into this face somewhat disturbing the structure of the surface. As the surface of polycrystalline electrode consists of a large amount of differently oriented single crystal faces the appearance of a sharp minimum can hardly be expected and the determination of the potential of zero charge on the basis of impedance measurements is complicated.

A minimum has also been exhibited at face (100) in the solutions of H_2SO_4. The minimum was also found to become deeper with the increase of the dilution of the solution but in contrast to that in $HClO_4$ solutions the potential of the

246

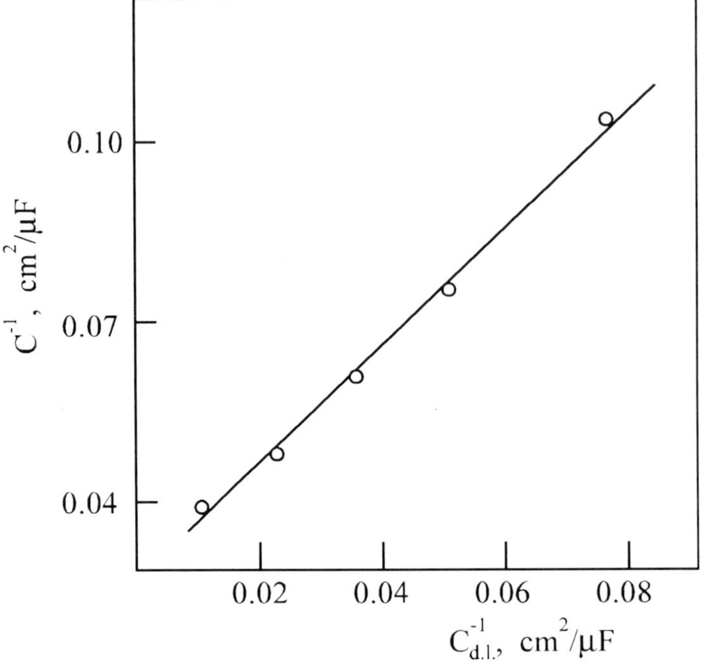

Fig. 13. Parsons - Zobel plot for face (100)

minimum somewhat depends on the acid concentration. The second significant difference observed is the appearance of a hump (maximum) on the C vs. E curves in H_2SO_4 solutions which has not been seen in $HClO_4$ solutions. These observations indicate the noteworthy specific adsorption of anions (HSO_4^-, SO_4^{2-}) on nickel surface. This may be the major reason for the dependence of $E_{\sigma=0}$ on the pH of the solution described in literature for sulphuric acid solutions [108]. The results of the determination of $E_{\sigma=0}$ on nickel confirm the assumption [63,70] that the potential of zero charge is within the region of potentials of the hydrogen overpotential measurements in acidic solutions. Consequently the ψ_1-potential for nickel is determined by the adsorption of components of solution rather than by the diffuseness of the electric double layer.

4.6. Coverage of nickel surface with adsorbed hydrogen.

The significance of the knowledge of the degree of coverage of nickel surface with atomic hydrogen θ_H especially the dependence of that on potential has been emphasised above already. In chapter 3 a short review of the existing experimental techniques worked out for the determination of adsorbed hydrogen has been given. The results obtained by these techniques for nickel electrodes will be discussed below:

(I) Acidic solutions.

As a remarkable corrosion of nickel electrodes proceeds in acidic solutions the open-circuit potential corresponds to the corrosion potential of nickel in the solution and the reversible hydrogen potential cannot be established. Therefore the measurements of θ_H^0 and even of θ_H^c at low overpotentials are not available. Thus anodic charging curves and potentiodynamic cycling are not applicable for the determination of θ_H in acidic solutions. In this case it is possible to determine only a part of θ_H^c by electrochemical methods.

The most convenient methods in these conditions are the overpotential decay and electrochemical impedance measurements [36,47,64,101]. At high overpotentials the influence of corrosion is not pronounced and the electrochemical behaviour of nickel is determined by the regularities of the HER only. As can be seen from Fig. 3 the capacitance obtained by overpotential decay measurements has a nearly constant value in the region of high overpotentials, which can be regarded as the capacitance of double layer. As no remarkable pseudocapacitance has been observed in the region studied it can be concluded that the change of θ_H^c at $\eta > 0.15\text{V}$ does not exceed 0.01.

(II) Alkaline solutions.

As the rate of corrosion of compact nickel is very low in alkaline solutions different methods for the study of adsorbed hydrogen are applicable, in principle.

The results of various measurements indicate that in the region of positive potentials not far from the potential of reversible hydrogen electrode in the same solution remarkable oxidation of the surface of nickel proceeds (the reversible potential of the reaction $Ni + 2\,OH^- \Leftrightarrow Ni(OH)_2 + 2\,e^-$ is 0.1 V [56]). The oxidation process can be observed as a maximum of current on potentiodynamic i,E - curves or pseudocapacitance on C,E - curves in the region of 0.2...0.5V (Fig. 14). A hump observed in some cases at the region of lower positive potentials can be attributed to the electrochemical formation of hydroxyl radicals and/or ionisation of adsorbed or absorbed hydrogen [51,114-117].

These possible processes cannot be distinguished easily and contrary interpretations of the experimental results have been presented.

An applicable method for the determination of θ_H seems to be the double charging method, used in [29] to determine the degree of coverage with hydrogen of nickel surface.

In our work [118] an additional attempt has been carried out to use the double charging method for the determination of θ_H in the case of nickel in alkaline solution. Two series of measurements have been carried out. The first one followed the technique described in [29], the normal curve has been measured starting from different steady state overpotentials, the compensation curve has been measured immediately after the normal curve starting from a small positive potential (without polarisation at cathodic potentials). The typical charging curves of this series have presented on Fig. 15 (curves 1 and 2) and the plots of $(i_H - i_{an}\theta)$ vs. time are shown in Fig. 16. The amount of charge determined on the basis of these curves has been found to be essentially dependent on the initial overpotential. These results are in good agreement with those in [29].

248

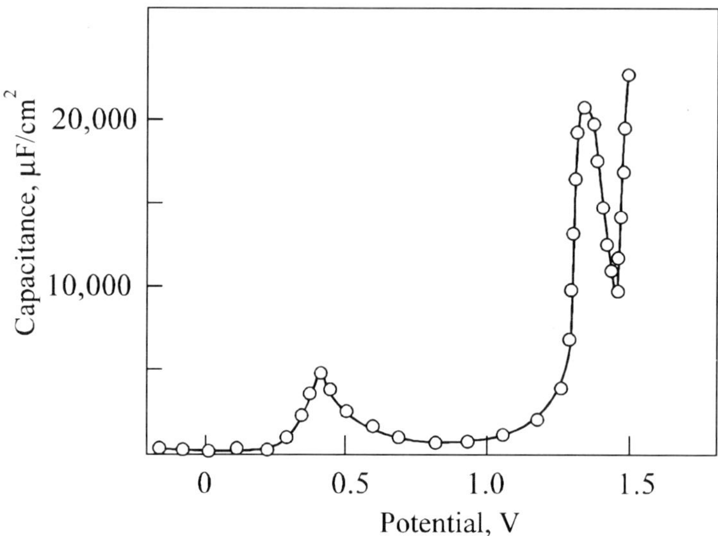

Fig. 14. Dependence of capacitance on potential for nickel in 0.5 M KOH solution
at anodic potentials.

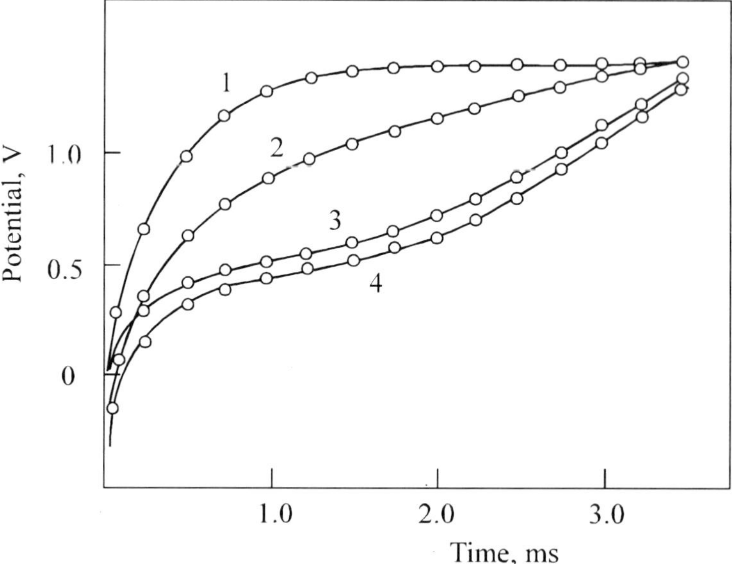

Fig. 15. Charging curves for nickel in 0.5 M KOH solution:
1,2 - the first series, 3,4 - the second series, 2,4 - normal curves, 1,3 -
compensation curves.

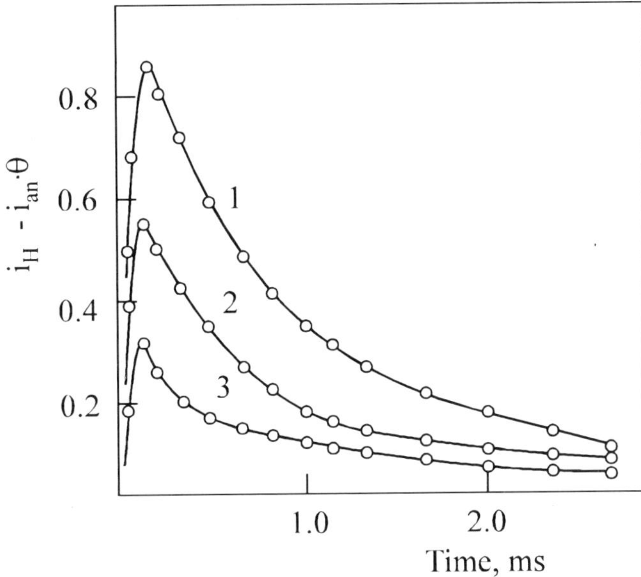

Fig. 16. ($i_H - i_{an}\theta$) vs. time plots of charging curves:
1 - E = -0.3 V, 2 - E = -0.2 V, 3 - E = -0.1 V.

The second series has measured by a changed technique. The electrode has been polarised before every charging curve (both normal and compensation curves) for 10 minutes at 0.35 V and after that 5 min. at the same initial potentials as in the first series. All the curves if measured by this technique had similar shapes (Fig. 15, curves 3 and 4).

Comparing the results of these two series it can be claimed that in the case of the first series not all of the oxide has been reduced during short cathodic polarisation at low overpotentials and thus even normal charging curves have been measured on partially oxidised surface, but the compensation curves on essentially oxidised surface. Consequently, the large amounts of charge obtained in the first series are consistent with the change of the amount of $Ni(OH)_2$ rather than of adsorbed hydrogen on the nickel surface. Although the technique used in the second series enables to carry out the measurements on reduced surface, the determination of the value of θ_H by this method is complicated as the difference between normal and compensation curves is very small (in both cases the main part of charge is required for the oxidation of the electrode). In view of the analysis presented above it must be concluded that the double charging method is not convenient to apply for the determination of θ_H in the case of nickel.

The change of θ_H^c can be successfully determined on nickel in alkaline solutions using overpotential decay measurements [36,40,41,42,49,119]. Remarkable pseudocapacitance has been observed in the range of comparatively low overpotentials dependent on the type of nickel electrode used. In some cases a maximum over 0.05 V has been noticed on C vs. η curves [41,119]. At low overpotentials the values of $d\eta/dt$ and the Faradaic current i_t are quite small and the error of their determination is considerably large if to compare with the case of higher overpotentials. Therefore the value of pseudocapacitance in this region of overpotentials can be obtained by overpotential decay curves with great error [120]. In order to overcome this source of error the switch-over curves have been used [42]. The initial current density was switched off and a quite small anodic current density i_a was switched on.

In this case the pseudocapacitance can be calculated by the relation:

$$C' = -\frac{i_t + i_a}{d\eta / dt} \tag{30}$$

The C vs. η curves determined by overpotential decay and over-switching transients are shown in Fig. 17 and 18.

It can be seen that if i_a is not too large C vs. η curves calculated from decay and over-switching curves are close and only at $\eta < 0.1$ the difference becomes remarkable. It is evident that at low overpotentials more reliable results can be obtained on the basis of switch-over curves. The results obtained and the theoretical values of θ_H^c calculated according to the mechanism of slow recombination [78] are presented in Table 5.

Table 5
Surface coverage of nickel electrodes with adsorbed hydrogen

Type of electrode	θ_H^c at η, V			
	0.15	0.20	0.25	0.30
mechanically polished	0.058	0.075	0.092	—
chemically polished	0.025	0.033	0.043	0.05
theoretically calculated value	0.13	0.17	0.23	0.26

It can be seen that θ_H^c for mechanically polished nickel is approximately two times larger than for chemically polished nickel, but the values of θ_H^c in both cases are essentially lower than that predicted by the recombination theory of the HER.

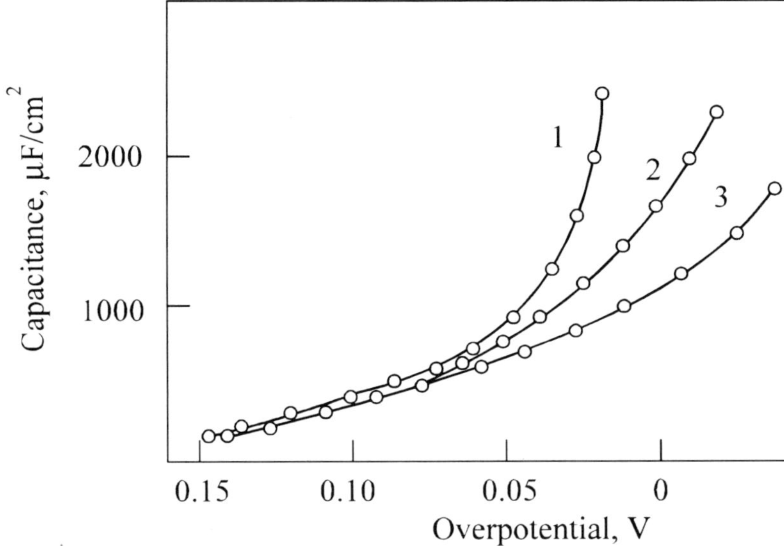

Fig. 17. C vs. η curves for mechanically polished nickel in 0.5 M KOH:
1 - i_a = 0, 2 - i_a = 0.05 mA/cm^2, 3 - i_a = 0.2 mA/cm^2

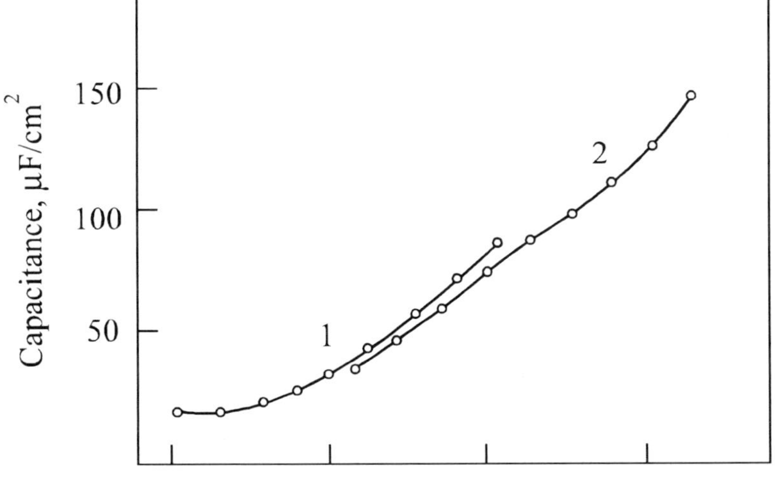

Fig. 18. C vs. η curves for chemically polished nickel in 0.5 M KOH:
1 - obtained by overpotential decay, 2 - obtained by a switch-over curve.

252

4.7. Influence of halide ions.

Halide ions belong to surface active ions having essential influence on the regularities of the hydrogen evolution reaction. It has been established that, as a rule, halide ions decrease the value of overpotential (at constant current density) on the metals of high overpotential and increase the value of η on the metals of medium overpotential [19,93].

According to the conventional model the effect of halide ions on the value of hydrogen overpotential can be divided into two parts:

$$\Delta\eta = \Delta\eta_{\Psi_1} + \Delta\eta_{ads} \quad (31)$$

The first term of equation (31) $\Delta\eta_{\Psi_1}$ reflects the influence of the adsorption of halide ions on the overpotential due to the change of Ψ_1-potential, the second term $\Delta\eta_{ads}$ is the shift of overpotential due to the change of the value of the energy of the metal-hydrogen bond. This interpretation is a little simplified as the term $\Delta\eta_{ads}$ is related to the influence of several factors [93]:

$$\Delta\eta_{ads} = \Delta\left\{ \frac{RT}{F}\ln\left[H_{ads}\right] + \frac{g_H}{F} + \frac{g_{H_2O}}{F} + \frac{1-\alpha}{\alpha}\frac{g_{H_3O^+}}{F} \right\} \quad (32)$$

Thus the adsorption of halide ions may change the energies of adsorption of all the reactants taking part of the HER.

The two terms of the expression (31) have contrary influence on hydrogen overpotential. The first term has a negative value and decreases the overpotential, the second one has a positive value and increases the hydrogen overpotential. It has been shown that in the case of nickel the second term is the dominating one.

The effect of halide ions on hydrogen overpotential on nickel depends on the nature and concentration of halide ions and on the potential of the electrode. The shift of overpotential for mechanically polished nickel caused by the influence of different halide ions of different concentration at constant current density is shown in Fig. 19. The polarisation curves for nickel measured in the presence of iodide ions in the solution are presented in Fig. 20. It has been established that, in general, Tafel slope b decreases with the increase of the concentration of halide ions. This tends to be essentially remarkable in the presence of bromide ions in the solution at low overpotentials (i.e. the region of potentials corresponding to small negative or even positive charge of nickel surface) where the adsorption of halide ions is stronger and consequently its effect on overpotential is more essential (Fig. 21, curve 3).

In order to get more detailed information about the effect of chemisorption of halide ions on the hydrogen overpotential the change of the overpotential after switching over the polarising current has been investigated. The electrode was polarised at high current density i_1. Then it was switched over to a lower

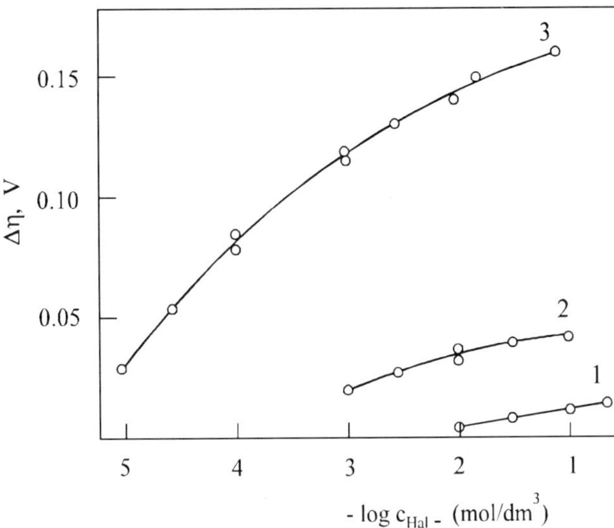

Fig. 19. Influence of the presence of halide ions of different concentration on hydrogen overpotential on mechanically polished nickel in 0.05 M H_2SO_4 solution:

1 - Cl^- - ions, 2 - Br^- - ions, 3 - I^- - ions.

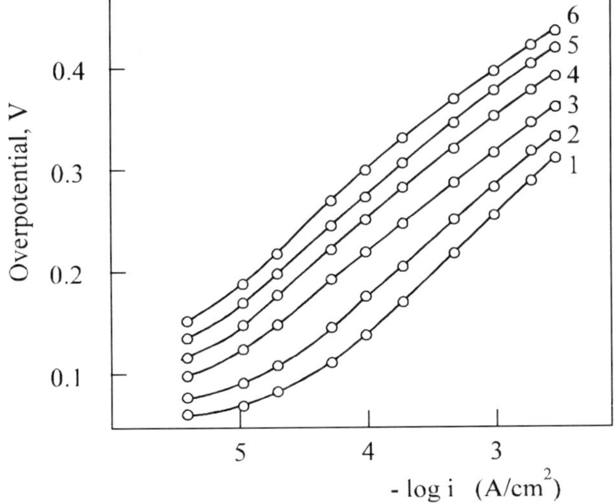

Fig. 20. Polarisation curves for mechanically polished nickel in 0.05 M H_2SO_4 solutions containing KI of different concentration:

1 - 0, 2 - 10^{-5}M, 3 - 10^{-4} M, 4 - 10^{-3} M, 5 - 10^{-2} M, 6 - 10^{-1} M.

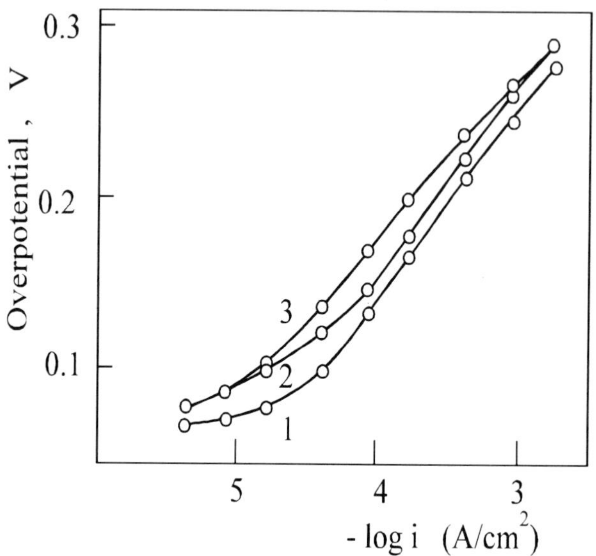

Fig. 21. Polarisation curves for mechanically polished nickel measured in:
1 - 0.05 M H_2SO_4, 2,3 - 0.05 M H_2SO_4 + 0.1 M KBr, 2 - built up on the basis of switch-over curves, 3 - usually measured.

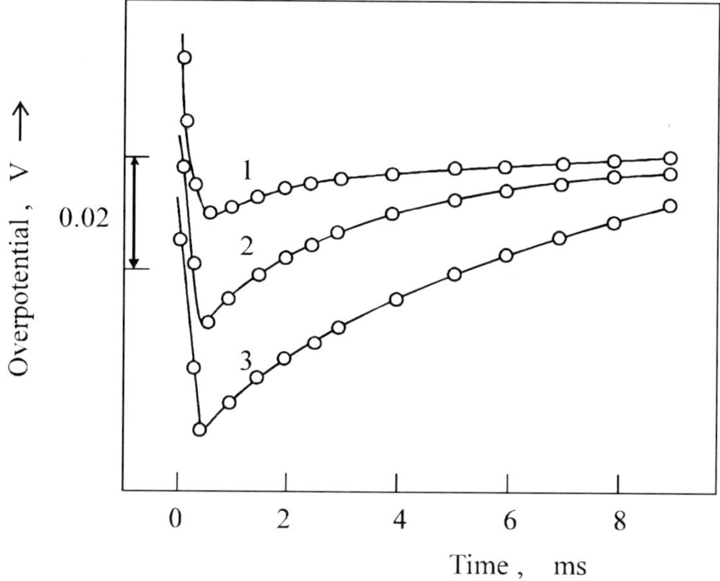

Fig. 22. Minima of switch-over curves in 0.05 M H_2SO_4 solution:
1 - 10^{-1} M KCl, 2 - 10^{-2} M KBr, 3 - 10^{-3} M KI.

value i_2 and the corresponding overpotential-time transient has been registered by oscillograph. It has been established that if the ratio of $i_1 : i_2$ is within the range of 5 - 1000, the switch-over curves exhibit a clear minimum (Fig. 22), the depth of which depends both on the value of $i_1 : i_2$ and on the nature and concentration of halide ions in the solution. The switch-over curves measured in the absence of halide ions in the solution have a smooth shape - the overpotential falls monotonously with time according to the discharge of the double layer of the electrode by the reduction of hydrogen ions. Therefore the appearance of a minimum on overpotential-time curves can be related to the change of adsorption of halide ions on the electrode surface. The change of the potential in the direction of less negative potentials during switch-over curves favours the adsorption of halide ions on nickel surface and consequently increases hydrogen overpotential.

The results of the switch-over measurements show clearly that the adsorption of halide ions is not a very quick process [90]. It is possible to divide switch-over curves into two branches. In the first one the overpotential decreases rapidly due to the discharge of double layer and then begins to increase slowly in the second branch. At the minimum point the double layer is almost discharged already but the adsorption of halide ions has not changed remarkably yet. Using the values of overpotential at minimum points of the switch-over curves to different current densities it is possible to build up the polarisation curves corresponding to the nearly constant degree of the surface coverage with chemisorbed halide ions.

A polarisation curve built up by the described above method starting from the initial overpotential 0.3V for nickel electrodes in the presence of bromide ions in the solution is presented in Fig. 21. As one can see such a non-stationary polarisation curve is shifted towards lower overpotentials in comparison with the one measured by the usual method in steady-state conditions and has a remarkably higher Tafel slope being similar to that of the curve measured in the absence of bromide ions. The decrease of the Tafel slope of non-stationary polarisation curve in the region of lower overpotentials can be explained assuming that in this case the adsorption of halide ions has already somewhat changed before the minimum point on the η vs. time curve. Consequently the degree of the surface coverage with strongly adsorbed halide ions does not correspond any longer to the constant initial value but has somewhat increased.

Minima of η vs. t curves in the presence of chloride ions in the solution are considerably less deep than in case of bromide and especially iodide ions. The upper part of the non-stationary polarisation curve has even a little higher value of Tafel slope than that in the absence of halide ions in the solution. This phenomenon supports the supposition of the remarkably more pronounced effect of the change of Ψ_1-potential on the hydrogen overpotential on nickel due to the comparatively weak adsorption of chloride ions differently from the strong adsorption of bromide and iodide ions on nickel .

Very interesting results have been obtained from overpotential decay and switch-over measurements for solutions containing bromide or iodide - ions. The essential pseudocapacitance has been established on C vs. η - curves (Fig. 23) [91], being dependent on the nature and concentration of halide ions and on the initial overpotential. The values of capacitance of the curves have been calculated by equations (14) and (30) using the current densities corresponding to non-stationary polarisation curves built up on the basis of switch-over curves. Such a pseudocapacitance may be caused either by the essential charge transfer during the chemisorption of halide ions or by the ionisation of some amount of the adsorbed on nickel atomic hydrogen displaced by halide-ions. It is reasonable to assume that both of the processes take place at the same time. The maximum value of total charge calculated from C vs. η - curves corresponds to 0.15 monolayer of atomic hydrogen (assuming that all the charge corresponds to the ionisation of atomic hydrogen).

It is not very easy to distinguish these processes by electrochemical methods. However, the charge transfer seems to be the dominating process at higher overpotentials and the contribution of adsorbed hydrogen may be more essential at lower overpotentials. It is reasonable to conclude that at least half of the total charge is connected with the ionisation of adsorbed atomic hydrogen and consequently $0.07 \leq \Delta\theta_H < 0.15$. This adsorbed hydrogen may correspond to θ_H^c at low overpotentials but the main part of it corresponds evidently to θ_H^0.

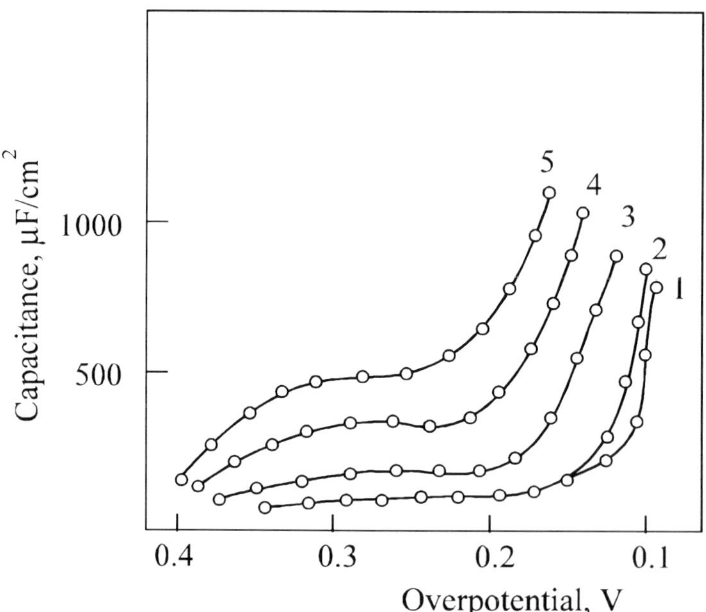

Fig. 23. C vs. η curves for mechanically polished nickel in 0.25 M H_2SO_4 solution in the presence of I^- - ions of different concentration:
1 - 0, 2 - 10^{-5} M, 3 - 10^{-4} M, 4 - 10^{-3} M, 5 - $2 \cdot 10^{-2}$ M.

5. CONCLUSION.

The most important data concerning hydrogen evolution reaction on nickel have been presented in this review. On the basis of these data it is possible to induce a conclusion of the mechanism of this process. Nevertheless, the experimental results do not form a whole, as some very important data are still not available yet and therefore the ambiguous conclusion cannot be done. The results presented above suggest that the recombination of adsorbed atomic hydrogen (the Tafel reaction) cannot be the rate-determining step of the overall process, at least at higher current densities. The strongest supports to this statement are the quite low values of θ_H^c and its weak dependence on overpotential at moderate and high current densities.

In the first approximation it is reasonable to suppose that the hydrogen evolution reaction proceeds according to the Volmer-Heyrovsky mechanism. In order to determine the rate-determining step to the Volmer-Heyrovsky mechanism the value of limiting coverage $\theta_{H,lim}$ is needed [119]:

$$\theta_{H,lim} = \frac{k_1}{k_1 + k_2} \quad , \tag{33}$$

where k_1 is the rate constant of discharge reaction and k_2 - the rate constant of electrochemical desorption reaction. Further the analysis will be carried out separately for acidic and alkaline solutions.

As it has been shown above, for acidic solutions it is impossible to determine the value of stoichiometric number and the coverage of the surface of nickel electrodes with adsorbed atomic hydrogen at the reversible hydrogen potential and even at low cathodic potentials due to the remarkable rate of the corrosion process of nickel at these potentials in this media. Thus the significant parameters of mechanism θ_H and its variation at low overpotentials are not known for acidic media, therefore we have no direct criteria to distinguish the slow discharge (the Volmer's mechanism) and the slow electrochemical desorption (the Heyrovsky's mechanism).

Some additional information is available from the studies of the influence of halide ions on hydrogen overpotential. It has been established that halide ions increase the hydrogen overpotential remarkably. It is reasonable to suppose that the term $\Delta\eta_{ads}$ in Eq. (31) is mainly determined by the decrease of the energy of adsorption of the hydrogen g_H. Consequently, the increase of overpotential due to the adsorption of halide ions suggests that the rate-determining step has to be the discharge. Contrary to this, the decrease of g_H has to accelerate the electrochemical desorption of adsorbed hydrogen and therefore to decrease hydrogen overpotential if this step is limiting the overall process. Nevertheless, it must be noted that the rate of the electrochemical desorption cannot be much higher than that of the discharge reaction as the coverage of nickel surface with adsorbed hydrogen is not extremely low that is confirmed by the ionisation of remarkable amount of hydrogen due to the adsorption of halide ions.

More experimental data are available for alkaline media . The most significant of them are the data of θ_H^c at cathodic potentials. It has been shown above that θ_H^c does not exceed 0.1 part of monolayer. Much higher values of θ_H^c obtained in [41] are probably connected with the specific properties of the nickel electrodes used in this work. The catalytic activity of the nickel film evaporated on glass used in [41] seems to be much higher than the activity of the surface of usual compact nickel.

As the exact value of θ_H^c is still not available we can only consider that $\theta_{H,lim}$ is within the limits $0.05 \leq \theta_{H,lim} \leq 0.95$ and consequently the ratio of constants $k_1:k_2$ varies approximately from 0.05 to 20. The same values of the ratio of constants of discharge and the electrochemical desorption have been obtained from the analysis of galvanostatic switch-in curves [121] and a similar ratio has been calculated on the basis of overpotential decay transients [119,120].

In the case of some types of nickel electrodes especially for anodically activated ones two slopes on Tafel plots have been observed [49,120,122]. The appearance of two different regions on polarisation curves has been explained by the influence of another electrode process at lower overpotentials e.g. of electrode corrosion. Such an essential influence of corrosion on the regularities of hydrogen evolution in alkaline media is not probable. The model calculations of polarisation curves for various combinations of the values of the rate constants of different steps have been carried out [123]. It has been established that there cannot be found a combination of rate constants and θ_H values to derive a polarisation curve with two remarkably different slopes having the higher slope at lower overpotentials.

Such a shape of polarisation curves obtained experimentally can be explained supposing that there are places of two different type on the surface of nickel electrode. It is possible that on the larger part of the surface the Volmer-Heyrovsky mechanism is operative but on the smaller part the rate-determining step is the recombination of hydrogen atoms (the Tafel's reaction) [14]. Almost the same results can be obtained if to suppose that the process is occurring by the Volmer-Heyrovsky mechanism but with essentially different transfer coefficients on different parts of surface [122].

The approach to the mechanism of the hydrogen evolution reaction presented above has been developed, in principle, for metals of high overpotential. In the case of metals of medium overpotential such as nickel it is very probable that significant interactions between the metal surface and as ions as water molecules take place, especially if the metal surface has positive charge. It is reliable that such interactions have an essential effect on the distribution of potential on the metal/solution interface causing significant differences in the regularities of hydrogen evolution reaction. Additional experimental and theoretical studies should be carried out before any novel approach can be developed.

REFERENCES

1. J. Tafel, Z. Phys. Chem., 50 (1905) 641.
2. J.A.V. Butler, Trans. Faraday Soc., 19 (1924) 729.
3. J. Heyrovsky, Rec. Trav. Chim., 46 (1927) 582.
4. T. Erdey-Gruz and M. Volmer, Z. Phys. Chem., A 150 (1930) 203.
5. N. Kobozev and N. Nekrasov, Z. Elektrochem., 36 (1930) 528.
6. R. Gurney, Proc. Roy. Soc., A 134 (1931) 137.
7. A. Frumkin, Z. Phys. Chem., A 160 (1932) 116.
8. J.A.V. Butler, Trans. Faraday Soc., 28 (1932) 379.
9. T. Erdey-Gruz and M. Volmer, Z. Phys. Chem., A 162 (1932) 53.
10. A. Frumkin, Z. Phys. Chem., A 164 (1933) 121.
11. J. Horiuti and M. Polanyi, Acta Physicochim. U.R.S.S., 2 (1935) 505.
12. J.A.V. Butler, Proc. Roy. Soc., A 157 (1936) 423.
13. G. Okamoto, J. Horiuti and K. Hirota, Sci. Pap. Inst. Phys. and Chem. Res., 29 (1936) 223.
14. A. Frumkin, Discuss. Faraday Soc., No. 1 (1947) 57.
15. J.O'M. Bockris, in Modern Aspects of Electrochem., Vol. 1, Eds. J.O'M. Bockris and B.E. Conway, Butterforths, London, 1954, p. 201.
16. K. Vetter, Elektrochemische Kinetic, Springer, Berlin, 1961.
17. J.O'M Bockris and E.C. Potter, J. Electrochem. Soc., 99 (1952) 169.
18. J.O'M. Bockris and R.G.H. Watson, J.Chem. Phys., 46 (1952) C 70.
19. A.N. Frumkin, in Advances in Electrochem. and Electrochem. Eng., Vol. 1, Ed. P. Delahay, Interscience, NY, London, 1961, p. 65.
20. A.N. Frumkin, in Advances in Electrochem. and Electrochem. Eng., Vol. 3, Ed. P. Delahay, Interscience, NY, London, 1963, p. 287.
21. S. Trasatti, in Advances in Electrochem. Science and Eng., Vol. 2, Eds. H. Gerischer and C.W. Tobias, VCH, Weinheim, 1992, p. 2.
22. E. Gileadi and B.E. Conway, in Modern Aspects of Electrochemistry, Vol. 3, Eds. J.O'M. Bockris and B.E. Conway, Butterworths, London, 1964, p. 347.
23. B.E. Conway, Theory and Principles of Electrode Processes, Ronald Press, NY, 1965.
24. A. Slygin and A. Frumkin, Acta Physicochim. U.R.S.S., 3 (1935) 79.
25. A.N. Frumkin, O.A. Petri and R.V. Marvet, J. Eletroanal. Chem., 12(1966) 504.
26. J.D. Pearson and J.A.V. Butler, Trans. Faraday Soc., 34 (1938) 1163.
27. M. Breiter, C.A. Knorr and W. Völkl, Z. Elektrochem., 59 (1955) 681.
28. M.A.V. Devanathan, J.O'M Bockris and W. Mehl, J. Electroanal. Chem., 1 (1959/60) 143.
29. M.A.V. Devanathan and M Selvaratnam, Trans. Faraday Soc., 56 (1960) 1820.
30. H. Gerischer and W. Mehl, Z. Elektrochem., 59 (1955) 1049.
31. M.W. Breiter, J. Electroanal. Chem., 19 (1968) 131.
32. G.A. Armstrong and I.A.V. Butler, Trans. Faraday Soc., 29 (1933) 1261.
33. A.N. Frumkin, Acta Physicochim. URSS, 18 (1943) 23.
34. A.N. Frumkin and N. Aladzhalova, Acta Physicochim. URSS, 19 (1944) 1.
35. M.B. Morley and F.E.W. Wetmore, Can. J. Chem., 34 (1956) 359.

36. V.E. Past and Z.A. Iofa, Dokl. AN SSSR, 106 (1956) 1052.

37. B.E. Conway and P.L. Bourgault, Trans. Faraday Soc., 58 (1962) 593.

38. U.V. Palm and V.E. Past, Dokl. AH SSSR, 146 (1962) 1374.

39. B.E. Conway and P.L. Bourgault, Can. J. Chem., 37 (1959) 292.

40. V.E. Past and Z.A. Iofa, Zh. Fiz. Khim., 33 (1959) 1230.

41. A. Matsuda and T. Ohmori, J. Res. Inst. Catalysis, Hokkaido Univ. 10 (1962) 215.

42. L. Tohver, J. Tamm and V. Past, Acta Comm. Univ. Tartuensis, No. 289 (1971) 3.

43. R.H. Burshtein, A.G. Pchenitchnikov, V.D. Kovalevskaja and M.E. Belyaeva, Elektrokhimiya 6 (1970) 1756.

44. D.C. Grahame, Chem. Rev., 41 (1947) 441.

45. P. Delahay, Double Layer and Electrode Kinetics, Wiley, NY, 1965.

46. A. Hamelin, in Modern Aspects of Electochemistry, Vol. 16, Eds. B.E. Conway and J.O'M. Bockris, Plenum Press, NY. 1985, p. 1

47. L. Tamm, J. Tamm and V. Past, Acta Comm. Univ. Tartuensis, No. 332 (1974) 3.

48. J. Tamm and V. Past, in Double Layer and Adsorption on Solid Electrodes, Vol. 1, Tartu University, 1968, p. 158.

49. L. Tohver and J. Tamm, Acta Comm. Univ. Tartuensis, No. 265 (1970) 14.

50. A.C. Makrides, J. Electrochem. Soc., 109 (1962) 256.

51. J.L. Weininger and M.W. Breiter, J. Electrochem. Soc., 111 (1964) 707.

52. G. Armstrong, F. Himsworth and J.A.V. Butler, Proc. Roy. Soc., A 143 (1933) 89.

53. F. Will and C. Knorr, Z. Elektrochem., 64 (1960) 258, 270.

54. T. Biegler, J. Electrochem. Soc., 116 (1969) 1131.

55. B.G. Pound, in Modern Aspects of Electrochem., Vol. 25, Eds. J.O'M. Bockris and B.E. Conway, 1993, p. 63.

56. M. Pourbaix, Atlas of Electrochem. Equilibria in Aqueous Solutions, Pergamon, Oxford, 1966.

57. J.O'M. Bockris, J. McBreen and L. Nanis, J. Electrochem. Soc., 112 (1965) 1025.

58. I.A. Bagotskaya, Zh. Fiz. Khim., 36 (1962) 2667.

59. P.A. Dowden and P.W. Reynolds, Discuss. Faraday Soc., 8 (1950) 184.

60. B.E. Conway, E.M. Beatty and P.A.D. De Maine, Electrochim. Acta, 7 (1962) 39.

61. Ya.B. Skuratnik, A.E. Kozachinskii, A.P. Pchelnikov and V.V. Losev, J. Electroanal. Chem., 366 (1994) 311.

62. A.E. Kozachinskii, A.P. Pchelnikov, Ya.B. Skuratnik and V.V. Losev, Elektrokhimiya, 29 (1993) 508.

63. P. Lukovtsev, S. Levina and A. Frumkin, Zh. Fiz. Khim., 13 (1939) 916.

64. A. Legran and S. Levina, Zh. Fiz. Khim., 14 (1940) 211.

65. P. Lukovtsev and S. Levina, Zh. Fiz. Khim, 21 (1947) 599.

66. J. O'M. Bockris and E.C. Potter, J. Chem. Phys., 20 (1958) 614.

67. R. Piontelli, L. Peraldo Bicelli and A. La. Vecchia, Atti Acad. Naz. Lincei Rend., Cl. Sci. Fis. Mat. Nat., 27 (1959) 312.

68. J. Mieluch, Bull. Acad. Pol. Sci., Ser. Sci. Chim., 17 (1969) 43.
69. V.A. Lavrenko, L.N. Yagupolskaya and V.L. Tikush, Elektrokhimiya, 6 (1970) 887.
70. J. Arold and J. Tamm, Elektrokhimiya, 25 (1989) 418.
71. S. Trasatti, Electrochim. Acta, 28 (1983) 1083.
72. J. Pielaszek, Bull. Acad. Pol. Sci., Ser. Sci. Chim., 20 (1972) 487, 611.
73. L. Tamm, J. Tamm and V. Past, Acta Comm. Univ. Tartuensis, No. 378 (1976) 3.
74. J. Horiuti, J. Res. Inst. Catalysis, Hokkaido Univ., 1 (1948) 8.
75. R. Parsons, Trans. Faraday Soc., 47 (1951) 1332.
76. K.J. Vetter, Z. Elektrochem., 59 (1955) 435.
77. A.N. Frumkin, Dokl. AN SSSR, 119 (1958) 318.
78. J. Horiuti, J. Res. Inst. Catalysis, Hokkaido Univ., 4 (1955) 55.
79. J. O'M. Bockris and A.M. Azzam, Trans. Faraday Soc., 48 (1952) 145
80. H. Kita and T. Yamazaki, J. Res. Inst. Catalysis, Hokkaido Univ., 11 (1963) 10.
81. H. Kita and O. Nomura, J. Res. Inst. Catalysis, Hokkaido Univ., 12 (1964) 107.
82. B.N. Kabanov, Acta Physicochim U.R.S.S., 4 (1936) 846.
83. T. Silk, J. Tamm and V. Past, Acta Comm. Univ. Tartuensis, No. 332 (1974) 18.
84. R.M. Wightman and D.O. Wipf, in Electroanal. Chem., Vol. 15, Ed. A.J.Bard, Dekker, NY, 1989, p. 267.
85. E.I. Michailova and Z.A. Iofa, Elektrokhimiya, 1 (1965) 107.
86. A.T.Skljarov and Ya.M. Kolotyrkin, Elektrokhimiya, 3 (1967) 1462.
87. N.D. Tomashov, N.M. Strukov and L.P. Vershina, Elektrokhimiya, 5 (1969) 26.
88. S. Rashkov, J. Hristova and N. Pangarov, Compt. Rend. Acad. Bulg. Sci., 23 (1970) 387.
89. E.I. Mikhailova and Z.A. Iofa, Elektrokhimiya, 6 (1970) 231.
90. L. Tamm, J. Tamm and V. Past, Electrokhimiya, 9 (1973) 1382.
91. L. Tamm, J. Tamm and V. Past, Elektrokhimiya, 11 (1975) 1581.
92. L. Tamm, J Tamm and V. Past, Acta Comm. Univ. Tartuensis, No. 378 (1976) 17.
93. J. Tamm, L. Tamm, P. Vares and J. Arold, Acta Comm. Univ. Tartuensis, No. 905 (1990) 107.
94. L.I. Krishtalik, Elektrokhimiya, 6 (1970) 1165.
95. L.I. Krishtalik, J. Electroanal. Chem., 35 (1972) 157.
96. J. Tamm, L. Tamm and P. Vares, Acta Comm. Univ. Tartuensis, No. 757 (1986) 34.
97. A.N. Frumkin, V.S. Bagotskii, Z.A. Iofa and B.N. Kabanov, Kinetika Electrodnyh Processov, Moscow Univ. Press, Moscow, 1952.
98. I.A. Ammar and S.A. Awad, J. Phys. Chem., 60 (1956) 837.
99. A. Matsuda and T Ohmori, J. Res. Inst. Catalysis, Hokkaido Univ., 10 (1962) 203.
100. L. Tamm, J. Tamm and V. Past, Elektrokhimiya, 10 (1974) 83.

262

101. L. Tamm, J. Tamm and V. Past, Acta Comm. Univ. Tartuensis, No. 332 (1974) 12.
102. J. Tamm and L. Tamm, Elektrokhimiya, 12 (1976) 955.
103. J. Tamm, L. Tamm and A. Pikat, Acta Comm. Univ. Tartuensis, No. 378 (1976) 11.
104. L.I. Antropov, Zh. Fiz. Khim., 26 (1952) 1688.
105. L.I. Antropov, Zh. Fiz. Khim., 28 (1954) 1336.
106. A.N. Frumkin, Potential of Zero Charge, 2nd ed. Nauka, Moscow, 1982.
107. V.L. Kheifets and L.S. Reishakrit, Uch. Zap. Leningr. Gos. Univ., Ser. Khim., No. 13 (1953) 173.
108. V.L. Kheifets and B.S. Krasikov, Zh. Fiz. Khim., 31(1957) 1992.
109. J.O'M. Bockris, S.D. Argade and E. Gileadi, Electrochim. Acta, 14 (1969).
110. E.M. Lazarova, Elektrokhimiya, 14 (1978) 1300.
111. T. Ohmori, J. Electroanal. Chem., 157(1983) 159.
112. J. Arold and J. Tamm, Elektrokhimiya, 25 (1989) 1411.
113. R. Parsons and F.G.R. Zobel, J. Electroanal. Chem., 9 (1965) 333.
114. Gu Lin-in, N.A. Shumilova and V.S. Bagotskii, Elektrokhimiya, 3 (1967).
115. N.A. Shumilova and V.S. Bagotskii, Electrochim. Acta, 13 (1968) 286.
116. V.R. Loodmaa, M.E. Haga and V.E. Past, Isv. Vuzov, Khim i Khim. Tekhnol., 9 (1966) 794.
117. L.A. Burkaltseva and A.G. Pshenichnikov, Elektrokhimiya, 12 (1976) 42.
118. V.Past, J. Tamm and T. Tenno, Acta Comm. Univ. Tartuensis, No. 219 (1968) 30.
119. B. Conway and L. Bai, Chem. Soc., Faraday Trans. I, 81 (1985) 1841.
120. A. Lasia and A Rami, J. Electroanal. Chem., 294 (1990) 123.
121. T. Silk, J. Tamm and V. Past, Acta Comm. Univ. Tartuensis, No. 378 (1976) 121.
122. J. Tamm and V. Past, Acta Comm. Univ. Tartucnsis, No. 235 (1969) 20.

Research in Chemical Kinetics, Volume 3
R.G. Compton and G. Hancock (editors)

Kinetics & mechanisms of silylene reactions : A prototype for gas-phase acid/base chemistry

R. Becerra[a] and R. Walsh[b]

[a]Instituto Quimica Fisica "Rocasolano", C.S.I.C., C/Serrano, 119, 28006, Madrid, Spain

[b]Department of Chemistry, University of Reading, Whiteknights, P.O.Box 224, Reading RG6 2AD., U.K

Contents

1. BACKGROUND

1.1 Introduction

Gas-phase chemistry is understandably dominated by studies of hydrocarbon reactions. This situation has come about because of the importance of fossil fuels in the world economy. Gas-phase hydrocarbon reactions, in the main, fall into the categories of pyrolysis, or combustion including atmospheric oxidation (so-called 'slow-combustion'). The mechanisms of such processes almost inevitably involve free-radical intermediates and so gas-phase chemistry has grown up largely illustrated by examples of free-radical reactions.

By contrast, other areas of chemistry are poorly represented. One such is that of silane and organosilicon chemistry. Organosilicon chemistry has a considerable importance in polymer production, the best-known examples of which are the silicones [1]. Silane chemistry is important to the vapour deposition processes (CVD) which are used to produce silicon-containing electronic materials [1]. The characteristic mechanisms of silane and organosilicon compound breakdown are, in general, not free radical processes, but rather they involve the intermediacy of silylenes. Silylene reactions represent a different type of chemistry than free radical processes and therefore this gives silane and organosilicon chemistry a distinctive character. It is true that silylenes are the silicon analogues of carbenes and there is a considerably developed chemistry of carbenes. However, as we shall see, studies of carbene reactions are hampered by certain difficulties which do not affect silylene studies. Thus in a variety of ways progress has been greater in understanding the behaviour of silylenes than that of carbenes. A caveat should be entered. All silylenes thus far studied possess singlet ground states, whereas many carbenes exist in triplet ground states. It is the chemistry and reactivity of the singlet states of silylenes which has developed. There is as yet no experimental knowledge of the chemical behaviour of triplet (excited) states of silylenes.

Our knowledge of silylenes has grown steadily since the first review of their occurrence and behaviour by Atwell and Weyenberg in 1969 [2,3]. The accumulating knowledge of their chemical behaviour has been particularly well described in a series of notable reviews by Gaspar [4,5], as well as in articles by others [6-10]. Most recently Strausz and his colleagues [11] have documented the increasing data base of experimental rate constants for silylene reactions. It is not the purpose of the present article to repeat these surveys in full detail. Rather our objective is to describe the insight gained into the mechanisms of prototype silylene processes as a result of the application of modern, time-resolved, techniques to measurements of their rate constants. The first experiment of this kind was performed by Inoue and Suzuki [12] in 1985, who used laser induced fluorescence to detect and monitor SiH_2 [13,14]. Since 1985 a steadily accumulating number of measurements of absolute rate constants have been published. The majority, but not all, of this work has been undertaken in the laboratories of Gaspar, Strausz, Jasinski and in our own. Prior to these investigations, information about the formation, intermediacy and reactivity of silylenes was inferred indirectly from final product distributions in numerous extensive experiments of a more classical kind using thermal or photochemical decomposition of appropriate organosilicon precursors to generate the silylenes. In these earlier studies direct measurements of rate constants of the intermediate silylenes were not possible, although relative rate constants could be obtained from carefully designed competitive studies, or inferred from kinetic modelling. Thus a picture emerged of silylenes as fairly reactive intermediates (not as reactive as carbenes) participating in reactions of certain types (insertions, additions) with certain bond types (but not others). The information was substantial but the picture lacked sharp detail. In a notable statement in their 1987 article [15] Gaspar and colleagues provided a focus for progress by framing the questions - What are the transition-state structures? What is the balance between electrophilic and nucleophilic character of the attacking silylene species in the transition state? What role do steric and electronic effects play in silylene reactions? To these we may add the more basic question - How do the structures and energies of silylenes affect their reactivity? This article attempts to show how the absolute rate studies have begun to throw some light on these matters. In order to set the scene we describe in this section some basic knowledge about the structure and energetics of silylenes as well as of their reaction types. Also included here is some information provided by theory which has progressed in tandem with experimental measurements. A very useful account of recent theoretical calculations on silylenes has been given by Apeloig [16].

1.2 Silylene structure, bonding and stability

Symmetrical silylenes, such as SiH_2, have C_{2v} structures with $^1A_1$ ground states and $^3B_1$ excited states [16]. An orbital occupancy diagram is shown in Figure 1.

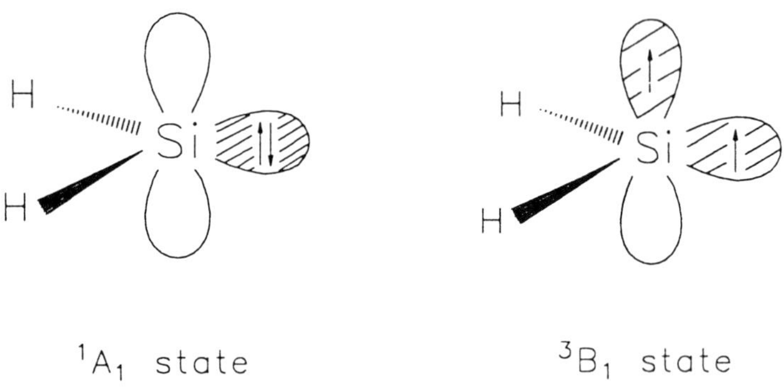

Figure 1. Orbital diagram for the lowest states of silylene

Experimental studies, based on analysis of the visible vibronic absorption spectrum [17], have given values for the Si-H bond length of 1.516Å and HSiH bond angle of 92.8°. Theoretical studies [16] are in excellent agreement.

A comparison with the angles of methylene, CH_2, is shown in Table 1, which indicates that in general the bond angles are *ca* 10° less in SiH_2 than CH_2.

Table 1
Bond angle values for lower electronic states of methylene and silylene

Species	HCH	Species	HSiH
CH_2 ($^1A_1$)	102.4[a]	SiH_2 ($^1A_1$)	92.8[a]
CH_2 ($^3B_1$)	130[a]	SiH_2 ($^3B_1$)	118.5[b]

a. Experimental values.   b. Theoretical value.

This is consistent with the general observation that silicon prefers to have non-bonding electrons in atomic orbitals with a high percentage of s-character. Luke et al [18] have calculated that in the $^1A_1$ state the lone pair is in a hybrid orbital with 89% s-character in SiH_2 compared with 60% s-character in CH_2. This reluctance of s orbitals to become involved in the bonding in silylene, reflects the decreased screening of valence s orbitals (compared with p orbitals) from the atomic nucleus in higher row elements of the periodic table. It is connected with the inert pair effect and general increasing prevalence of lower valence states in higher rows of the periodic table. This phenomenon has been quantified by Walsh for silylenes in the concept of Divalent State Stabilisation Energy or DSSE [19-21]. This is defined, for the silylene SiX_2, as follows:

$$DSSE = D(X_3Si-X) - D(X_2Si-X)$$

where $D(X_3Si-X)$ and $D(X_2Si-X)$ correspond to the first and second dissociation energies of the silane SiX_4. This equation reflects the fact that whereas breaking the first bond effectively produces little or no rehybridisation (the Si atomic orbitals being essentially sp^3 in both SiX_4 and SiX_3) breaking the second bond causes the rehybridisation of the Si atomic orbitals to become mainly p in the Si–X bonds, and s in the lone pair. Walsh [21] has estimated values for $DSSE$ for a number of silylenes and these are shown in Table 2.

Table 2
Thermodynamic data for silylenes [21]

Species	ΔH_f^o / kJ mol^{-1}	$DSSE$ / kJ mol^{-1}
SiH_2	273 ± 2	93 ± 3
$SiMe_2$	135 ± 8	121 ± 12[a]
SiF_2	-590 ± 8	205 ± 42
$SiCl_2$	-169 ± 8	159 ± 17
$SiBr_2$	-46 ± 8	142 ± 50
SiI_2	-92 ± 8	134 ± 54

a. Revised in this review.

The numbers on which these values depend are taken from the most recent figures for bond dissociation energies [21,22] in combination with enthalpies of formation, ΔH_f^o, of the silylenes themselves [21,23]. Although the errors are in some cases still large, it is nevertheless clear that silylenes are stabilised by electronegative substituents. The small but significant difference for $SiMe_2$ and SiH_2 shows that, as expected, methyl groups act as weakly electronegative substituents. The expectation in this case derives from Pauling electronegativity values (2.7 for C *vs* 2.2 for H relative to 1.8 for Si). The usefulness and significance of the idea of DSSE has been expanded and extended to other elements by Grev [24].

In a theoretical examination of substituent effects in silylenes, Luke *et al* [18], have reached similar conclusions for the monosubstituted silylenes RSiH, where R represent groups -Li, -BeH, $-BH_2$, $-CH_3$, $-NH_2$, -OH and -F. These results show that the most effective stabilisation is achieved with π-donors (eg $-NH_2$). This suggests the importance of change transfer to the empty $3p$ orbital and involvement of resonance structures shown in figure 2.

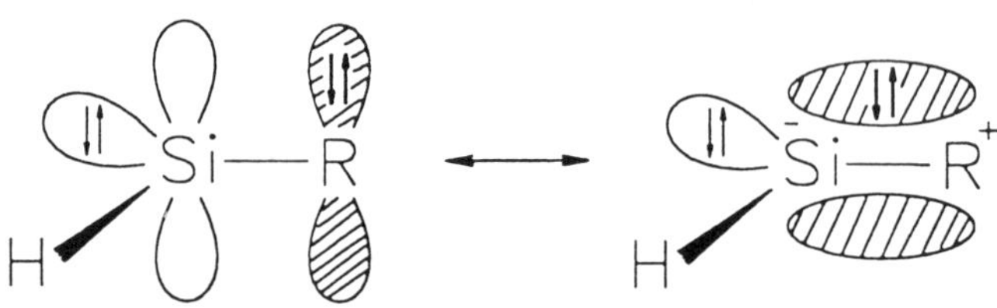

Figure 2. Contributing structures for silylenes with lone-pair carrying substituents

It is important to stress the point that stabilisation of silylenes, as discussed here, is essentially thermodynamic in nature, and does not necessarily have any bearing on the kinetics and general reactivity of the silylenes (as we shall see).

1.3 Nature of silylene reaction processes

It is clear from Figure 1 that $(^1A_1)$silylenes have two potential sites of reactivity, *viz.* an empty p orbital and a filled lone-pair orbital. All their reactions involve the use of both of these orbitals. Since these reactions all involve acceptance of a lone pair (by the Si $3p$ orbital) and the donation of a long pair (the Si s^2 pair) to the creation of a new bond (with a substrate) these processes may be considered examples of acid/base behaviour. More commonly the terminology "electrophilic" and "nucleophilic" has been used by theoreticians who have noted on the basis of *ab initio* calculations [25-27] that these processes appear to occur in stages. This may be illustrated for one of the most widely studied processes, the Si-H insertion reaction shown in Figure 3.

These ideas are very useful since they offer experimentalists a conceptual framework within which to probe the nature of the processes. If the stages of interaction do not occur simultaneously the question arises of whether other intermediates with only <u>one</u> lone pair interaction, may be involved. Thus questions which naturally arise are.. Does the "electrophilic" stage precede the "nucleophilic" stage? .. or vice versa? .. Or, indeed, do they occur synchronously?   In fact all of the major types of silylene process *viz.* σ-bond insertion, π-bond addition and reaction with n-type donors, can be considered in this manner.   These ideas are further illustrated during the discussion of specific reaction examples.

1.4 Comparison of silylenes with carbenes

Although a detailed discussion of carbene ($^1A_1$) reactions is beyond the scope of this article, in principle they may be equally well discussed within the same framework of approach adopted here.   However, despite intensive investigation [28] progress has been hampered by certain practical difficulties

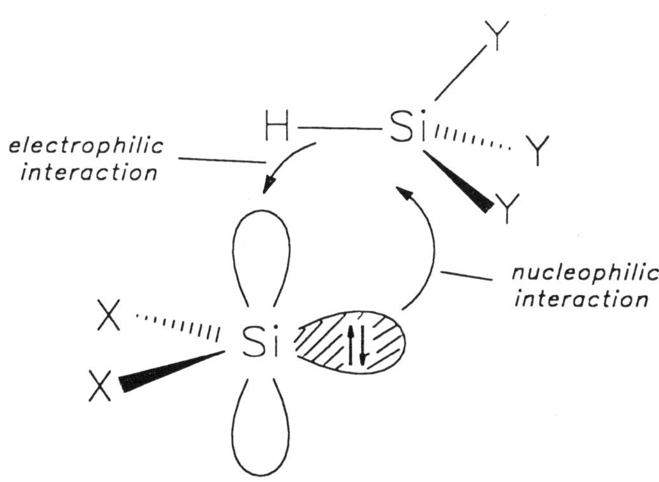

Figure 3. Interactions involved during the Si-H insertion process of silylenes

with their study. There are two main reasons for this. The first arises because, for many carbenes, the ground state is $^3B_1$ and even if $^1A_1$ can be prepared free from $^3B_1$, it is extremely difficult in kinetic studies of the $^1A_1$ state with reactive substrate molecules to separate the component of removal of $^1A_1$ by reaction from that due to deactivation (collisionally induced intersystem crossing) [29,30]. The second reason for difficulty arises with substituted alkyl carbenes. Alkyl substituents (such as methyl) are useful probes as modifiers of reactivity but bimolecular studies of alkylcarbenes are made difficult (if not impossible) by the very rapid intramolecular 1,2 H-shift process leading to alkenes [31], for example, dimethylcarbene to propene:

$$CMe_2 \rightarrow MeCH = CH_2$$

The analogous rearrangement for dimethylsilylene, *viz*:

$$SiMe_2 \rightarrow MeSiH = CH_2$$

has a substantial activation energy barrier [32,33] and is therefore very slow, thus permitting the study of $SiMe_2$ kinetics, untroubled by intramolecular reaction. This does not mean that a great deal of understanding of carbene reactions has not been achieved [34,35] but rather the potentially fruitful route of direct measurement of absolute rate constants for a variety of carbenes is considerably hindered.

A specific point that is worth mentioning is that of energy release. This may be illustrated by the comparison of the prototype reactions of CH_2 and SiH_2 as shown in Table 3.

Table 3
Energy release for prototype methylene and silylene reactions

Reaction	$-\Delta H^{\circ}$ / kJ mol^{-1}	Ref.
$^1CH_2 + CH_4 \rightarrow C_2H_6$	439	[36]
$^1CH_2 + C_2H_4 \rightarrow c\text{-}C_3H_6$	429	[36]
$^1CH_2 + H_2O \rightarrow CH_3OH$	390	[36]
$^1SiH_2 + SiH_4 \rightarrow Si_2H_6$	227	[21]
$^1SiH_2 + C_2H_4 \rightarrow c\text{-}SiC_2H_6$	201	[37]
$^1SiH_2 + H_2O \rightarrow H_3SiOH$	311	estd.

All reactions are highly exothermic, since each process involves the creation of two bonds at the expense of one. However the CH_2 reactions produce *ca* double the energy release of the SiH_2 reactions. This is because in the SiH_2 case the reactions involve giving up the DSSE of silylene. The practical outcome is that the products of an SiH_2 reaction are initially much less vibrationally excited than those of a CH_2 reaction.

One of the well known phenomena of CH_2 reactions is that of "chemical activation" in which initially formed products rearrange further to secondary products [38]. This characteristic is less evident in SiH_2 reactions although, as will be seen, the consequences of energy release manifest themselves in the different general phenomenon of the third-body stabilisation requirement for association processes. This is discussed in more detail in specific examples.

1.5 Advantages of the gas-phase

One advantage of the gas-phase is very straightforward *viz.* that the reactions can be studied free of the complicating effects of solvent. It is clear that species with the ambiphilic character of silylenes will be particularly prone to interaction with solvents. This may be illustrated by time-resolved studies of reactions of $SiMe_2$ with alcohols and ethers carried out by Levin *et al* [39]. The reactivity of $SiMe_2$ is modified in the presence of these O-donor species, as discussed in section 5, almost certainly due to complexation with the solvent. However in non-polar solvents, free of impurities, there is no reason why solution studies should not yield essentially similar results to the gas-phase. If differences occur they should be instructive. Moreover there are many silylenes which are difficult to generate in the gas-phase because they have involatile precursors. In these cases solution studies are the only possibility.

Another advantage of the gas-phase is that it allows us to probe the consequences of energy release. The pressure dependences associated with third-body assisted association reactions, permit us to gain information about the thermal stability of the products (as we shall see). This is not possible in solution.

Of course most real-life chemistry involves the use of solvents and therefore solution reactions unquestionably need detailed kinetic investigation in their own right. But the gas-phase serves as the vital reference point of the solvent-free system. Thus this review is for the most part limited to gas-phase data and solution studies are only discussed where they throw extra light on the mechanism in question.

2. REQUIREMENTS FOR DIRECT STUDY

2.1 General

Silylenes are among the most reactive transient species known and therefore their study via direct monitoring requires appropriate fast response techniques [40]. The most widely used method is that of laser flash photolysis with laser absorption detection used in our laboratories [41] and those of Jasinski [42] (so-called LRAFKS technique). Typically, the most reactive silylenes, like SiH_2 itself, when created in a transient high concentration pulse, decay away via reaction on microsecond timescales. This imposes two requirements on an experimental monitoring system (i) a rapid silylene creation source and (ii) a fast response detection system. The first is generally met by use of excimer (sometimes more correctly called exciplex) lasers with typically intense UV pulses (of energy *ca* 100 mJ/pulse and with *ca* 10 nsec linewidths) in combination with a suitable UV absorbing molecular silylene precursor. Conventional UV flash lamp sources can also be used for silylenes such as $SiCl_2$, which are less reactive [43]. Other methods for creating silylenes include the use of infrared multiphoton decomposition (IRMPD) [44], although this has not been much used in direct kinetic studies. The second requirement is met by use of optical techniques in combination with fast response electronic recording. Optical absorption can be used where the combination of (a) the silylene extinction coefficient (or oscillator strength), (b) the transient concentration and (c) the absorbing path length are sufficient to obtain good signals. This has been used for the silylenes, SiH_2, SiD_2, MeSiH, $SiMe_2$, PhSiH, $SiCl_2$, $SiBr_2$. The alternative is to use optical emission as in laser induced fluorescence (LIF). This is in principle more sensitive but is limited to those silylenes which emit (SiH_2, SiHCl, SiF_2). These points are expanded in the following sections. They are illustrated mainly by our own techniques and experience but the general principles apply to all such studies.

2.2 Silylene photochemical sources

The UV photochemistry of silicon containing molecules is not a well developed subject. However, it is known that both aromatic silanes and non-aromatic disilanes have pronounced UV absorptions. It turns out that these molecules do provide most of the known photochemical sources of silylenes. A list of these precursor species is shown in Table 4.

Table 4
Silylene photochemical sources

Silylene	Precursor	λ / nm	Ref.
SiH_2	$PhSiH_3$	193	[12] [42] [45]
SiH_2	SiH_3I	248	[42]
SiH_2	Si_2H_6 Si_3H_8	193	[46]
SiD_2	$PhSiD_3$	193	[47]
$SiMe_2$	PMDS[a]	193	[48]
MeSiH	DMDS[b]	193	[49]
PhSiH	$PhSiH_3$	193	[50]
ClSiH	$ClSiH_3$ Cl_2SiH_2	193	[51] [52]
$SiCl_2$	Si_2Cl_6	UV[c]	[43]
$SiBr_2$	$SiBr_4$	UV[c]	[53]

a. Pentamethyldisilane.   b. 1,2-Dimethyldisilane.   c. Broad band.

There is almost certainly scope for the development of further molecular sources. One of the problems with existing sources is the occurrence of side reactions. With the more hydrogenated disilanes, such as $MeSiH_2SiH_2Me$ dust is formed simultaneously with MeSiH [49]. This can be minimised with sufficiently low precursor pressures. With $PhSiH_3$ both SiH_2 and PhSiH are formed simultaneously [54] but since these absorb in different places, they do not interfere with one another. A good quantum yield of silylene is important since it is advantageous to keep the precursor pressure as low as possible in order to minimise kinetic losses of silylene through reaction with the precursor itself.

A particular problem with current sources is the use of 193nm radiation. Many molecules of potential interest as reactive substrates for the silylenes generated, themselves absorb and undergo photochemistry at this wavelength. This makes them difficult or impossible to study at present.

2.3 Detection and monitoring

Silylenes possess strong absorptions in the visible and UV regions arising from electronic transitions of the type $^1B_1 \leftarrow {}^1A_1$. The spectroscopy of silylenes has been studied by both gaseous flash photolysis methods and in low temperature matrices. The first silylene spectrum to be characterised was that of SiH_2 itself observed in the flash photolysis of $PhSiH_3$ [55]. Subsequently, much improved spectra were obtained by Dubois, Herzberg and Verma [56] in the 480-650 nm region, from silylene produced by the flash discharge of silane in excess hydrogen. The spectrum consisted of seven bands, with the strongest bands at 610.0, 579.6 and 552.6 nm. A detailed rotational analysis of these bands was carried out by Dubois [17]. Jasinski and Chu [57] have recorded the most highly resolved spectrum to date using laser absorption flash

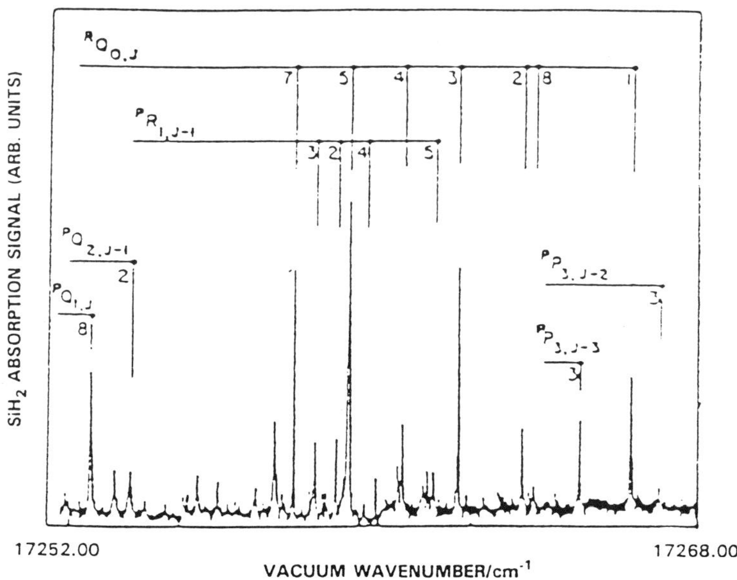

Figure 4. A portion of the visible rovibronic absorption spectrum of SiH_2 [57].

kinetic spectroscopy. They scanned the spectral region from 17242 to 17352 cm^{-1}, around the origin of the $(0,2,0) \leftarrow (0,0,0)$ vibronic band of the $^1B_1 \leftarrow {}^1A_1$ transition. Their spectrum consisted of approximately 100 single rotational lines showing that it was considerably richer than previously believed. About half the lines were readily assigned using known SiH_2 transitions [17]. A portion of the spectrum is shown in Figure 4 including the intense $^RQ_{0,5}$ line which has been used by Jasinski [42] and ourselves [45] to monitor SiH_2.

The narrow line widths of ca 0.05 cm^{-1} (FWHM) of this highly resolved (small molecule) spectrum means that a narrow line optical probe is required for its monitoring. This is provided by a single-mode (ring) dye laser which can be tuned to the appropriate wavenumber (17259.50 cm^{-1}). For SiD_2 an assigned high resolution rovibronic spectrum is not available but from a point-by-point scan, Mason, Frey and Walsh [47] have obtained a spectrum suitable for kinetic studies using laser flash photolysis of $PhSiD_3$.

Other silylenes have broader band spectra many of which have been recorded by matrix isolation and documented by West $et\ al$ [58]. An example of a spectrum of this type is that for MeSiH shown in Figure 5. The gaseous spectrum is typically slightly red-shifted relative to that of the matrix. Table 5 shows a list of absorbing regions for silylenes which have been investigated in the gas phase. There may be an error in the published PhSiH spectrum [50]. This is discussed in section 3. These species place less demand on the optical probe system, and an argon ion laser has proven to be a convenient absorption

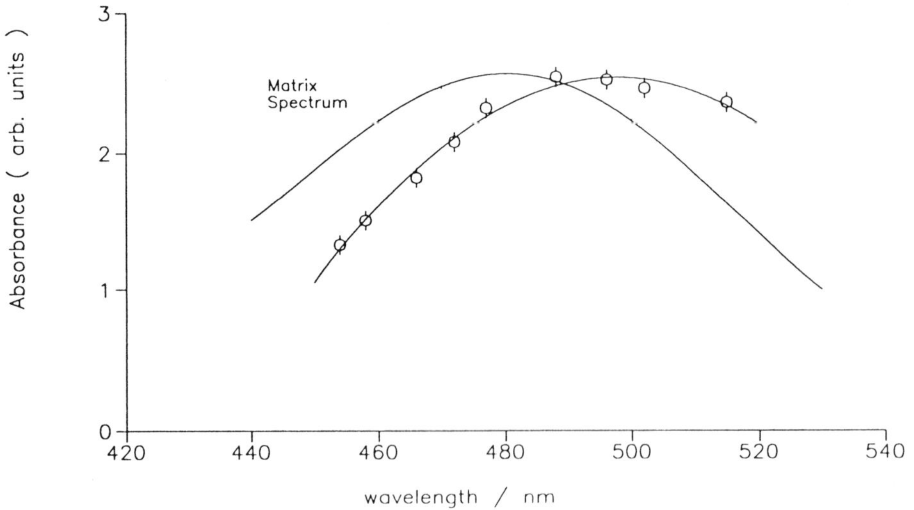

Figure 5. The visible absorption spectrum of MeSiH

monitor. Several silylenes (MeSiH, $SiMe_2$, PhSiH) possess absorptions conveniently within the range (450-515 nm) of the several lines available for study. Another convenient feature of many of the silylene absorption spectra is their occurrence in the visible region, which is uncluttered by absorptions of many of the desirable reactant molecules. The use of lasers as detectors offers the benefit of increasing the absorption path length by means of multipassing the beam (using Whites optics) which increases path lengths in practice to well over 1 metre [48]. An important experimental check, is to verify that the kinetics are independent of the absorption wavelength.

Table 5
Approximate absorption (monitoring) wavelengths for silylenes ($^1B_1 \leftarrow ^1A_1$ band)

Silylene	λ / nm	Ref
SiH_2	480-650	[42]
	(579.39)	[45]
SiD_2	480-650	[47]
	(575.14)	
$SiMe_2$	420-500	[48]
$MeSiH$	430-520	[49]
$PhSiH$	460-520	[50]
$SiCl_2$	308-328	[43]
$SiBr_2$	340-400	[53]

A schematic diagram of the laser flash photolysis apparatus and experimental setup for such experiments in our laboratories is shown in Figure 6. With modern fast response photodiodes and electronic signal detection, amplification and storage are relatively straightforward. RF noise from the excimer laser pulse can be a problem but with suitable screening of the electronics can be kept to a minimum. The raw experimental result consists of the time evolution of the transient silylene absorption over a suitable time scale (usually microseconds). Signal-to-noise is improved by the accumulation of the traces from several photolysis runs (excimer shots) using a transient digitizer. It is important to check that such experimental repetition does not seriously perturb the gaseous reaction mixtures during a series of photolysis shots. To avoid reactant depletion or product interference gaseous flowing systems can be employed. However if materials are costly or of limited availability a static reaction vessel can be used provided the irradiated volume is a relatively small proportion of the total (< 5%). In this way, with shot repetition rates of *ca* 1Hz, there is plenty of time for replenishment of the reaction zone by diffusion between photolysis laser pulses.

Under typical operating conditions, excimer laser pulses of *ca* 100 mJ can create transient concentrations in the reaction zone of *ca* 0.1-1 mTorr (10^{13} molecule cm^{-3}) which give detectable absorbances of *ca* 0.05. Precise

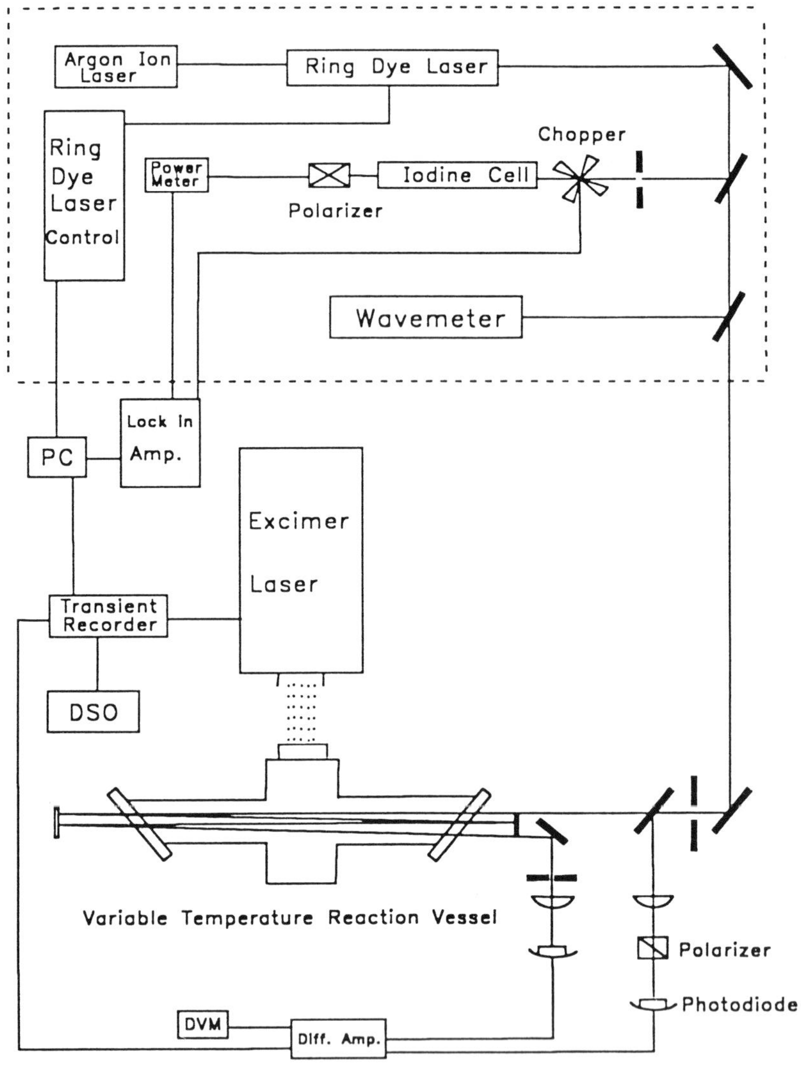

Figure 6. Schematic diagram of the laser flash photolysis apparatus used to study kinetics of SiH_2 reactions [45]

knowledge of the silylene concentrations are not usually required since the decay kinetics are almost invariably quasi-first order.

Gaseous reaction mixtures usually consist of a few mTorr of precursor (sufficient to give a silylene signal), an appropriate pressure of reactive substrate (which can be from a few mTorr up to several Torr, depending on reactivity) and an inert diluent gas. The diluent serves two purposes. The first function is to thermalize the products of laser photolysis so that the silylene is present with a (relaxed) Boltzmann energy distribution. Usually 5 Torr (and often less) are sufficient for this purpose. Polyatomic bath gases, such as SF_6 which does not interfere with the chemistry, are best for this. Secondly, the inert diluent serves as a pressure control for reaction when the overall kinetics are pressure dependent. In practice the maximum pressure has turned out to be usually *ca* 100 Torr for many silylene studies. Signals appear to be significantly quenched at higher pressures (although the reasons are not understood). At higher pressures care has to be taken to ensure complete mixing of reagents.

2.4 Data processing

The extremely high reactivities of silylenes mean that even in the presence of only a few mTorr of substrates, silylene decay is dominated by bimolecular reaction with the substrate, *viz:*

$$SiX_2 + substrate \rightarrow product$$

Since under operating conditions, substrate concentrations are always greater than those of the silylene, SiX_2, this means that the SiX_2 decay kinetics are quasi-first order with a rate constant, k_{obs}, linearly dependent on [substrate]. The true second order rate constant is obtainable from studies of this dependence. This is standard kinetic treatment, easily programable with microcomputers with standard error packages.

Occasionally (for example in the reactions of $SiMe_2$ with ethers [59]) more complex kinetics arise and this can be handled through alternative computer fitting routines or numerical integration packages, for a hypothesised mechanistic model.

For further experimental details of silylene studies, readers are referred to individual papers.

3. THE INSERTION PROCESS

It is often stated that silylenes characteristically insert into Si–H, Si–OR and O–H bonds. While this is formally correct, in fact it is believed that the reactions with Si–OR and O–H bonds proceed *via* initial silylene co-ordination to oxygen. Therefore these latter reactions are treated separately (in section 5). This section, therefore, is devoted largely to the Si–H insertion process, although the important reaction with H_2 and its isotopic variants is also included.

3.1 SiH$_2$ insertion reactions

<u>(a) SiH$_2$ + H$_2$/D$_2$/HD (and SiD$_2$ + H$_2$/D$_2$)</u>

The first absolute rate constant for the reaction of SiH$_2$ with H$_2$ was reported by Inoue and Suzuki [12] using the laser-induced fluorescence method (LIF) to monitor the concentration of SiH$_2$. At the total pressure of 1.8 Torr (He) they found a rate constant of (1.0 ± 0.4) x 10^{-13}cm^{3}molec^{-1}s^{-1}. Nearly at the same time Jasinski [60] measured the rate constant for SiH$_2$ with D$_2$ using the laser resonance absorption flash kinetic spectroscopy (LRAFKS) technique. The rate constant obtained at 2-3 Torr total pressure (He) was (2.7 ± 0.7) x 10^{-12}cm^3 molec^{-1}s^{-1}. This value is much higher than that obtained by Inoue and Suzuki for reaction with H$_2$.

Jasinski reasoned that his rate constant more nearly represented the true rate of insertion because only one of the six possible channels for decomposition of the vibrationally highly excited adduct, SiH$_2$D$_2$*, leads back to SiH$_2$, whereas SiH$_4$*, formed from SiH$_2$+H$_2$ will revert to reactants unless collisionally stabilised thus making it a much more pressure dependent reaction.

The immediate result of these measurements, however, was to dispel the view that the reaction of SiH$_2$ + H$_2$ had a small but positive activation energy variously estimated at 23.1 kJ mol^{-1} [61] and 43 kJ mol^{-1} [62]. These values were derived from relative rate measurements which were based (in hindsight) on unreliable assumptions. If correct the room temperature rate constant for SiH$_2$ + H$_2$ would have had to be at least four orders of magnitude lower than observed. Interestingly at this time theoretical calculations showed a drop in barrier from *ca* 30 kJ mol^{-1} [27,63-65] to 7 kJ mol^{-1} [66]. The magnitude of the barrier was measured, using the same technique as Jasinski, by Baggott *et al* [67] who obtained a temperature independent rate constant of (1.8 ± 0.2) x 10^{-12} cm^3 molec^{-1}s^{-1} for SiH$_2$+D$_2$ over the range 268-330 K (at 5 Torr total pressure in SF$_6$). This corresponds to zero activation energy and the authors estimated a maximum possible value of 1.3 kJ mol^{-1}. More recently Al-Rubaiey and Walsh [68] have extended the temperature range for this reaction up to 553 K and now found a negative activation energy of -2.03 kJ mol^{-1}.

Jasinski and Chu [57] measured absolute rate constants for the SiH$_2$ + H$_2$ reaction in the 1-100 Torr pressure range (He). The pressure dependence of the rate constants for this reaction was reproduced by RRKM calculations and the high pressure limiting value derived of k^∞ = 3.2 x 10^{-12} cm^{-3}molec^{-1}s^{-1}, in good agreement with the value obtained from the SiH$_2$ + D$_2$ experiments. The rate constant measured at 2 Torr total pressure was 2.7x10^{-13}cm^{3}molec^{-1}s^{-1}, in rough agreement with the value obtained at 1.8 Torr by Inoue and Suzuki [12]. These studies also confirmed that the reactions of SiH$_2$ with HD and D$_2$ show much less pressure dependence and the isotope effect appears to be small [57].

Table 6 summarises these results and also includes a recent value for the rate constant for SiD$_2$+H$_2$, which indicates that the isotope effects, although small are certainly not zero [47]. Just as for the SiH$_2$+D$_2$ system, this reaction is almost pressure independent and should represent the true rate of insertion (Jasinski's argument).

The study of SiD$_2$ + H$_2$ has been extended up to higher temperatures [69]. The Arrhenius parameters for these two processes are shown in Table 7 along with those obtained in a comprehensive (unpublished) study of the (pressure dependent) $SiH_2 + H_2$ and $SiD_2 + D_2$ reactions [68].

These most recent investigations [68,69] extend the previous studies, but are consistent with them. They include transition state theory calculations of isotope effects which are beyond the scope of this review, but which provide a distinction between two published transition state structures for the $SiH_2 + H_2$ reaction. These are shown in the Figure 7.

Table 6
Rate constants for $SiH_2 + H_2$ and isotopic variants at room temperature

Reaction	$k/10^{-12}$ cm^{3}molec^{-1}s^{-1}	Ref.
$SiH_2 + H_2$	3.2[a]	[57]
$SiH_2 + D_2$	2.6 ± 0.7	[60]
	1.9 ± 0.7	[67]
$SiH_2 + HD$	2.0 ± 0.4	[57]
$SiD_2 + H_2$	3.8 ± 0.1	[47]

a. Infinite pressure value.

Table 7
Arrhenius parameters for $SiH_2 + H_2$ and isotopic variants

Reaction	T range/ K	$\log(A/$cm^{3}molec^{-1}s$^{-1})$	$E_a/$kJ mol^{-1}
$SiH_2 + D_2$	300-553	-12.07	-2.03
$SiD_2 + H_2$	294-528	-11.87	-2.54
$SiH_2 + H_2$	300-513	-12.01[a]	-2.34[a]
		-11.75[b]	-1.84[b]
$SiD_2 + D_2$	298-498	-12.35[a]	-3.14[a]
		-12.09[b]	-3.52[b]

a. Infinite pressure values obtained from extrapolated rate constants.
b. Values obtained *via* transition state theory of isotope effects.

The isotope effects support the *ab initio* structure (C_s symmetry) of Gordon *et al* [66] rather than the semi empirical one (C_{2v} symmetry) of Roenijk *et al* [70]. However RRKM calculations based on the *ab initio* structure do not reproduce the measured pressure dependence of the rate constants very well [68]. Nevertheless the *ab initio* structure seems more likely since it represents a geometry of approach of H_2 to SiH_2 in which the electrophilic interaction between the H_2 bonding electrons and empty *p* orbital of SiH_2 is favoured thus strongly suggesting an electrophilically-led reaction.

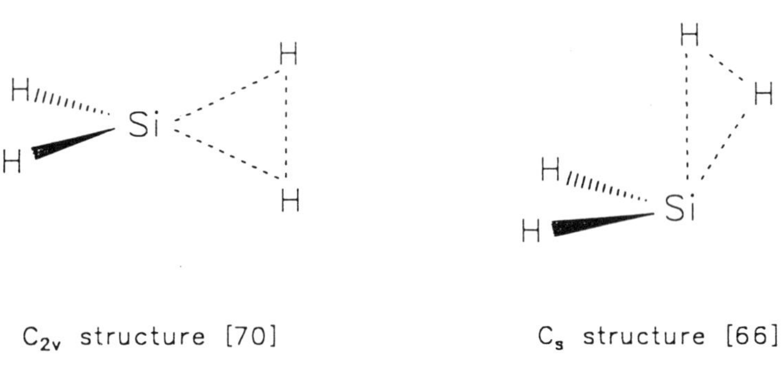

C_{2v} structure [70] C_s structure [66]

Figure 7. Transition state structures for reaction of SiH_2 + H_2

(b) SiH_2 + SiH_4

Insertion of SiH_2 into the Si–H bond is one of the key processes in chemical vapour deposition. Insertion into the Si–H bonds of SiH_4 is the most fundamental of these processes. Once again direct measurements have supplanted the older, much lower estimates of the rate constant [61,62]. Inoue and Suzuki [12] measured the absolute rate constant, using the LIF technique, for this reaction at 1 Torr total pressure (He). At 298 K they found a value of (1.1 ± 0.2) x $10^{-10} cm^3$ molec^{-1}s^{-1}. Jasinski and Chu [57], using the LRAFKS technique, determined the absolute rate constant to be (6.7 ± 0.7) x 10^{-11} cm^3 molec^{-1}s^{-1} at 1 Torr (He) in reasonable agreement. These authors also studied the effect of pressure up to 95 Torr and performed RRKM calculations to reproduce the observed dependence. This yielded the high pressure limiting rate constant, k^{∞} = 2.2 x 10^{-10} cm^3 molec^{-1}s^{-1}.

Using the same technique Baggott *et al* [71] measured absolute rate constants for this reaction and obtained values *ca* 1.7 times higher over the same pressure range (He bath gas), thus leading to a value for k^{∞} = 4.0 x 10^{-10} cm^3 molec^{-1}s^{-1}. The cause of this discrepancy is not obvious.

Another study by Dietrich *et al* [72] using IRMPD to generate SiH_2 from SiH_4 yields pressure dependent rate constants leading to k^{∞} = (1.1 ± 0.2) x 10^{-10} cm^3 molec^{-1}s^{-1}. However these experiments were carried out under solid deposition conditions and it is not clear that this corresponds to comparable (298 K) temperatures. A more complete temperature and pressure study has been carried out by Becerra *et al* [73], yielding the rate constants a selection of which are shown in Table 8.

Table 8
Rate constants for SiH_2 + SiH_4 at two pressures and six temperatures

T/K	k / 10^{-10} cm^{3}molec^{-1}s^{-1}	
	P = 10 Torr	P = ∞[a]
296	4.1 ± 0.3	4.6 ± 0.3
363	3.2 ± 0.3	3.9 ± 0.3
432	2.3 ± 0.3	3.2 ± 0.3
488	1.6 ± 0.3	2.6 ± 0.3
578	1.2 ± 0.4	2.6 ± 0.4
658	0.8 ± 0.4	2.2 ± 0.4

a. Values obtained by extrapolation

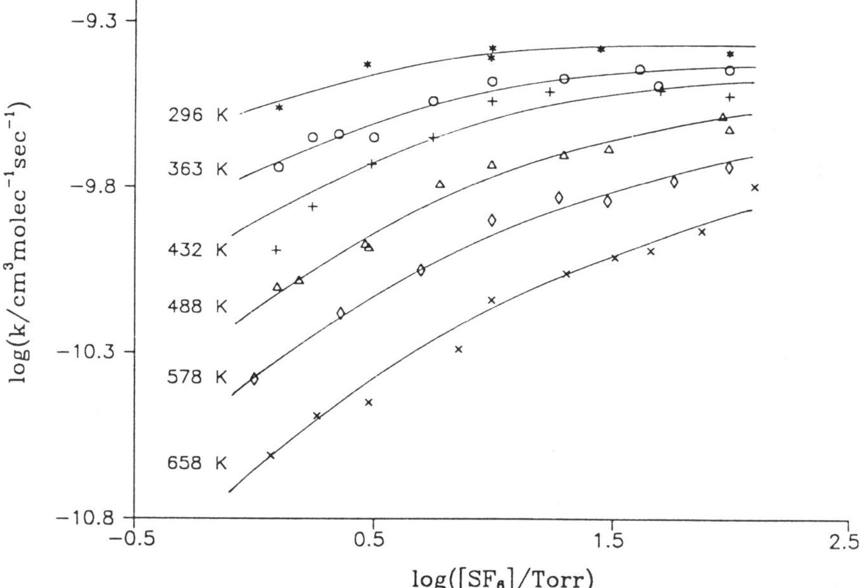

Figure 8. Pressure and temperature dependence of rate constants for SiH_2 + SiH_4

In this table, k^∞ values were obtained by fitting RRKM calculated fall-off curves as shown in Figure 8. The modelling of these data used a variational transition state. The 298 K value of k^∞ confirms the earlier measurements in our laboratory and points to the high collisional efficiency of this reaction. Arrhenius treatment of the temperature dependence of k^∞ leads to $A^\infty = 1.3 \times 10^{-10}$ cm^3 molec^{-1} s^{-1} and $E_a = -3.0$ kJ mol^{-1} although the plot is slightly curved. The negative activation energy is consistent with the idea of a weak intermediate complex on the approach potential for this reaction. *Ab initio* calculations [73] support the presence of such a complex at *ca* 52 kJ mol^{-1} below the reaction threshold although the barrier to rearrangement of this complex to Si_2H_6 is only 6.5 kJ mol^{-1}. The structure of the complex is shown in Figure 9.

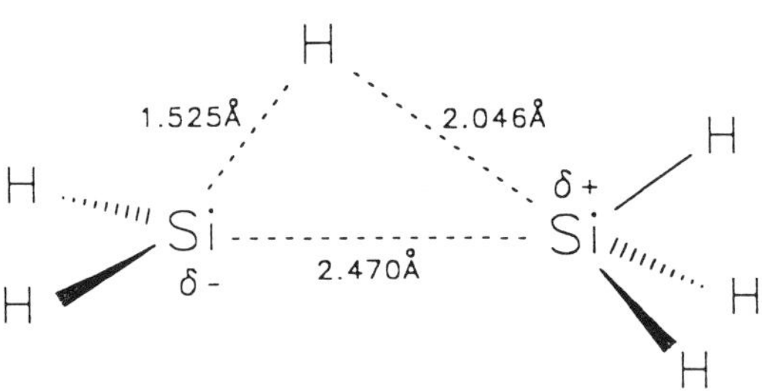

Figure 9. Structure of the intermediate complex in the reaction of SiH_2 + SiH_4

The geometry of the complex suggests substantial transfer of H from SiH_4 to SiH_2 again supporting the idea of an electrophilically led insertion process.

Although thermochemical considerations are not the focus of this review, the rate constants of this reaction system and its reverse [74] are the basis of the currently most precise value for $\Delta H_f^o(SiH_2) = 273 \pm 2$ kJ mol^{-1} [75].

<u>(c) SiH_2 + Si_2H_6</u>

Once again direct measurements have supplanted older values [61,62]. Absolute rate constants have been obtained in the same laboratories as for SiH_2 + SiH_4 [12,57,71,75]. Values are shown in Table 9.

The value of Dietrich *et al* [72] (2.7±0.4) x 10^{-10} cm^3 molec^{-1} s^{-1}) was obtained at an uncertain temperature and is not included. The main discrepancy is between Jasinski and Chu [57] and the most recent values found in our laboratories [75]. Additionally we have found a pressure independent reaction between 3 and 30 Torr in SF_6 and only a slight dependence on bath gas (Ar, C_3H_8, SF_6). Inoue and Suzuki's value [12] is in line with this. Jasinski and Chu measured a rate constant variation of more than a factor of 2 between 1 and 9.5 Torr (He). Our feeling is that there is some error here because the energised $Si_3H_8^*$ product has an alternative pathway to decomposition which is certainly predominant at higher temperatures [76] *viz:*

$$SiH_2 + Si_2H_6 \rightleftarrows Si_3H_8^* \rightarrow SiH_3SiH + SiH_4$$

This means that pressure dependence in this system should be relatively unimportant.

Table 9
Comparison of room temperature rate constants for $SiH_2 + Si_2H_6$

$k/10^{-10}$ cm^{3}molec^{-1}s^{-1}	P / Torr	Reference
5.7 ± 0.2	1 (He)	[12]
1.5 ± 0.2	1 (He)	[57]
2.8 ± 0.3	5 (He)	[57]
3.4 ± 0.3	9.5 (He)	[57]
4.6 ± 0.7	5 (Ar)	[71]
5.8 ± 0.6	10 (Ar)	[75]
6.6 ± 0.3	10 (SF$_6$)	[75]
6.4 ± 0.2	10 (C$_3$H$_8$)	[75]

The temperature dependence (295-590 K) of this reaction, studied by Becerra *et al* [75], gives the average Arrhenius parameters A = 3.1 x 10^{-10} cm^3 molec^{-1} s^{-1}, E_a = -1.9 kJ mol^{-1}, although the Arrhenius plot is slightly curved. The negative temperature dependence and high collisional efficiency of this reaction suggest that similar considerations determine its mechanism as for $SiH_2 + SiH_4$.

<u>(d) $SiH_2 + Si_3H_8$</u>
No published kinetic data exists for this reaction. Its kinetics have been investigated by Becerra *et al* [75], in an unpublished study, for a temperature range of 297-595 K in C_3H_8 and Ar buffer gases, and at six temperatures over the range 295 to 578 K in SF_6. Si_3H_8 was in this case both the precursor of SiH_2 and the substrate for the reaction. The reaction was tested for pressure dependence over the range 1 - 30 Torr. None was found.

Rate constants at room temperature were (7.8 ± 0.9) x 10^{-10} cm^3 molec^{-1}s^{-1} in SF_6 and (8.0 ± 0.2) x 10^{-10} cm^3 molec^{-1} s^{-1} in C_3H_8. In Ar, values were somewhat lower (between 50 and 80% of those in SF_6) over the temperature range. We suspect an undetected experimental error in the Ar experiments. The increased molecular complexity of this reaction (relative to $SiH_2 + SiH_4$ and $SiH_2 + Si_2H_6$) combined with the probability of alternative decomposition pathways from the energised product (*cf* SiH_2 + Si_2H_6) explain the lack of a pressure dependence in this reaction. The temperature dependence gives rise to the average Arrhenius parameters, A = 3.7 x 10^{-10} cm^3 $molec^{-1}$ s^{-1} and E_a = -2.0 kJ mol^{-1}, although the Arrhenius plot is slightly curved. These curved Arrhenius plots for the reactions of SiH_2 with SiH_4, Si_2H_6 and Si_3H_8 are shown in Figure 10. The magnitude of the rate constants and the almost statistical relationship between them (Table 10) attests to the high collisional efficiency of these reactions, and the looseness of the activated complexes.

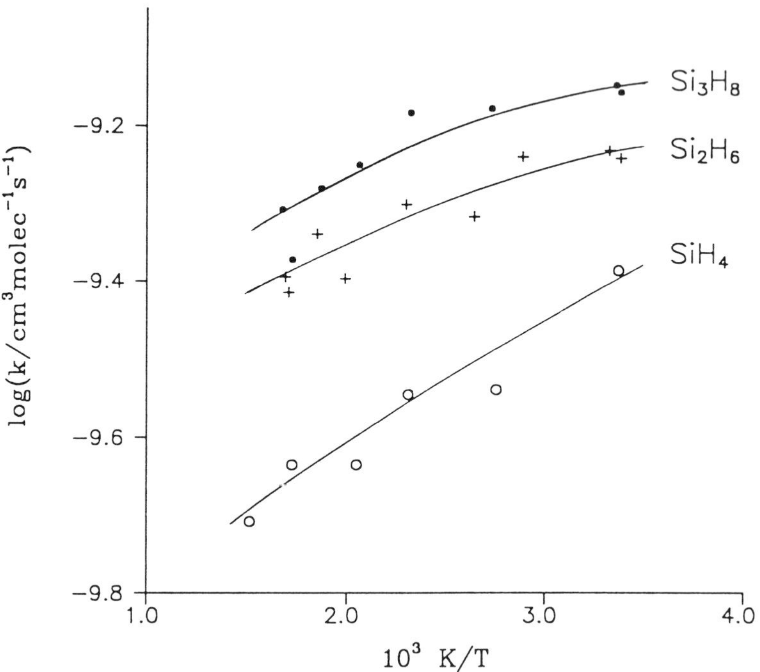

Figure 10. Arrhenius plot for reactions of SiH_2 with silanes.

Table 10
Comparison of rate constants for SiH_2 insertion (per Si–H bond)

Substrate	$k/10^{-10}$ cm^{3}molec^{-1}s^{-1}	
	$T \approx 298$ K	$T \approx 588$ K
SiH_4	1.15	0.59
Si_2H_6	1.08	0.72
Si_3H_8	0.99	0.72

(e) SiH_2 + Me_nSiH_{4-n} (n = 1-4)

Baggott *et al* [71] measured the first rate constants for reaction of SiH_2 with the methylsilanes ($MeSiH_3$, Me_2SiH_2, Me_3SiH and Me_4Si). The study was carried out at room temperature and the results are shown in Table 11.

Table 11
Rate constants for SiH_2 reactions with the methylsilanes

Species	$k/10^{-10}$ cm^{3}molec^{-1}s^{-1}
$MeSiH_3$	3.7 ± 0.2
Me_2SiH_2	3.3 ± 0.3
Me_3SiH	2.5 ± 0.1
Me_4Si	≤ 0.0027

No pressure dependence was found for SiH_2 + $MeSiH_3$ and it was assumed therefore that the other reactions were pressure independent.

Carpenter and Walsh [77] have extended these studies for $MeSiH_3$, Me_2SiH_2 and Me_3SiH over a temperature range of 295-625 K and a pressure range of 3 to 100 Torr (SF_6). No pressure dependencies were found except for SiH_2 + $MeSiH_3$ at 600 K. The measured Arrhenius parameters can be seen in Table 12. The mechanism of these reactions is discussed later. The study of SiH_2 + Me_4Si has been repeated by Becerra and Walsh [78] giving an (apparent) rate constant of $(1.7 \pm 0.2) \times 10^{-13}$ cm^3 molec^{-1} s^{-1}. The rate constant value was shown to reduce with further purification of the already 99.9% pure Me_4Si. However even 0.1% of a remaining reactive impurity would still account for this rate constant. There are reasons for believing that the reaction of SiH_2 with Me_4Si is very much slower (see below).

Table 12
Arrhenius parameters for SiH_2 reactions with the methylsilanes

Species	$\log(A/cm^3 molec^{-1}s^{-1})$	$E_a/kJ\ mol^{-1}$
$MeSiH_3$	-9.82	-2.39
Me_2SiH_2	-10.00	-3.12
Me_3SiH	-10.14	-3.20

<u>(f) SiH_2 + alkanes</u>

Inoue and Suzuki [12] reported the absolute rate constant, $k = (1.0 \pm 0.3)$ x 10^{-12} cm^3 $molec^{-1}s^{-1}$ for the reaction of SiH_2 with CH_4 measured by the LIF technique at 1 Torr total pressure. Chu *et al* [42], using the LRAFKS technique, determined upper limits for the rate constants of the reactions with CH_4 and C_2H_6 at 5 Torr total pressure. They found substantially lower values of (2.5 ± 0.5) x 10^{-14} and (1.2 ± 0.5) x 10^{-14} cm^3 $molec^{-1}s^{-1}$ respectively. The authors explained that a 0.01% impurity in the methane or ethane, reacting on every collision with SiH_2, could account for all of the observed silylene removal. This interpretation is also consistent with the observation that ethane appears to react slightly slower even though more CH bonds are available. Eley *et al* [79] found no observable reaction of silylene with methane in their competitive rate study. These results are consistent with previous indirect evidence [80] and *ab initio* calculations [27] all of which conclude that there is a substantial energy barrier to silylene insertion into C–H bonds in saturated hydrocarbons. In order to test the reactivity with strained carbon-carbon bonds Chu *et al* [42] examined the reaction of silylene with cyclopropane. The attempt was unsuccessful in the sense that all observed reactivity could be accounted for by the $\approx$ 1% propylene impurity in the commercially available sample of cyclopropane.

3.2 Insertion reactions of substituted silylenes

This section begins with dimethylsilylene, $SiMe_2$. Time-resolved experimental studies on $SiMe_2$ in the gas-phase began in 1987 with the work of Baggott *et al* [48]. At the same time Gaspar *et al* [15] published their account of solution studies of the direct kinetics of silylenes, including $SiMe_2$. Prior to this there had been a good deal of controversy about the absorption spectrum of $SiMe_2$ a summary of which is given in Apeloig's article [16]. This was resolved with agreement that the absorption spectrum of $SiMe_2$ possesses a broad maximum at 450 nm [81]. The gas-phase spectrum is similar [48] but with a small red shift. The solution spectrum obtained by Gaspar *et al* [15] was also similar but it appears probable that it was (at least in part) due to another species since subsequent experiments by Shizuka *et al* [82] in solution using a different source of $SiMe_2$ give significantly different values for some of the rate constants. These later values were more consistent with those from the gas-phase, described below. As before absolute measurements replaced earlier estimates [83], which were much closer for $SiMe_2$ than for SiH_2.

(a) $\underline{SiMe_2 + Me_nSiH_{4-n}\ (n = 0\text{-}4)}$

The results obtained by Baggott *et al* [48] on this reaction system are shown in Table 13. These were landmark results; the first gas-phase directly measured rate constants for an alkyl-substituted silylene.

Table 13
Rate constants for $SiMe_2$ reactions with silane and the methylsilanes

Substrate	$k/10^{-13}\ cm^3molec^{-1}s^{-1}$
SiH_4	2.0 ± 0.3
$MeSiH_3$	19 ± 2
Me_2SiH_2	55 ± 5
Me_3SiH	45 ± 5
Me_4Si	< 0.5

They showed (i) the significant activating effect of methyl substituents in the substrate silane, and (ii) the significant reduction in silylene reactivity, resulting from methyl substitution (by comparison of the rate constants with those for SiH_2 [12,57], subsequently confirmed in similar experiments [71,73]). They reconfirmed the extremely low reactivity of silylenes for both C–H and Si–C bonds, from the essential lack of reaction with Me_4Si. Product studies confirmed the reactions were Si–H bond insertions. Amongst alternative explanations for the methyl substituent effect was that of changes in Si–H bond polarity with increasing methyl substitution. This is discussed later. Further studies by Baggott *et al* [84] over the temperature range 295-608 K established the Arrhenius parameters shown in Table 14.

Table 14
Arrhenius parameters for $SiMe_2$ insertion reactions

Substrate	$\log(A/cm^3molec^{-1}s^{-1})$	$E_a/kJ\ mol^{-1}$
SiH_4	-12.54 ± 0.16	$+1.25 \pm 1.23$
$MeSiH_3$	-12.90 ± 0.09	-6.73 ± 0.68
Me_2SiH_2	-13.15 ± 0.05	-10.55 ± 0.38
Me_3SiH	-13.41 ± 0.14	-11.20 ± 1.03

These results showed that the methyl substituent effects in the substrate, persist from room temperature to higher temperatures leading to increasingly negative activation energies with increased methyl substitution. These studies

288

reinforced an emerging characteristic of silylene kinetics, *viz.* that of *decreasing* rates with *increasing* temperature. This is seen very clearly in the graphical representation of these results in Figure 11. The magnitudes of the rate constants show that these $SiMe_2$ reactions (unlike those of SiH_2) do not occur close to the collision rate. The explanation for this and the mechanism of these reactions is discussed later. The kinetic results for $SiMe_2$ insertion may be combined with those for reverse decomposition of the (product) disilanes to give ΔH_f^o ($SiMe_2$). This is another story of twists and turns, with the key items of uncertainty having been, until recently, the enthalpies of formation of the methylated disilanes. This appears to have been settled by a recent solution calorimetric study by Pilcher *et al* [23] leading to $\Delta H_f^o(SiMe_2)$ = 140 ± 6 kJ mol^{-1} in reasonable agreement with other values [85]. This corresponds to a DSSE value of 121 ± 12 kJ mol^{-1} (see Table 2). Readers are referred to original papers [39,82] for discussion of the solution phase studies of $SiMe_2$.

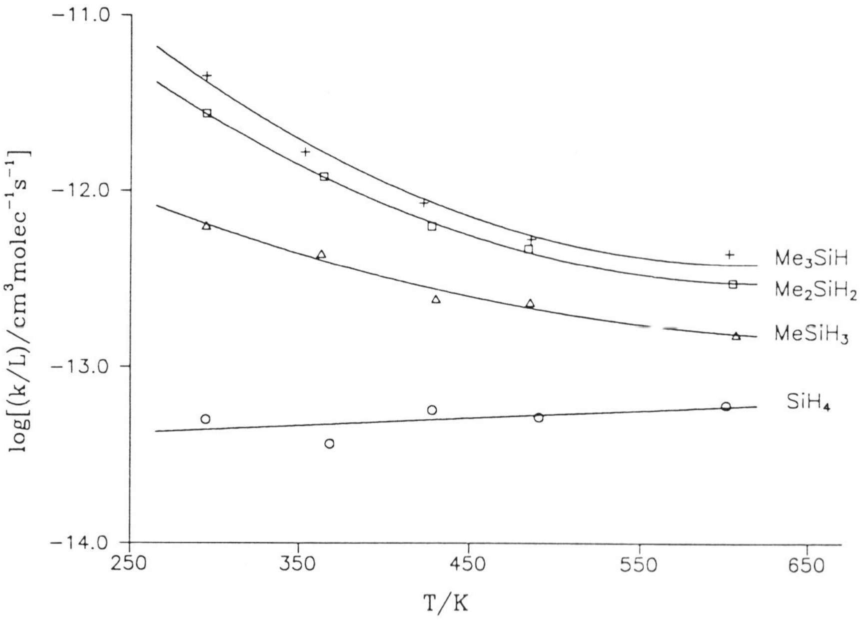

Figure 11. Temperature dependence of rate constants (per Si-H bond) for reactions of $SiMe_2$ with methylsilanes.

(b) MeSiH + Me$_n$SiH$_{4-n}$ (n = 0-3)

The first, and at the present time only, direct kinetic study of MeSiH has been carried out by Becerra *et al* [49]. Following after earlier studies with SiH_2

[71] and $SiMe_2$ [48] this was limited to SiH_4, $MeSiH_3$, Me_2SiH_2 and Me_3SiH, since Me_4Si appeared unreactive to the other silylenes. Products were confirmed by gc analysis, and reactions checked for the absence of total pressure effects (in both Ar and SF_6). These studies were more difficult than those of SiH_2 and $SiMe_2$, because of the combination of low sensitivity of detection for MeSiH, combined with its high reactivity. This gives rise to more scatter in the individual rate constants and consequently a more erratic temperature dependence, as shown in the Arrhenius parameters of Table 15. Because of the scatter the authors also give adjusted E_a values based on an assumed A factor (per Si–H bond) of $10^{-12.4}$ cm^3 $molec^{-1}s^{-1}$. These values are shown in parentheses.

Table 15
Arrhenius parameters for MeSiH insertion reactions

Substrate	$\log(A/cm^3molec^{-1}s^{-1})$	$E_a/kJ\ mol^{-1}$
SiH_4	-11.4 ± 0.2 (-11.8)	-7.5 ± 1.7 (-10.9)
$MeSiH_3$	-12.1 ± 0.3 (-11.9)	-14.5 ± 2.7 (-12.5)
Me_2SiH_2	-11.8 ± 0.2 (-12.1)	-11.5 ± 2.1 (-14.3)
Me_3SiH	-12.8 ± 0.1 (-12.4)	-18.4 ± 0.9 (-14.9)

The trends with substrate methyl substitution are, however, the same as for the insertion reactions of $SiMe_2$ [84]. Negative activation energies are again found in these reactions. The rate constants are much closer in magnitude to those of SiH_2 than to those of $SiMe_2$. The mechanism of these reactions is discussed later.

The kinetic results for MeSiH insertion have again been combined with those of the reverse (decomposition) reactions by Becerra *et al* [49] to give ΔH_f^o(MeSiH) = 202 ± 6 kJ mol^{-1} in agreement with other values [85].

(c) PhSiH + Me_nSiH_{4-n} (n = 0-3)
The first, and again only, study of PhSiH has been carried out by Blitz *et al* [50]. The general findings are very similar to those for MeSiH. PhSiH is close in reactivity to MeSiH. It is worth mentioning that the original spectrum attributed to PhSiH between 540 and 600 nm is probably due to SiH_2 (the precursor or source, $PhSiH_3$, is common to both). However since PhSiH was monitored at 515 nm (where SiH_2 does not absorb) for the time-resolved studies the reported kinetics are not affected. The results in terms of Arrhenius parameters are given in Table 16.

Because of the scatter the authors also give adjusted E_a values based on an assumed A factor (per Si–H bond) of $10^{-12.0}cm^3$ $molec^{-1}$ s^{-1}. These values are shown in parentheses. The trends are similar to those for MeSiH [49]. The mechanism of these reactions is discussed later.

Table 16
Arrhenius parameters for PhSiH insertion reactions

Substrate	$\log(A/\mathrm{cm}^3\mathrm{molec}^{-1}\mathrm{s}^{-1})$	$E_a/\mathrm{kJ\ mol}^{-1}$
SiH_4	-11.54 ± 0.50 (-11.4)	-7.1 ± 4.1 (-6.0)
$MeSiH_3$	-11.60 ± 0.69 (-11.5)	-12.5 ± 5.7 (-12.0)
Me_2SiH_2	-11.31 ± 0.16 (-11.7)	-9.3 ± 1.2 (-12.6)
Me_3SiH	-12.14 ± 0.28 (-12.0)	-15.6 ± 2.3 (-15.0)

(d) ClSiH, SiCl$_2$, SiF$_2$

Ho *et al* [52], using LIF to monitor ClSiH, have obtained the rate constants listed in the Table 17 at total pressures of 500 Torr (N_2).

Table 17
Rate constants for ClSiH insertion reactions

Substrate	$k/10^{-13}\ \mathrm{cm}^3\mathrm{molec}^{-1}\mathrm{s}^{-1}$
H_2	$\leq .003$
SiH_4	0.053 ± 0.001
Cl_2SiH_2	4.1 ± 0.3

These values are considerably less than those of MeSiH of PhSiH thus indicating the substantial deactivating effect of the chlorine substituent *in the silylene*. It is worth noting, however, that like methyl-substitution, chlorine-substitution *in the silane* is considerably activating.

There are no absolute rate constants for insertion reactions of $SiCl_2$ although Safarik *et al* [86] did explore its reaction with i-C_4H_{10}, for which reaction is either very slow or non-existent. No Si–H insertion process has been studied.

Freedman *et al* [87] found no reaction of SiF_2 with H_2 even up to 1400 K, setting an upper limit of $10^{-15}\ \mathrm{cm}^3\ \mathrm{molec}^{-1}\ \mathrm{s}^{-1}$ for the rate constant even at the highest temperature. This is consistent with *ab initio* calculations of Sosa and Schlegel [88], which predict a large increase in activation barrier from the reaction of $SiH_2 + H_2$ to that of $SiF_2 + H_2$.

Relative rate studies [89] for silylene and halogenated silylenes have established the following relative order of reactivity for insertion into SiH bonds: $SiH_2 > ClSiH > FSiH >> SiCl_2, SiF_2$.

3.3 General discussion of the Si–H insertion process

Although direct silylene kinetics is still a relatively new field, the published studies on the Si–H insertion process have already supplied a considerable amount of detail. Substituent effects have been probed and activation energies obtained. These have in turn provided insight into the nature of the process and its detailed mechanism which are discussed in more detail here.

(a) Comparison of substituent effects

The differing reactivities of SiH_2, $SiMe_2$, $MeSiH$ and $PhSiH$ are brought out clearly by the comparisons in Tables 18 and 19. The two temperatures selected, 298 K and 600 K, represent approximately the two extremes of the experimental range of study.

Table 18
Comparison of insertion rate constants (10^{-10} $cm^3molec^{-1}s^{-1}$) at 298 K

silane	silylene			
	SiH_2	$SiMe_2$	$MeSiH$	$PhSiH$
SiH_4	4.0	0.0020	0.81	0.51
$MeSiH_3$	3.7	0.019	2.55	3.9
Me_2SiH_2	3.3	0.055	1.69	2.1
Me_3SiH	2.5	0.045	2.78	3.9

Table 19
Comparison of insertion rate constants (10^{-10} $cm^3molec^{-1}s^{-1}$) at 600 K

silane	silylene			
	SiH_2	$SiMe_2$	$MeSiH$	$PhSiH$
SiH_4	2.2	0.0024	0.179	0.12
$MeSiH_3$	2.4	0.0046	0.133	0.31
Me_2SiH_2	1.9	0.0060	0.164	0.32
Me_3SiH	1.4	0.0045	0.060	0.165

The tables bring out a number of factors. The silylene reactivity sequence is: SiH_2 > $PhSiH$ ~ $MeSiH$ > $SiMe_2$. However replacing Me-for-H in the silylene does not produce a constant effect. The differences in reactivity between SiH_2 and $MeSiH$ are much less than those between $MeSiH$ and $SiMe_2$, although the relative rate factors get closer as temperature increases, because of the sharper drop in values for $MeSiH$ than for $SiMe_2$. The $PhSiH$ is slightly more reactive than $MeSiH$ except with SiH_4 itself. The substrate reactivity sequence is less obvious, but on a *per Si–H* basis and allowing for a slight scatter, is: Me_3SiH > Me_2SiH_2 > $MeSiH_3$ > SiH_4 for all silylenes at both temperatures. Table 10 shows that on a per Si–H basis SiH_4, Si_2H_6 and Si_3H_8 have almost identical reactivities for SiH_2. This, incidentally, tends to argue against any pathway involving Si–Si insertion, which has been ruled out in other cases [90].

<u>(b) Nature of the Si–H insertion process</u>

This process, as pointed out in the introduction, can be discussed in terms of the coupling of electrophilic and nucleophilic interactions involved [25-27]. In the general reaction:

$$SiX_2 + HSiY_3 \rightarrow HX_2SiSiY_3$$

three possibilities have been considered [84], *viz.*
(i) a fully synchronous process in which both electrophilic, *e*, and nucleophilic, *n*, stages occur simultaneously;
(ii) a process led by the electrophilic, *e*, interaction *viz.*

(iii) a process led by the nucleophilic, *n*, interaction *viz.*

In processes (ii) and (iii) distinct but different intermediate complexes are involved. The clue about which is the best description comes from calculations by Gordon [91] referred to in our recent discussion of this question [49]. The geometry of the intermediate complex in the SiH_2 + SiH_4 reaction (Figure 9) shows almost complete transfer of H from SiH_4 to SiH_2, thus favouring (ii) with the electrophilic interaction leading. Figure 9 suggests some degree of Si–Si bond formation. However the looseness of structure (high entropy requirement of the transition state) argues against the completely synchronous process. This is supported by more recent theoretical calculations by Sakai and Nakamura [92] suggesting a hydride anion transfer almost corresponding to ion pair formation. This is not dissimilar to Hoffmann's finding [93], that in the CH_2 + CH_4 reaction the initial approach is abstraction-like. This harks back to an old suggestion of Benson [94] about radical pair formation in the CH_2 + CH_4 reaction and the involvement of ionic states. The H-bridged structure for the process (ii) is supported by the calculations of Trinquier [95] who has shown that the structures of the type $H_2M..H..MH_3$ become increasingly stable as M changes from C to Pb down group 14 of the periodic table. The methyl group substituent effects can now be understood in terms of process (ii) although on their own, the effects do not provide an unambiguous interpretation.

Since carbon is more electronegative than silicon it is plausible that methyl substituents can function as weakly electron-withdrawing groups. They tend to draw off negative charge inductively from silicon centres. Silylenes are so electrophilic that we must presume they have little difficulty in accepting an electron pair regardless of substituent (X $\equiv$ H or Me). The donor character of methylsilanes (Y $\equiv$ Me) is also probably not affected significantly by methyl groups (Si–H bond dissociation energies are virtually constant in the methylsilanes and independent of methyl group substitution [96,97]). It is the second, nucleophilic, stage where the methyl groups exert their influence. If the methyl groups are present in the silylene (X $\equiv$ Me) then it will be more reluctant to donate its lone pair. This suggests the second step should be slower for $SiMe_2$ than SiH_2. If the methyl groups are present on the silane (Y $\equiv$ Me) then they will help draw off charge and facilitate acceptance of the lone pair. Thus if the first (e) stage is rate determining, the Y $\equiv$ Me substituent effect will be negligible. This closely corresponds to the SiH_2 insertion. If, on the other hand, the second (n) stage is rate determining then the Y $\equiv$ Me substituent effects will become significant. This corresponds to the $SiMe_2$ insertion. With MeSiH the situation is in between, suggesting a possible switch with the e stage rate determining at lower temperatures and n stage at higher temperatures.

The effects of Ph-for-Me substitution on the silylene are very small and so are compatible with this explanation of the insertion process. The phenyl group, in spite of its increased size and different electronic character from methyl, clearly plays no special role. There appears to be no special interaction (π-type or otherwise) with the empty or filled silylene orbitals. This is explained by the poor overlap between $Si(3p)$ orbitals and $C(2p)$ orbitals and is supported by the observation of an insignificant energetic interaction in the phenylsilyl radical [98].

The effects of Cl-for-Me substitution on the silylene are large. Cl may deactivate the silylene in two ways. Back donation of a chlorine atom lone pair of electrons into the Si empty p orbital (known to stabilise the silylene [16], as shown in Figure 2) will reduce the electrophilic character and may slow the e stage. The electronegativity of Cl will cause orbital contraction of the silylene, making it a less effective donor and thus slowing the n stage.

294

<u>(c) Intermediate complex mechanism</u>

The negative activation energies of these reactions, in some cases quite large, and the probable two stage character of this reaction, discussed above, have led us to consider a general kinetic treatment of a two stage process involving an intermediate complex. This has been outlined for both $SiMe_2$ [84] and MeSiH [49]. Part of this treatment for the latter case is reproduced here. The insertion process is written as follows:

$$MeSiH + H\text{--}SiR_3 \overset{1}{\underset{-1}{\rightleftharpoons}} MeSiH\text{---}HSiR_3 \overset{2}{\rightarrow} MeH_2SiSiR_3$$

where $R \equiv H$ and/or Me.

Step (1) may be viewed as a loose association of the methylsilylene with the substrate to form the H-bridged complex. Step (2) involves the reorganization of the intermediate complex, requiring restriction of its motion via a much tighter transition state in which the two Si atoms approach whilst the H still bridges before completing its migration to give the product disilane. The observation of single-exponential decays supports the assumption of a steady-state population of intermediate complexes. The phenomenological second-order rate constant, k, is given by:

$$k = k_1 k_2/(k_{-1} + k_2) = k_1/(1 + k_{-1}/k_2)$$

Since step (1) is a loose association process the authors assume that it will have at most a weak temperature dependence, i.e. E_1 is approximately zero. This suggests the major temperature dependence is associated with k_{-1}/k_2. For MeSiH the observed rate constants, k, are sufficiently high, that they approach the collisional rate especially at the lower temperatures. This implies that at sufficiently low temperatures $k \approx k_1$ and k_{-1}/k_2 is small. Then as the temperature increases k_{-1}/k_2 becomes more significant. The mechanism switches from one in which step (1) is mainly rate determining to one in which step (2) becomes rate determining. The same analysis of $SiMe_2$ data [84] suggested step (2) was rate determining over the whole temperature range of study (300-600 K).

In order to carry out the analysis the authors made the assumption that $k_1 = L \times 10^{-10}$ cm^3 $molec^{-1}$ s^{-1} (where L is the path degeneracy). The values for k_{-1}/k_2 were then calculated from the observed values for k. For the MeSiH insertion reactions the following general expression was found:

$$k_{-1}/k_2 \approx 10^{2.6} \exp((E_2 - E_{-1})/RT)$$

The derived values for $E_2 - E_{-1}$ are shown in Table 20 as well as those derived for $SiMe_2$.

Table 20
Activation energy differences for rearrangement (E_2) and redissociation (E_{-1}) of intermediate complexes in the Si–H insertion reactions of MeSiH and SiMe$_2$.

Substrate	$(E_2 - E_{-1})$ / kJ mol^{-1}	
	MeSiH	SiMe$_2$
SiH$_4$	-12.6	-0.3
MeSiH$_3$	-14.7	-6.0
Me$_2$SiH$_2$	-16.7	-9.4
Me$_3$SiH	-17.3	-10.5

These values represent the lowering of the secondary barrier (step (2)) compared with that for redissociation of the intermediate complex (step (-1)). This can be seen in Figure 12. The A factor ratio, $A_{-1}/A_2 = 10^{2.6}$, gives support to the view that the first transition state (TS1) is significantly looser in structure than the second (TS2). Figure 12 shows two representative (qualitative) potential energy surfaces with an indication of the positions of the two transition states involved in the process.

The differences between the Si–H insertion reactions of MeSiH and SiMe$_2$ over the temperature range 300 to 600 K may be rationalised. For both silylenes their kinetic behaviour is determined by the relative tendencies of the intermediate complexes to redissociate (k_{-1}) or rearrange (k_2). For all the insertion reactions of SiMe$_2$ redissociation of the complexes is more favourable than rearrangement, because the large entropic advantages outweigh the small energetic disadvantages. For the insertion reactions of MeSiH, the redissociation and rearrangement of the complexes are more closely competitive with one another. This is because the entropic advantage of redissociation is partially offset by the energetic disadvantage. The effect of energy is to favour rearrangement at lower temperatures. It also favours rearrangement of the complexes formed with the higher methyl-substituted silanes.

An examination of the figures of Table 20 shows that for a given substrate molecule $|E_2 - E_{-1}|$ is larger for reactions of MeSiH than for those of SiMe$_2$. The values for $|E_2 - E_{-1}|$ for reactions of SiH$_2$ cannot be derived from experiment because the reactions are so fast that their rate constants only show slight decreases over the temperature range 300-600 K. Effectively this means that step (1) is rate controlling under all conditions. This must be because E_2 is small compared with E_{-1}. In other words, even though they cannot be measured, $|E_2 - E_{-1}|$ values are large for SiH$_2$, consistent with trends of Table 20. This is supported by the *ab initio* calculations for SiH$_2$ + SiH$_4$ [91]. The increases of $|E_2 - E_{-1}|$ with methyl substitution on the silane, could reflect *either* an increasing stabilisation of the intermediate complex (increasing E_{-1}) *or*

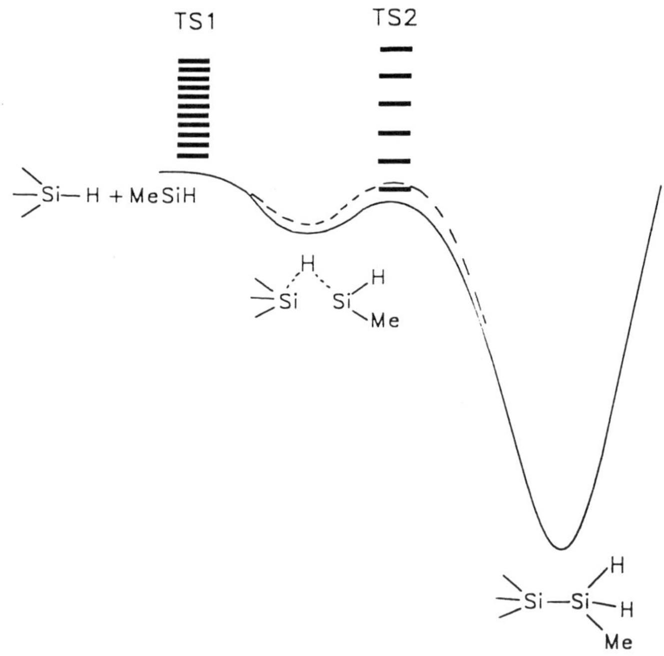

Figure 12. Schematic potential energy surface for the Si-H insertion process of MeSiH. Solid line applies to the more favourable reaction (with Me_3SiH); dashed line applies to less favourable reaction (with SiH_4).

a decreasing stabilisation of the transition state, TS2, for rearrangement (decreasing E_2). It is not obvious which of these is more likely. However the well depths associated with these complexes cannot be too great or non-stationary state kinetic effects would be observed. An estimate of ≤ 41 kJ mol^{-1} has been made [84] for the well depth of the complex formed from $SiMe_2$ + $HSiMe_3$.

Baggott *et al* [84] have noted that other theories such as that developed by Houk *et al* [99] based on the idea of entropy control are not inconsistent with the kinetic data on silylene insertion reactions. The secondary barrier (TS2) in the intermediate complex model represents an entropy bottleneck. Baggott *et al* have discussed in detail possible differences between these theories and tests to distinguish them.

4. THE π-TYPE ADDITION PROCESS

After the Si–H bond insertion reaction this is probably the most widely studied process. One problem of this reaction is that the initial products of addition are the strained ring siliranes (addition to alkenes) [100] and silirenes (addition to alkynes) [101] which are relatively unstable particularly in the gas phase. This instability is a combination of handling (analytical) difficulty and thermal instability. Although this may seem a disadvantage, study of the reactions by direct monitoring of the silylene decay not only avoids this difficulty but in some cases provides information on the thermal stability of the products. This section covers the addition reaction to alkenes, alkynes and dienes, including, in the cases of ethene and ethyne, their isotopic variants C_2D_4 and C_2D_2.

4.1. SiH$_2$ addition reactions
(a) $SiH_2 + C_2H_4/C_2D_4$
Inoue and Suzuki [12] measured the first absolute rate constant for SiH_2 with C_2H_4 at room temperature, using the LIF technique. They found a value of k = $(9.7 \pm 1.2) \times 10^{-11}$ cm^3 molec^{-1} s^{-1} at 1 Torr total pressure in He. Using the LRAFKS technique, Chu *et al* [42] obtained a value of $(2.7 \pm 0.3) \times 10^{-11}$ cm^3 molec^{-1} s^{-1} under the same conditions, and additionally observed a pressure dependence. Al Rubaiey *et al* [102] have repeated these measurements in Ar and SF$_6$ bath gases. The values are compared in Table 21.

Table 21
Rate constants for $SiH_2 + C_2H_4$ in the presence of different bath gases

	k / 10^{-10} cm^{3}molec^{-1}s^{-1}		
P / Torr	He	Ar[c]	SF$_6$[c]
1	0.97[a] 0.27[b]	0.39	0.67
5	0.55[b]	0.79	1.30
10	0.80[b]	1.30	1.58

a. Ref. [12]   b. Ref. [42]   c. Ref. [102]

Al-Rubaiey *et al* [102] have extended their measurements over the pressure range 1-50 Torr and shown that, apart from the Inoue and Suzuki value, the results can be explained in terms of a third body assisted association process with differing collisional efficiencies for He, Ar and SF$_6$, *viz:*

$$SiH_2 + H_2C=CH_2 \rightleftarrows H_2Si\underset{CH_2}{\overset{CH_2}{\diagup}}{}^{*}$$

$$\downarrow +M$$

$$H_2Si\underset{CH_2}{\overset{CH_2}{\diagup}}$$

Additionally Al Rubaiey *et al* [102] have shown that the pressure dependence can be modelled *via* an RRKM calculation. To obtain a good fit to the data they required a value for the activation energy for silirane decomposition of *ca* 197 kJ mol^{-1} consistent with theoretical calculations [103,104], but in disagreement with a measured value of 130 kJ mol^{-1} for the analogous hexamethylsilirane decomposition [105] (the only silirane decomposition for which at present there is an experimental activation energy). In a more extensive study Al-Rubaiey and Walsh [37] have studied the reaction over the pressure range 1-100 Torr (in SF_6) and at five temperatures in the range 298-595 K. Some of the rate constants are shown in Table 22.

The infinite pressure values were found by extrapolation of the pressure dependent curves with the assistence of RRKM modelling. The temperature dependent rate constants correspond to Arrhenius parameters of $\log(A^{\infty}/cm^3 \text{ molec}^{-1}\text{ s}^{-1}) = -9.97$ and $E_a^{\infty} = -2.9$ kJ mol^{-1}. These indicate a fast collisionally-controlled association process.

Table 22
Rate constants for $SiH_2 + C_2H_4$ at two pressures and five temperatures

	k / 10^{-10} cm^{3}molec^{-1}s^{-1}	
T/K	P = 10 Torr	P = ∞[a]
298	1.58 ± 0.03	3.5 ± 1.2
355	1.15 ± 0.02	3.0 ± 1.2
415	0.731 ± 0.017	2.5 ± 1.0
515	0.357 ± 0.005	2.2 ± 0.9
595	0.148 ± 0.007	1.9 ± 0.9

a. Values obtained by extrapolation.

In a disagreement over product formation Al-Rubaiey and Walsh [37] searched, using GC analysis, but found no evidence for vinylsilane formation in this reaction, as claimed by Fisher and Lampe [106]. Other analytical evidence [62] is against vinylsilane formation and Al-Rubaiey and Walsh have suggested that Fisher and Lampe may have seen the mass spectrum of silirane. The latter have recently measured relative rate constants for reaction of SiH_2 with C_2H_4 and with SiH_4. It seems [37] as if these are not in good agreement with absolute values. This may be the result of non-thermalised SiH_2 produced by the IR multiphoton method. *Ab initio* theoretical calculations by Anwari and Gordon [107] of the reaction of SiH_2 with C_2H_4 indicate a reaction with no energy barrier, consistent with the high rate constants. In similar fashion to the insertion reaction (see sections 1 and 3) the reaction may be described as occuring in two stages *viz.* an initial π attack ("electrophilic stage" - donation of $C=C$ π electrons into the Si $3p$ orbital), follow by σ attack ("nucleophilic stage" - donation of the silicon lone pair electrons into the $C=C$ antibonding π^* orbital). Figure 13 shows these processes and gives an approximate indication of the geometry of the transition state.

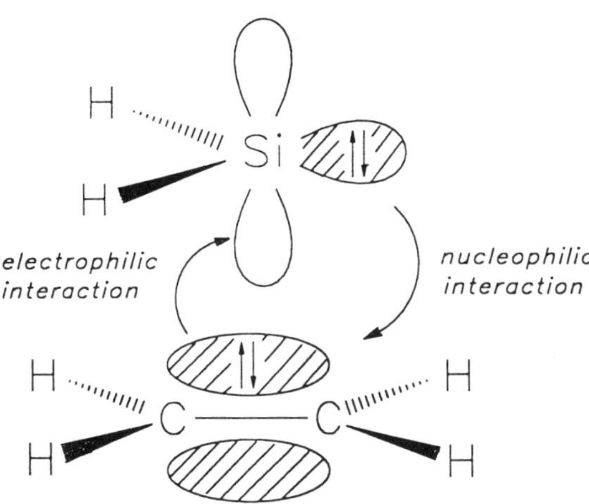

Figure 13. Orbital interactions involved during the process of π-addition of SiH_2 to C_2H_4

An unpublished study of the reaction of SiH_2 + C_2D_4 by Al-Rubaiey and Walsh [108] shows that there is a significant and pressure dependent isotope effect which is more important at low than high pressures as shown in Figure 14.

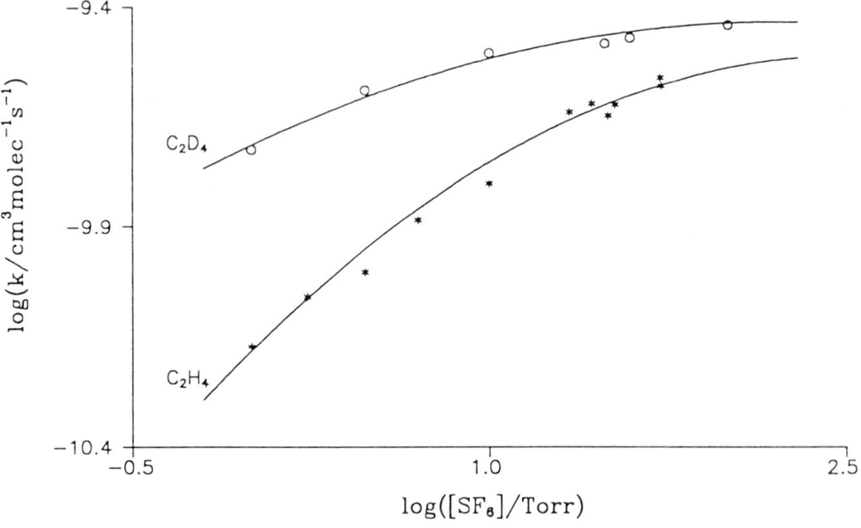

Figure 14. Comparison of pressure dependent rate constants for reactions of SiH_2 with C_2H_4 and C_2D_4 at 298 K.

This has been investigated at five temperatures in the range 291-595 K. The high pressure limiting rate constants are, within experimental error, the same as those for SiH_2 + C_2H_4, but RRKM fitting of the fall-off curves on the assumption of involvement of 2,2,3,3-tetradeuterosilirane alone did not fit the data. At low pressures the reaction was too fast. This suggests the involvement of a rapid isotopic scrambling mechanism as occurs for the SiH_2 + C_2D_2 reaction [109,110] (see subsection (d)). A mechanism for this, based on higher temperature studies [62,111] plausibly involves the intermediacy of ethylsilylene-d_4, *viz:*

The propensity for such rapid processes has been documented by Davidson [112] and the energetic considerations are discussed later.

(b) $SiH_2 + C_3H_6$

The only published rate constant for this reaction is a value of (1.2 ± 0.1) x 10^{-10} cm^3 molec^{-1} s^{-1} obtained by Chu et al [42] at room temperature and 5 Torr (He). In an unpublished study in our laboratories, Al-Rubaiey [113] has found the reaction to be weakly pressure dependent (1-100 Torr, SF_6) in the temperature range 294-520 K. At 5 Torr (SF_6) a rate constant of ca 2.0×10^{-10} cm^3 molec^{-1} s^{-1} can be interpolated in reasonable consistency with value in He [42], allowing for different collision efficiencies. The high pressure limiting rate constants give the Arrhenius parameters $\log(A^{\infty}/cm^3$ molec^{-1} s$^{-1}) = -10.38$ and $E_a^{\infty} = -5.1$ kJ mol^{-1}. These rate constants are ca 20% smaller than those for $SiH_2 + C_2H_4$, although because of the extrapolation they could be equal (within error limits). Thus the methyl substituent effect on this reaction is slightly negative. This contrasts with the analogous results for $SiMe_2$ [114] (discussed in section 4.2). RRKM modelling of the pressure dependence can be made to fit provided the activation energy for decomposition of the product, 2-methylsilirane (back to $SiH_2 + C_3H_6$) is ca 160 kJ mol^{-1}. This is significantly less than the activation energy for decomposition of silirane itself (obtained from the $SiH_2 + C_2H_4$ system). Unless there is an undiscovered source of error in this study, this means that there is a surprising methyl substituent effect on silirane ring stability (and presumably strain energy). This finding, however, is consistent with the difference already noted, with hexamethylsilirane [105].

(c) $SiH_2 + C_4H_6$

The only published rate constant for this reaction has a value of (1.9 ± 0.2) x 10^{-10} cm^3 molec^{-1} s^{-1}, obtained by Chu et al [42] at room temperature and 5 Torr (He). This is close to the collision number and consistent with values for the reactions of SiH_2 with C_2H_4 and C_3H_6. Pressure and temperature variation studies have not yet been carried out. This experimental value is ca 10^5 greater than that proposed by Rogers et al [115] derived from relative rate studies. It is also not very consistent with relative rate studies of Gaspar et al [116], who found that SiH_2 apparently reacts 12 ± 4 times faster with SiH_4 than C_4H_6 (at 298 K) from chemical product analysis (and 9 ± 1 from nuclear recoil measurements [117]). It is possible that the high values for these ratios arise because of incomplete product recovery from the $SiH_2 + C_4H_6$ reaction (for which only 4-silacyclopentene, the rearranged product of 2-vinylsilirane, has ever been detected).

(d) $SiH_2 + C_2H_2/C_2D_2$

Chu et al [42] obtained a value of $(9.8 \pm 1.2) \times 10^{-11}$ cm^3 molec^{-1} s^{-1} for $SiH_2 + C_2H_2$ at 298 K and 5 Torr (He). Becerra et al [109,110] have made a comprehensive study of this reaction as well as its isotopic variant, $SiH_2 + C_2D_2$, over the pressure range 1-100 Torr (SF_6) and temperature range 291-613 K. A selection of rate constants is shown in Table 23.

Table 23
Rate constants (10^{-10} cm^{3}molec^{-1}s^{-1}) for $SiH_2 + C_2H_2/C_2D_2$ at 10 Torr total pressure (SF_6)

T/K	C_2H_2	T/K	C_2D_2
291	3.21 ± 0.33	291	3.74 ± 0.16
346	2.56 ± 0.04	346	2.94 ± 0.07
399	1.99 ± 0.05	395	2.63 ± 0.06
483	1.26 ± 0.04	481	2.02 ± 0.08
613	0.61 ± 0.05	613	1.44 ± 0.06

The room temperature value for $SiH_2 + C_2H_2$ is consistent with that of Chu *et al* [42] taking into account pressure and bath gas differences. Just as for the $SiH_2 + C_2H_4$ reaction, the rate constants show a negative temperature dependence and an isotope effect favouring $SiH_2 + C_2D_2$. The rate constants also show a pressure dependence as illustrated for $SiH_2 + C_2H_2$ in Figure 15.

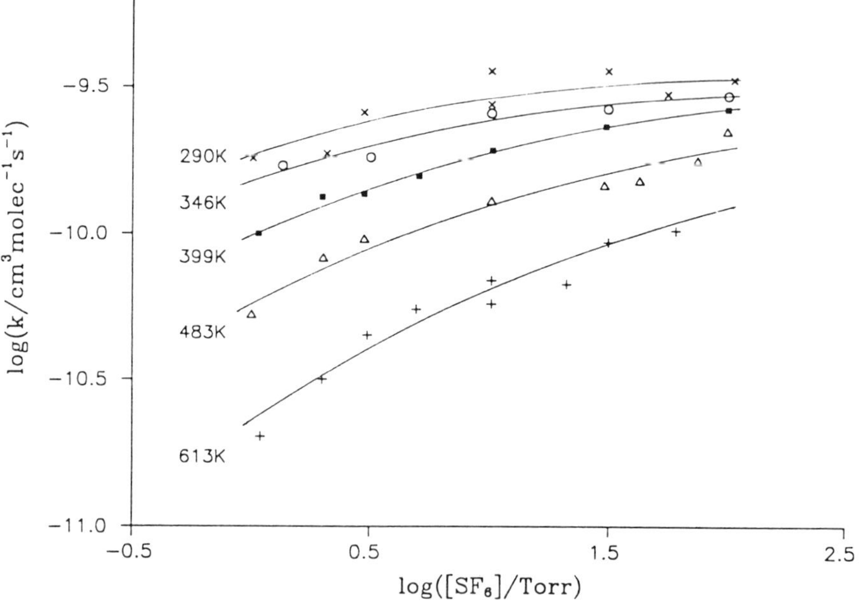

Figure 15. Pressure and temperature dependence of rate constants for $SiH_2 + C_2H_2$.

Extrapolation to infinite pressure gives rate constants corresponding to the Arrhenius parameters of $\log(A^{\infty}/cm^3 \text{ molec}^{-1} \text{ s}^{-1}) = -9.99$ and $E_a^{\infty} = -3.3$ kJ mol^{-1}. The curves for $SiH_2 + C_2D_2$ converge to the same limit within experimental error. These are again consistent with a fast collision-controlled association process, which is third-body-assisted at lower pressures. These studies bring together a total of four direct investigations of SiH_2 with $PhSiH_3$ [71], Me_3SiH [71], C_2H_4 [37] and C_2H_2 [110] which have been shown [37] to be consistent with relative rate measurements carried out in the Reading laboratories by Eley et al [79] prior to the first absolute rate measurements. Other estimates of the rate constants for $SiH_2 + C_2H_2$ based on relative rate constants at higher temperatures [62] are significantly in error.

The direct experimental results are supported by the ab $initio$ calculation of Boatz et al [118] which indicates a barrierless reaction. The mechanism of addition is similar to that for $SiH_2 + C_2H_4$ and is further discussed in section 4.2.

RRKM theoretical attempts to fit the pressure dependence of $SiH_2 + C_2H_2$, in contrast to the $SiH_2 + C_2H_4$ reaction, were only successful with an unrealistically high activation energy for decomposition of the supposed silirene product [109]. This led to the suggestion that another product $viz.$ ethynylsilane (silylacetylene) was probably formed according the scheme:

$$SiH_2 + HC \equiv CH \ \rightleftarrows \ H_2Si \diagdown^{\displaystyle CH^*}_{\displaystyle CH} \ \|\ \rightarrow \ H_3SiC \equiv CH$$

$$\downarrow +M$$

$$H_2Si \diagdown^{\displaystyle CH}_{\displaystyle CH} \ \|$$

Subsequent experiments [119] have confirmed the formation of ethynylsilane. Modelling of this system [110] gave a value for the activation energy for the isomerisation of silirene to ethynylsilane of ca 156 kJ mol^{-1}, assuming a fairly tight transition state. The isotope effect points to a similar isotopic scrambling mechanism for $SiH_2 + C_2D_2$ as observed for $SiH_2 + C_2D_4$. Becerra and Walsh [110] have proposed a mechanism involving vinylsilylene-d_2 $viz:$

$$\text{SiH}_2 + \text{C}_2\text{D}_2 \rightleftarrows \quad \overset{\text{H}}{\underset{\text{H}}{>}}\text{Si}\overset{\text{CD}^*}{\underset{\text{CD}}{<}}\|$$

$$\text{SiHD} + \text{C}_2\text{HD} \leftarrow \quad \overset{\text{H}}{\underset{\text{D}}{>}}\text{Si}\overset{\text{CD}^*}{\underset{\text{CH}}{<}}\|$$

$$\text{SiD}_2 + \text{C}_2\text{H}_2 \leftarrow \quad \overset{\text{D}}{\underset{\text{D}}{>}}\text{Si}\overset{\text{CH}^*}{\underset{\text{CH}}{<}}\|$$

$$\text{HSiCD} = \text{CDH}$$

$$\text{DSiCH} = \text{CDH}$$

This reaction has been incorporated into the model to fit the pressure dependence of $\text{SiH}_2 + \text{C}_2\text{D}_2$, from which an approximate fit gives an activation energy of *ca* 151 kJ mol^{-1} for vinylsilylene-d$_2$ formation from silirene-d$_2$. This complex mechanism is very similar to that for the $\text{SiH}_2 + \text{C}_2\text{H}_2$ reaction system suggested by Ring, O'Neal and coworkers [111,115] to operate at the higher temperatures of pyrolysis studies. It appears to occur as low as room temperature.

4.2 SiMe$_2$ addition reactions

Room temperature absolute rate constants for addition of SiMe$_2$ to alkenes, dienes and alkynes have been obtained by Baggott *et al* [114]. These are shown in Table 24.

These rate constants were not thought to be pressure dependent although the test was only for $\text{SiMe}_2 + \text{C}_2\text{H}_4$ and for a limited pressure range of 5-20 Torr (SF$_6$). There are no other gas phase data for comparison but rate constants of similar magnitude to those of Table 24 for other alkenes, dienes and alkynes have been obtained by Levin *et al* [39] in cyclohexane solution (using dodecamethylcyclohexasilane as precursor). The solution data of Gaspar *et al* [15] (already mentioned) give apparent rate constants for reaction of SiMe$_2$ with dienes in cyclohexane (using (PhMe$_2$Si)$_2$SiMe$_2$ as precursor) which are at least an order of magnitude lower than those of Levin *et al* [39] and probably should be attributed (as discussed) to another intermediate.

In none of the time-resolved studies did the authors report product detection but since these should have been siliranes or silirenes this is not too surprising. Baggott *et al* [114] failed to find identifiable products using gas chromatography. Even in the case of SiMe$_2$ with buta-1,3-diene, the known high temperature product 3,3-dimethyl-3-silacyclopentene was not found. They suggested that the initial vinylsilirane product does not rearrange at room temperature. The mechanism of reaction of SiMe$_2$ with dienes has been discussed in several papers [112,120,121]]. Baggott *et al* [114] have shown that the rate constants of Table 24 correlate with substrate ionisation energies, as shown in Figure 16.

Table 24
Rate constants for addition reactions of $SiMe_2$ at 298 K [114]

Reactant	k / 10^{-11} $cm^3 molec^{-1}s^{-1}$
$CH_2{=}CH_2$	2.21 ± 0.12
$CH_3CH{=}CH_2$	3.72 ± 0.27
$(CH_3)_2C{=}CH_2$	6.82 ± 0.18
t-$CH_3CH{=}CHCH_3$	4.43 ± 0.19
$CH_3CH{=}C(CH_3)_2$	5.46 ± 0.22
$(CH_3)_2C{=}C(CH_3)_2$	4.89 ± 0.22
$ClCH{=}CH_2$	1.25 ± 0.04
$FCH{=}CH_2$	0.88 ± 0.05
t-$FCH{=}CHF$	0.072 ± 0.009
$CH_2{=}CHCH{=}CH_2$	7.45 ± 0.45
$CH{\equiv}CH$	4.63 ± 0.22
$CH{\equiv}CCH_3$	9.37 ± 0.39
$CH_3C{\equiv}CCH_3$	17.0 ± 0.9

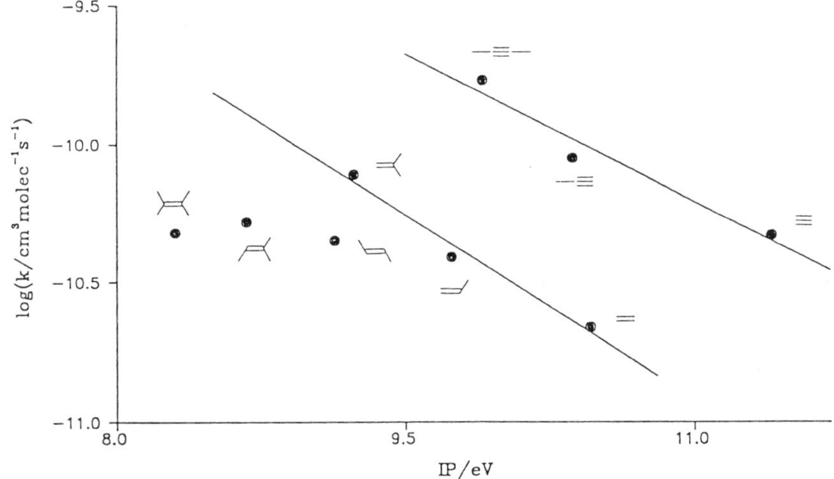

Figure 16. Correlations between $SiMe_2$ addition rate constants and ionisation energies of alkenes/alkynes.

The data for alkenes and alkynes fall on different but parallel lines. The more substituted alkenes fall below the appropriate correlation line and the authors suggested that the explanation for this might be a steric effect associated with crowding between the methyl groups in $SiMe_2$ and those of the alkene. Figure 17 shows, therefore, the preferred approach geometry in the $SiMe_2$ + i-C_4H_8 reaction. This also provides indirect evidence for the non-least motion approach pathway indicated by theoretical calculations [107].

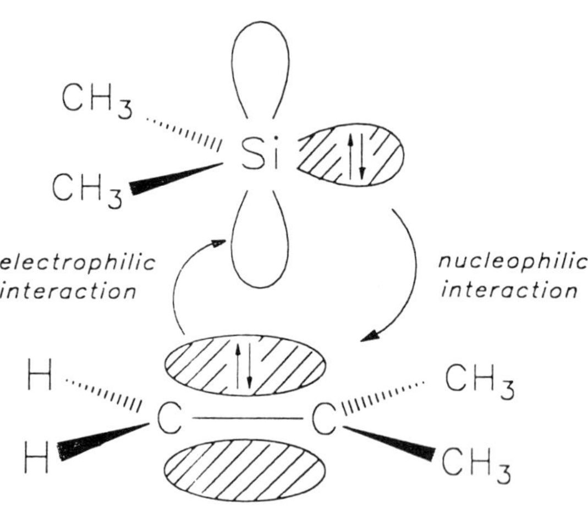

Figure 17. Orbital interactions and preferred geometry of approach during the process of π-addition of $SiMe_2$ to isobutene.

Temperature variation studies have been undertaken on a selected number of alkenes/alkynes by Blitz [122]. These results show that rate constants decrease sharply (by factors of between 3.5 and 15 over the temperature range 298 to *ca* 580 K) for the five compounds studied (C_2H_4, C_3H_6, i-C_4H_8, trans-$C_2H_2F_2$ and $CH_3C \equiv CCH_3$). Arrhenius plots of the data are slightly curved. These results are not analysed in detail here but appear to suggest the possible involvement of intermediate complexes in this reaction, although the only theoretical evidence for this seems to be in case of alkynes [118].

4.3 $SiCl_2$ addition reactions

After the discovery of the UV absorption spectrum of $SiCl_2$ [43], Safarik *et al* [86] measured the absolute rate constants for the reactions of $SiCl_2$ with a series of unsaturated hydrocarbons using the flash photolysis-kinetic absorption spectroscopy technique. The results are shown in Table 25.

From this table we can see the rate constant values increasing with the substitution on both double and triple bonds. This suggests again that the electrophilic interaction is dominant in influencing the rate of reaction even though $SiCl_2$ is a less reactive silylene than either SiH_2 or $SiMe_2$. No products were positively identified although mass spectral evidence [43] pointed to formation of a compound thought to have arisen by rearrangement of the initially formed dichlorosilirane (in the reaction of $SiCl_2$ + i-C_4H_8).

Table 25
Rate constants for $SiCl_2$ addition reactions at 298 K [86]

Reactant	k / 10^{-13} $cm^3molec^{-1}s^{-1}$
C_2H_2	0.71 ± 0.18
C_2H_4	1.29 ± 0.33
C_3H_6	3.82 ± 0.10
t-C_4H_8	5.15 ± 0.13
1-C_4H_6	13.1 ± 0.31

4.4 General discussion of the π-type addition process

The kinetic (and theoretical) data clearly support a process involving an initial electrophilic interaction, as already indicated. The question of whether intermediate complexes are involved is an open one and needs further investigation. However, the mechanism of addition can be usefully illuminated first by comparisons of silylene reactivity and secondly by discussion of the energetics of the prototype silylene addition reactions.

(a) <u>Comparisons of silylene reactivity</u>

The growing data base of absolute rate constants collected in this review enable us to compare silylene reactivities both with one another and with methylene. The values of rate constants for their reactions with selected unsaturated hydrocarbons are shown in Table 26.

It should be noted that the data for $CH_2(^1A_1)$ have been measured and corrected for non-reactive collisionally-induced intersystem crossing in the same studies and also that the SiH_2 data represent the limiting high pressure values. So far as is known the reactions of CH_2, $SiMe_2$ and $SiCl_2$ are pressure independent processes. The rate constants for CH_2 and SiH_2 are close to the

Table 26
Comparison of rate constants for addition reactions of methylene and several silylenes ($^1A_1$ states)

Species	k / 10^{-12} cm^{3}molec^{-1}s^{-1}			
	$^1CH_2$	SiH_2	$SiMe_2$	$SiCl_2$
C_2H_2	280[a]	400	47	0.071
C_2H_4	190[b]	350	22	0.130
C_3H_6	250[b]	340	39	0.38
i-C_4H_8	250[b]	-	78	-
1,3-C_4H_6	260[b]	190	75	-

a. Ref. [123]   b. Ref. [35]

collisional maximum. SiH_2 is about twice as reactive as CH_2 in accordance with the simple idea of an electrophilic reaction in which the initial interaction is dominated by the size of the receptor orbital ($3p$ for SiH_2, $2p$ for CH_2). Me and Cl substitution on the silylene deactivate it, just as was seen for the insertion reaction. The reasons for this, *viz.*, orbital contraction by electronegative substituents, and back donation from the Cl lone pair into the empty Si $3p$ orbital are almost certainly the same.

(b) <u>Energy considerations</u>

The mechanisms of the prototype π-addition processes are examined here in terms of energy surfaces calculated by Al-Rubaiey and Walsh [37] (SiH_2 + C_2H_4) and Becerra and Walsh [110] (SiH_2 + C_2H_2). These are shown in Figures 18 and 19.

Figure 18 shows that silirane formed from SiH_2 + C_2H_4 can either revert to SiH_2 + C_2H_4 or rearrange to vinylsilane (and other isomers) with approximately equal energies. Since Al-Rubaiey and Walsh [37] found no evidence for isomerisation but only for reversion (inferred from the pressure dependence of the kinetics), they suggested that A factors (or relative looseness of transition states) was the explanation for this. The energy of ethylsilylene shows that it is a plausible species to explain the isotopic scrambling mechanism suggested [108] in the SiH_2 + C_2D_4 studies. An estimate has been made of the activation barrier to its formation from silirane in higher temperature studies [124]. The value is dependent on the ring strain in silirane but in any case is fairly low. What is clear is that, because ethylsilylene is endothermic relative to silirane, if ethylsilylene can be readily formed it will rapidly revert to silirane, thus accounting for the proposed label scrambling, in the isotopically substituted silirane species.

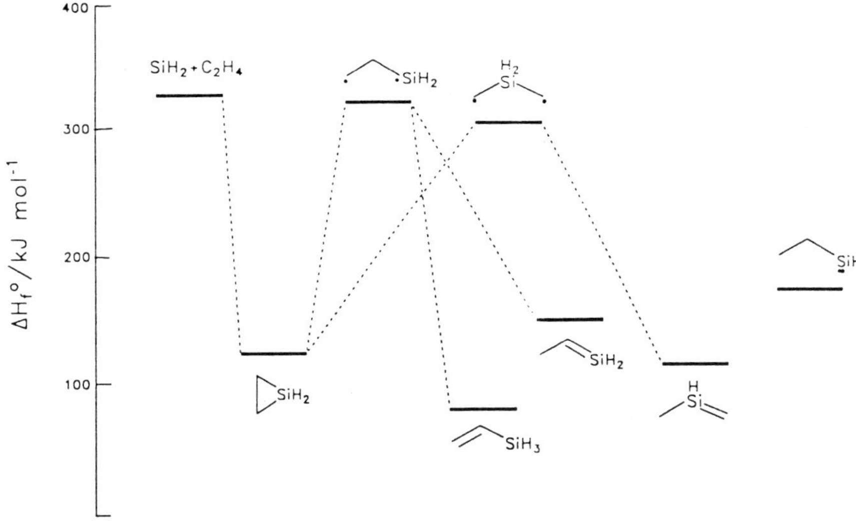

Figure 18. Enthalpy surface for the reaction of $SiH_2 + C_2H_4$

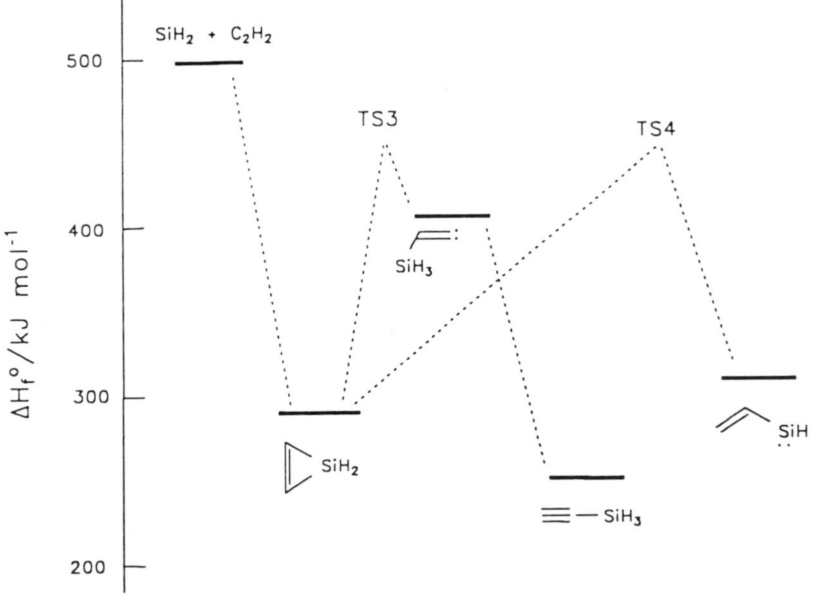

Figure 19. Enthalpy surface for the reaction of $SiH_2 + C_2H_2$

Examination of Figure 19 shows that, in the $SiH_2 + C_2H_2$ reaction, silirene (in contrast to silirane) can isomerise more easily than revert to $SiH_2 + C_2H_2$. This can be explained by the low energy accessibility of the silylvinylidene intermediate leading to ethynylsilane (as found by Becerra and Walsh [119]), so that even if the transition state for its formation is tight, the energetic advantage is sufficient to offset the benefit of the loose transition state for reversion of silirene to $SiH_2 + C_2H_2$.

The energies of vinylsilylene and the transition state for its formation show that they can explain the isotopic scrambling mechanism suggested [109,110] in the $SiH_2 + C_2D_2$ studies. Because vinylsilylene is endothermic relative to silirene, when vinylsilylene is formed it will rapidly revert to silirene, again explaining the proposed label scrambling mechanism in the isotopic silirene species.

As explained in section 1, SiH_2 reactions are considerably less exothermic than those of CH_2. It is the lowering of energy associated with silylenes (DSSE) which significantly determines the energy surface diagram and gives rise to their special chemistry. The facility of the three membered rings to open to silylenes (or decompose to give SiH_2) shows that ring opening to biradicals (as for cyclopropane and cyclobutane [125]) is by no means the only or obvious fate in the thermolysis of small strained rings.

5. REACTIONS WITH O-DONOR MOLECULES

That there is a high propensity for reaction of silylenes with lone-pair donor molecules is most vividly illustrated by the matrix isolation studies of the groups of West [126,127] and Ando [128,129] which have revealed the existence of molecular complexes, stabilised at low temperatures, between silylenes and a variety of O-, N- and S- containing molecules. The formation of such species is strongly supported by theoretical calculations [130] which show the existence of energy minima on the potential surfaces for reactions of SiH_2 with NH_3, H_2O, HF, PH_3, H_2S and HCl. This is illustrated in Figure 20 for the reaction of SiH_2 with H_2O.

The H_2Si--OH_2 complex has a significant barrier to rearrangement (via 1,2 H-migration) to the final product H_3SiOH. Similar surfaces, but with differing well depths and rearrangement barriers for the donor-acceptor complexes, were found for the other reactions. Prior to the matrix-isolation and theoretical work, Weber and co-workers had already proposed the idea of the formation of an initial zwitterionic complex in the reactions $SiMe_2$ with alcohols and ethers [131-133] based on product studies and isotope effect measurements in solution. There is very little direct kinetic data on these reactions and in virtually all that exists is on reactions of $SiMe_2$ with alcohols and ethers, either in solution [39,82] or the gas phase [59,134]. The kinetics are considered within the mechanistic framework implied by involvement of the donor-acceptor (zwitterionic) intermediate complexes.

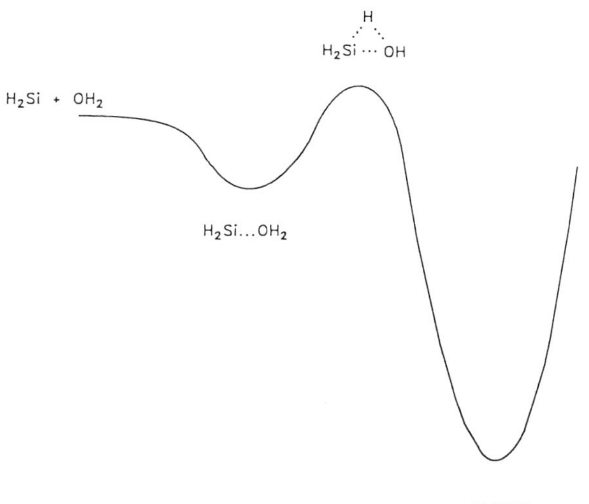

Figure 20. Potential energy surface for reaction of SiH$_2$ with H$_2$O

5.1. SiMe$_2$ + water, alcohols, ethers

Gas-phase data at 295K have been obtained by Baggott *et al* [59,134]. The rate constants obtained are shown in Table 27. This is the only gas-phase

Table 27
Gas-phase rate constants for reactions of SiMe$_2$ with water, alcohols and ethers at 298 K

Reactant	$k/10^{-12}$ cm^{3}molec^{-1}s^{-1}	P / Torr (Ar)
H$_2$O	0.12 ± 0.05	10
D$_2$O	0.077 ± 0.005	10
MeOH	1.18 ± 0.09	5
MeOD	1.03 ± 0.05	5
Me$_2$O	2.41[a]	5-15
Oxirane	3.29 ± 0.06	1
	3.59 ± 0.08	2
	3.80 ± 0.14	3.5
	4.16 ± 0.09	5
	4.60 ± 0.09	10
Oxetane	33.4 ± 1.17	5
3,3-Dimethyloxetane	46.6 ± 2.4	5
Tetrahydrofuran (THF)	3.92[a]	2,5

a. Obtained from modelling biexponential decays.

312

data. The solution phase data is discussed later. Two rate constants in Table 27 were obtained by modelling bi-exponential decay traces. These are the only two cases thus far uncovered of such kinetic behaviour for silylene reactions. An example of these decay traces is shown in Figure 21.

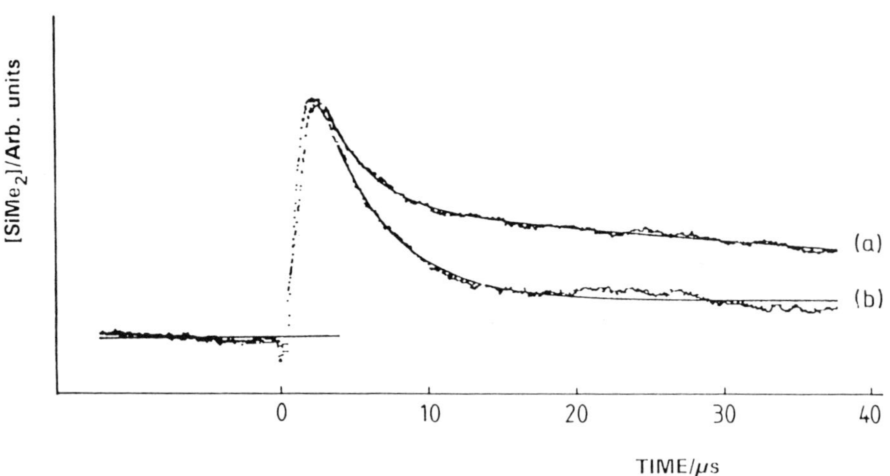

Figure 21   SiMe$_2$ time decay profiles for reaction with (a) Me$_2$O (3.5 Torr) and (b) THF (1.6 Torr). Solid lines correspond to a four step mechanism in each case

Baggott, Blitz and Lightfoot [59] considered a number of mechanisms to account for this behaviour for the reaction with Me$_2$O. All of them involved reversible formation of the donor-acceptor complex, *viz:*

$$Me_2Si + OMe_2 \rightleftarrows Me_2Si\text{---}OMe_2$$

The preferred fit mechanism then involved a scheme in which uncomplexed (i.e. residual) SiMe$_2$ decayed slowly by reaction with precursor augmented by reaction with photolysis products of Me$_2$O. This and the study of SiMe$_2$ with TMF thus provide kinetic confirmation of the involvement of donor-acceptor complexes. From the kinetic data the authors estimated potential well depths (ΔH_f^o/kJ mol^{-1}) of -37 $\pm$ 3 and -45 $\pm$ 3 for the two complexes using a thermodynamic estimate of the entropy change involved in their formation. This is broadly consistent with the theoretical estimate of the stability of H$_2$Si--OH$_2$ [130,135]. Even more convincing was the study, in cyclohexane solution, carried out by Levin *et al* [39]. They observed directly, *via* its absorption at 310 nm, the formation of the Me$_2$Si--THF complex concomitent with SiMe$_2$ decay in the presence of THF. The growth and decay constants of the two species were the same. However there are two differences with the gas phase. First biexponential behaviour was not observed. This is explained by the use of an 8-fold higher concentration of THF by Levin *et al* [39] which drives complex formation nearly to completion. The second difference is that the rate constant for $SiMe_2$ + THF is 2.2×10^{-11} cm^3 $molec^{-1}$ s^{-1}. This is a factor of 5.5 greater than that in the gas-phase. This is a large factor for a non-polar solvent like cyclohexane and there may be another explanation (see later).

The rate constants of Table 27 were correlated by Baggott *et al* [134] with the ionisation energies of the substrate molecules (both for the "in plane" (a_1) and "out of plane" (b_1) orbitals). The correlation was reasonable for H_2O, methanol and unstrained ethers but poor for the strained rings (particularly oxetane). Thus this provided some support for the rate influencing effect of the electrophilic interaction. The following mechanism was proposed:

This is simply the extension of the complex forming reaction to allow for its rearrangement to final product (via Y-group migration) and corresponds to the two stage, "electrophilic" followed by "nucleophilic", process proposed for the Si–H insertion reaction.

However the $SiMe_2$ + MeOH/MeOD results pose some difficulties. The isotope effect in the gas-phase is *ca* 1.15, in contrast to a value of 2.14 for $SiMe_2$ + EtOH/EtOD in cyclohexane [133]. The latter value led Steele & Weber to suggest the second step in the process was rate determining, whilst the gas-phase data suggest a less clear cut situation. There are several values for absolute rate constants for reaction with alcohols (in non polar solvents) which are shown in Table 28. The earlier questionable [39] values [15] are not included.

The data show no great effect of alkyl groups on the rate constants, but the value for MeOH exceeds the gas-phase value by more than a factor of 10. Unless there is an unexpected solvent effect there may be undetected pressure dependences of the gas-phase rate constants for some of the reactions (particularly the small molecules) in Table 27. At the present time until this question is resolved, there will remain a question mark about which step is rate determining in the $SiMe_2$ + alcohol systems. The high values for the rate constants of Table 28 tend to support the first step (zwitterion formation). Further discussions of the mechanism of the $SiMe_2$ + MeOH system may be found in the original paper [134].

Table 28
Rate constants for reactions of $SiMe_2$ with alcohols in solution at 298 K

Reactant	$k/10^{-11}$ $cm^3molec^{-1}s^{-1}$	Solvent	Reference
MeOH	1.5	methylcyclohexane	[82]
EtOH	1.5	cyclohexane	[39]
Me_2CHOH	1.7	cyclohexane	[39]
Me_3COH	1.7	cyclohexane	[39]

For $SiMe_2$ + H_2O, the most recent theoretical calculation indicates a secondary barrier of *ca* 26 kJ mol^{-1}, which would certainly make the second step rate determining. The measured gas-phase rate constant however would only accommodate a maximum (overall) barrier of 17 kJ mol^{-1} [134]. The uncertainties of the data on $SiMe_2$ + H_2O/D_2O unfortunately do not allow any conclusions based on the isotope effect.

More recently Blitz [122] has studied the temperature dependence of the $SiMe_2$ + MeOH reaction at pressures of 10 Torr (Ar) over a temperature range 295-421 K. The rate constants decreased strongly with temperature corresponding to Arrhenius parameters of $\log(A/cm^3$ $molec^{-1}$ $s^{-1}) = -14.7$ and $E_a = -15.7$ kJ mol^{-1}. A brief check apparently showed the reaction to be independent of pressure. At temperatures above 421 K the $SiMe_2$ decay plots were not single exponential. These data are broadly consistent with the complex mechanism, although the question of the rate determining step (at 295 K) remains.

In the ether systems at 295 K [134] a pressure dependence was not observed for $SiMe_2$ + Me_2O, but was for $SiMe_2$ + oxirane. However in the former case the pressure range was limited to 5-15 Torr (Ar). Higher rate constants (at infinite pressure) for these two reactions would certainly be more consistent with those $SiMe_2$ + oxetane and $SiMe_2$ + 3,3-dimethyloxetane. The proposed mechanism for the oxirane reaction was:

$$Me_2Si + O\underset{CH_2}{\overset{CH_2}{\diagdown \, |}} \rightleftarrows Me_2Si\text{---}O\underset{CH_2}{\overset{CH_2}{\diagdown \, |}} \rightarrow Me_2Si=O + C_2H_4$$

Product C_2H_4 was detected, although $Me_2Si=O$ (or its oligomers) were not. Since the overall reaction is not an association process the pressure dependence was argued to be related to the formation of the intermediate complex. This has been further supported by the unpublished results of Blitz [122] which show that the pressure dependence persists up to 430 K but diminishes in magnitude. The rate constants also decrease (at a given pressure). At 5 Torr (Ar) the Arrhenius parameters are $\log(A/cm^3$ $molec^{-1}$ $s^{-1}) = -12.84$ and $E_a = -8.2$ kJ mol^{-1}. The pressure dependence remains to be modelled but its temperature dependence is contrary to expectations for a third body assisted association process. If confirmed this would be the first evidence for pressure stabilisation of a donor-acceptor complex.

5.2 SiH$_2$ + Me$_2$O

SiH$_2$ kinetic studies with either H$_2$O or MeOH have yet to be undertaken. The reaction with Me$_2$O is the only reaction of SiH$_2$ with an O-donor for which there are absolute rate constants. King, Lawrance and Staker [136] have obtained values of between 6.5 x 10^{-12} cm^3 molec^{-1} s^{-1} at 3 Torr and 1.25 x 10^{-11} cm^3 molec^{-1} s^{-1} at 11 Torr (Ar), both at 298 K. Additionally they have studied the temperature dependence from 298-434 K from which they have obtained the Arrhenius parameters (at Ar, 5 Torr) of log(A/cm^3 molec^{-1} s^{-1}) = -13.71 and E_a = -15.3 kJ mol^{-1}. These however do not represent the limiting high pressure values since the rate constants are clearly pressure dependent.

Becerra, Carpenter and Walsh [137] have confirmed the pressure dependence of the reaction in SF$_6$ and obtained values of 1.35, 1.72, 4.80 and 7.54 x 10^{-11} cm^3 molec^{-1} s^{-1} at pressures of 3, 5, 20 and 50 Torr. The results are clearly consistent with a third body assisted association reaction with SF$_6$ a more efficient collider than Ar. The surprise is that this reaction shows no evidence (biexponential decays) of reaching equilibrium as observed for SiMe$_2$ + Me$_2$O. An independent GC search failed to find evidence for MeSiH$_2$OMe [138]. This appears to suggest that the reaction proceeds irreversibly to the zwitterionic complex as shown below, and that there is no further reaction.

$$H_2Si + OMe_2 \rightarrow H_2Si\text{---}OMe_2$$

The implication is that the pressure dependence reflects this association and that the complex is more strongly bound than Me$_2$Si--OMe$_2$. It is not clear why this should be so. RRKM calculations (not yet undertaken) should give information about the binding energy.

5.3 Reactivity of complexed SiMe$_2$

Studies of SiMe$_2$ in the presence of several complexing agents in cyclohexane solution have been carried out by Levin et al [39]. Although not a gas-phase experiment this study offers the first direct kinetic information about "solvated" silylenes ("silylenoids" would be another name) and we thought this important enough for inclusion. Table 29 shows the data for a number of reactions of the SiMe$_2$--THF complex and also for reactions of several other SiMe$_2$ complexes with C$_2$H$_5$OH. These were obtained by direct monitoring of the absorptions of the complexes.

These results clearly show the reduced reactivity of the complexed SiMe$_2$ relative to the free silylene in the reactions with ethanol. Comparison with data in sections 3 and 4 also shows the reduced reactivity of the SiMe$_2$--THF complex relative to free SiMe$_2$ in the Si–H insertion process and the alkene addition reactions. The authors did not attempt to correlate the reactivities of different complexes with any solvent or other parameter of the complexing agent. However they did apply some checks (a) to show product ratios were consistent with the kinetics and (b) to rule out the possibility that these rates

Table 29
Rate constants for $SiMe_2$ complexes in solution [39]

Complexing agent	Reagent	$k/cm^3molec^{-1}s^{-1}$
None	C_2H_5OH	1.5×10^{-11}
THF	C_2H_5OH	3.2×10^{-13}
	$(CH_3)_3COH$	7.3×10^{-14}
	$CH_2{=}CH(CH_2)_3CH_3$	4.0×10^{-15}
	$(CH_2)_3SiCH{=}CH_2$	4.3×10^{-15}
	$(CH_3CH_2CH_2)_3SiH$	1.4×10^{-15}
Et_2O	C_2H_5OH	1.6×10^{-12}
Et_3N	C_2H_5OH	5.5×10^{-14}
CH_3CN	C_2H_5OH	1.8×10^{-13}

could be attributable to a low concentration of uncomplexed $SiMe_2$ in equilibrium with the complexes themselves. Presumably then the measured rate constants refer to nucleophilic displacement reactions of the complexing agent from the complex by the reactive substrate, *viz:*

$$Me_2Si\text{--}complex + substrate \rightarrow Me_2Si\text{--}substrate\ (product) + complex$$

These are therefore examples of S_N2 reactions on silicon.

6. REACTIONS WITH SMALL INORGANIC MOLECULES

Silylenes react with many small inorganic molecules. These reactions belie their apparent simplicity in the potential variety of secondary chemistry which can arise. Although very little of the chemistry is known with great confidence there has been considerable discussion and conjecture. Much of this has already been discussed in the excellent review of Safarik *et al* [11]. Since relatively little has been published since then we decided against a detailed repetition of the discussion of these reactions. For completeness however we include a tabulation of rate constant data. To provide a contrast with the earlier review this is done by substrate molecule. The data are shown in Table 30.

A number of these reactions are pressure dependent which makes comparison of their rate constants a tricky proposition. Where comparisons are possible they generally conform to the silylene reactivity sequence $SiH_2 >$ $SiMe_2 > SiCl_2 > SiBr_2$. Shortage of data makes the placement of SiF_2 in this sequence impossible. It is generally agreed that SiF_2 is fairly unreactive [5,7].

Exceptions to the general sequence, however, exist since with CO, SiH_2 and $SiMe_2$ do not react under experimental conditions whereas $SiCl_2$ does. This

Table 30
Rate constants for silylene reactions with small inorganic molecules at 298 K

Substrate	Silylene	k / $cm^3 molec^{-1} s^{-1}$	P / Torr	Ref.
O_2	SiH_2	$(7.5 \pm 0.8) \times 10^{-12}$	1 (He)	[139]
		$(1.4 \pm 0.2) \times 10^{-11}$	9.5 (He)	[139]
	$SiMe_2$	$(2.5 \pm 0.2) \times 10^{-13}$	5 (Ar)	[134]
		$(4.2 \pm 0.5) \times 10^{-12}$	(Solution)	[39]
	SiF_2	$< 2 \times 10^{-17}$	2 (Ar)	[87]
	$SiCl_2$	$(5.65 \pm 0.12) \times 10^{-12}$	100 (Ar)	[140]
	$SiBr_2$	$(9.3 \pm 0.07) \times 10^{-13}$	40 (Ar)	[141]
NO	SiH_2	$(1.5 \pm 0.2) \times 10^{-11}$	1 (He)	[139]
		$(2.1 \pm 0.2) \times 10^{-11}$	9.5 (He)	[139]
	$SiCl_2$	$(2.66 \pm 0.02) \times 10^{-12}$	100 (Ar)	[142]
	$SiBr_2$	$(4.65 \pm 0.07) \times 10^{-13}$	40 (Ar)	[141]
CO	SiH_2	$< 1 \times 10^{-13}$	5 (He)	[139]
	$SiMe_2$	$< 5 \times 10^{-14}$	5 (Ar)	[134]
	$SiCl_2$	1.05×10^{-12}	100 (Ar)	[143]
N_2O	SiH_2	$(1.90 \pm 0.09) \times 10^{-12}$	P indep.	[144]
	$SiCl_2$	$(9.47 \pm 0.05) \times 10^{-13}$	100 (Ar)	[143]
HCl	SiH_2	$(4.3 \pm 0.6) \times 10^{-12}$	1 (He)	[139]
		$(7.5 \pm 1.0) \times 10^{-12}$	9.5 (He)	[139]
Cl_2	SiH_2	$(1.4 \pm 0.2) \times 10^{-10}$	5 (He)	[139]
	SiF_2	$(5.1 \pm 0.6) \times 10^{-13}$	2.5 (Ar)	[145]
F_2	SiF_2	$(4.7 \pm 0.3) \times 10^{-13}$	≤ 10 (Ar)	[145]

may be a question of third body stabilisation (i.e. a pressure effect), which was not sufficient under experimental conditions for SiH_2 and $SiMe_2$ since stable adducts have been calculated to exist [146] and indeed observed in low temperature matrices [147,148]. It is probable that, as with the reactions already discussed, the electrophilic character of the silylene determines the initial interaction with these substrates since they are all capable of lone pair donation.

The only measured temperature dependences amongst these reactions are those for SiH_2 + N_2O [144], SiF_2 + Cl_2 [145] and SiF_2 + F_2 [145]. SiH_2 + N_2O has a negative temperature dependence (295-747 K) with E_a = -2.0±0.3 kJ mol^{-1} whilst SiF_2 + F_2 has a positive temperature dependence (282-617 K) with E_a = +7.7±0.8 kJ mol^{-1}. For SiF_2 + Cl_2, rate constants increase with temperature but the Arrhenius plot is curved above 700 K. Between 281-700 K, E_a = 3.3 kJ mol^{-1}.

The relative rate studies of the group of Lampe, using IRMPD generation of SiH_2, of the reactions of SiH_2 + HCl [149], and SiH_2 + NO [150], in each case relative to SiH_2 + SiH_4, do not seem to match the absolute values very closely.

7. GENERAL COMMENTS AND CONCLUSIONS

In conclusion to their survey of silylene rate constants in 1990, Safarik *et al* [11] highlighted the fact that the absolute rate constant measurements which only began in 1985, had conclusively demonstrated that silylene reactions were much faster than was estimated earlier from indirect or relative methods. More cautiously they stated that only a few broad correlations could be made of silylene reactivity and much more work was needed to derive a clear appreciation of structure-reactivity relationships. We may now be a little more confident. Significant patterns have begun to emerge. Four years ago, at the time of the earlier review, almost no temperature dependence studies had been carried out. During the intervening period a significant number have been undertaken and published. One finding, that of negative activation energies, seems to be almost universal. The magnitudes of some of these negative activation energies strongly support mechanisms involving intermediate complexes. In the case of reaction with O-donors, such complexes have been directly observed and their kinetics monitored. Equilibrium between the free silylene (in the case of $SiMe_2$) and the donor adducts have been established in two examples and the stabilities of the complexes (in terms of their binding energies) have been measured. For the Si–H insertion reaction the comprehensive investigation of methyl substituent effects has revealed a wealth of detail, not all of it yet fully understood, but consistent with a very polar, H-bonded, intermediate complex. By and large, theoretical calculations of potential energy surfaces have confirmed (or indeed anticipated) this picture.

The simple idea of electron pair donor-acceptor interactions seems to underpin the mechanisms of all these processes. It is remarkable that for all three major classes of reaction, the Si–H bond insertion, the π-type addition and the n-donor addition, they can be described as two stage processes involving an initial electrophilic stage followed by a subsequent nucleophilic stage. The coupling of these stages may be very close, as in the case of the SiH_2 + SiH_4 reaction, with a barely identifiable intermediate. Alternatively the stages may be well separated, as in the Me_2Si + OMe_2 reaction where the intermediate, Me_2Si--OMe_2 is sufficiently stable that there is no actual evidence for the second stage, *viz.* its unimolecular rearrangement to Me_3SiOMe, on the timescale of the experiments. The contrasting behaviour of the different silylenes has begun to take shape. The prototype, SiH_2, emerges as one of the most reactive species known, with rate constants at the collisional limit in several of its reactions. Substitution of H for other groups reduces the reactivity, the effect depending on the substituent. Methyl groups deactivate to a variable extent, depending on the reaction and whether one or both hydrogens in SiH_2 are replaced. This may be understood, generally, in terms of an electron withdrawal (inductive) effect causing shrinkage of the electron-pair-seeking $Si(3p)$ orbital on the silylene. Methyl groups, being marginally electronegative relative to silicon behave differently than in purely organic systems. Substitution of H by halogens causes greater deactivation still, explained by the back donation (π-type interaction) of lone pair electron density into the empty $Si(3p)$ orbital thus reducing its electrophilicity.

The movement of electron pairs and therefore centres of charge density in silylene reactions causes charge separation in the intermediates. It is therefore to be expected that silylene reactions will be particularly sensitive to polarisation effects. This can be vividly illustrated by the contrast in behaviour in the reaction of silylenes with Si–H and C–H bonds. In the former case the polarity of the bond is $Si^{\delta+}$–$H^{\delta-}$ and the electrophilic silylene has no difficulty approaching the negatively charged hydrogen in the initial interaction stage. By contrast carbon-hydrogen bonds are polarised $C^{\delta-}$–$H^{\delta+}$ and the silylene access to the desired electron pair is hampered by the positively charged hydrogen. Reactions of SiH_2 with C–H bonds are not observed, whereas with Si–H bonds they can occur on every collision! It is worth noting that C–H bonds are not totally immune from reaction with silylenes since, *when they are built into the same molecule*, reaction can occur. The rearrangement of ethylsilylene to silirane is an illustration of this. There are many other higher temperature examples of these intramomecular C–H insertion processes [5]. The process, however, looks as though it requires considerable activation. A lot of effort has been made, particularly by Davidson and coworkers [112,151] to estimate the activation energies involved.

We may address one final question concerning the behaviour of silylenes. How does their stability affect their reactivity? Here we face a paradox. It almost seems that the answer is not at all! There is no doubt that, in terms of thermodynamics and chemical bonding, silylenes are stabilised, the magnitude of this having been formalised in the DSSE scale. Yet silylenes can react on every collision. SiH_2 adds to C_2H_4 faster than CH_2 (which is not stabilised). It is true that the more stabilised silylenes, such as $SiCl_2$, are less reactive than the prototype SiH_2, but nevertheless the reactivity is remarkably high. Another interesting comparison would be $SiCl_2$ with CCl_2 but direct data are not available for the latter. Where the stabilisation energy effects impose themselves on the reactivity pattern is in the diminished energy release of silylene reactions (in comparison, say, to carbene reactions). This was mentioned at the beginning of this article. Its detailed manifestation emerges in the secondary behaviour of the initially-formed silylene reaction products. Products such as Si_2H_6 from $SiH_2 + SiH_4$, or silirane from $SiH_2 + C_2H_4$ only have enough energy to revert (partially) to reactants, as shown by the third body stabilisation requirement. In the analogous CH_2 reactions, with their greater energy release, extensive further reaction can occur through the effects of chemical activation. Other consequences of silylene stabilisation are the involvement of substituted silylenes such as ethyl- and vinyl-silylene in the rearrangements of silirane and silirene, respectively, as revealed by the isotopic scrambling studies. Thus the connection between reactivity and stability for silylenes is a rather loose one. Silylene reactions offer yet one more example to chemists, and indeed an excellent one, that thermodynamic and kinetic stability are two quite different matters.

7.1 The future for silylene kinetic studies

Although the pattern of kinetic and mechanistic behaviour of silylenes has begun to emerge a lot remains to be done. Identifying, characterising and learning about the intermediate complexes in silylene reactions needs a lot more work. Whether they even occur in the π-type addition reaction is an open question. Certainly the separation of the electrophilic and nucleophilic stages of these processes and their individual kinetic characterisation is a challenging prospect. Combined pressure and temperature variation studies remain to be carried out for many of the simplest silylene reactions even within the currently easily accessible range. Extension of studies outside of the current range, particularly to sub-ambient temperatures would be very useful. Product characterisation remains a problem in many reactions. Development of methods to handle and study some of the specially sensitive compounds involved such as siliranes and silirenes would be very useful. The powerful methods of reaction dynamics (study of state-selected species, beam orientation studies, examination of the details of energy partitioning amongst product quantum states) have not yet been applied at all in this area. Silylene reactions and behaviour could surely benefit enormously from use of these techniques.

7.2. The general prospect

Silylenes are both electron-pair acceptors and donors *at the same time*. They are both Lewis acids and Lewis bases. They are ambiphilic. Their reactions represent a class of acid/base processes in the gas-phase now sufficiently broad to provide models of behaviour. There are few other gas-phase examples, although solution chemistry abounds with them. Free of the complications of solvent effects, gas-phase silylene chemistry thus offers a blueprint which may be extended to other areas. Two which spring to mind are boron hydride chemistry, and carbenium ion-molecule reactions. An examination of these fields from a different perspective offers a stimulating prospect.

ACKNOWLEDGEMENT

The authors would like to express their thanks to many colleagues at the University of Reading who have contributed to the work described in this article, as well as to members of the wider organosilicon chemistry community who have provided stimulation, encouragement, advice and much helpful discussion on this subject, over the years.

REFERENCES

1.   E.G. Rochow, Silicon and Silicones, Springer-Verlag, Berlin, 1987.
2.   W.H. Atwell and D.R. Weyenberg, Angew. Chem. Int. Ed. Engl., 8 (1969) 469.
3.   Silylenes were discussed earlier than 1969, but as part of an overall review of Carbene Analogues by O.M. Nefedov and M.N. Manakov, Angew. Chem. Int. Ed. Engl., 5 (1966) 1021.
4.   P.P. Gaspar and B.J. Herold, Chap. 13 in Carbene Chemistry, W. Kirmse (ed.), Academic Press, New York, 1971, p. 504.
5.   P.P. Gaspar in Reactive Intermediates, M. Jones Jr. and R.A. Moss (eds.), Wiley, New York; 1 (1978) 229; 2 (1981) 335; 3 (1985) 333.
6.   E.A. Chernyshev, D.G. Komalenkova and S.A. Bashkirova, Russ. Chem. Rev., 45 (1976) 913.
7.   Y-N. Tang, Chap. 4 in Reactive Intermediates, R.A. Abramovitch (ed.), Plenum, 1982, Vol. 2, p. 297.
8.   C-S. Liu and T-L. Hwang, Chap. 1 in Adv. in Inorganic Chemistry and Radiochemistry, H.J. Emeléus and A.G. Sharpe (eds.), Academic Press, London, 1985, Vol. 29, p. 1.
9.   I.M.T. Davidson, Ann. Rep. Prog. Chem., Sect. C, 81 (1985) 47.
10.   M.J. Almond, Short Lived Molecules, Ellis Horwood, London, 1990, Chap. 5, p. 93.
11.   I. Safarik, V. Sandhu, E.M. Lown, O.P. Strausz and T.N. Bell, Res. Chem. Intermediates, 14 (1990) 105.
12.   G. Inoue and M. Suzuki, Chem. Phys. Lett., 122 (1985) 361.
13.   In fact the first claimed absolute rate study of a silylene was by Gaspar and colleagues [14], who reported time resolved kinetics in solution, attributed to PhSiMe. It is now thought, from the reported rate constant values, that the transient involved was probably not a silylene.
14.   P.P. Gaspar, B-H. Boo, S. Chari, A.K. Ghosh, D. Holten and S. Konieczny, Chem. Phys. Lett., 105 (1984) 153.
15.   P.P. Gaspar, D. Holten, S. Konieczny and J.Y. Corey, Acc. Chem. Res., 20 (1987) 329.
16.   Y. Apeloig, Chap. 2 in The Chemistry of Organic Silicon Compounds (Part 1), S. Patai and Z. Rappoport (eds.), Wiley, New York, 1989, p. 57.
17.   I. Dubois, Can. J. Phys., 46 (1968) 2485.
18.   B.T. Luke, J.A. Pople, M-B. Krogh-Jespersen, Y. Apeloig, M. Karni, J. Chandrasekhar and P.v.R. Schleyer, J. Am. Chem. Soc., 108 (1986) 270.

19. R. Walsh, J. Chem. Soc., Faraday Trans. 1, 79 (1983) 2233.
20. R. Walsh, Pure & Appl. Chem., 59 (1987) 69.
21. R. Walsh, Chap. 11 in Energetics of Organometallic Species. J.A. Martinho-Simôes (ed.), NATO-ASI Series C, Vol. 367, Kluwer, Dordrecht, 1992, p. 171.
22. W.J. Bullock, R. Walsh and K.D. King, J. Phys. Chem., 98 (1994) 2595.
23. G. Pilcher, M.P.L. Leitâo, Y. Meng-Yan and R. Walsh, J. Chem. Soc., Faraday Trans., 87 (1991) 841.
24. R. Grev, Adv. Organometallic Chem., 33 (1991) 125.
25. H. Kollmar, Tetrahedron, 28 (1972) 5893.
26. H. Kollmar and V. Staemmler, Theor. Chem. Acta, 51 (1979) 207.
27. M.S. Gordon and D.R. Gano, J. Am. Chem. Soc., 106 (1984) 5421.
28. Kinetics and Spectroscopy of Carbenes and Biradicals, M.S. Platz (ed.), Plenum, New York, 1990.
29. M.N.R. Ashfold, M.A.Fullstone, G. Hancock and G.W. Ketley, Chem. Phys., 55 (1981) 245.
30. A.O. Langford, H. Petek and C.B. Moore, J. Chem. Phys., 78 (1983) 6650.
31. H.M. Frey and I.D.R. Stevens, J. Chem. Soc., (1962) 3514.
32. H.F. Schaefer III, Acc. Chem. Res., 15 (1982) 283.
33. R. Walsh, J. Chem. Soc., Chem. Comm., (1982) 1415.
34. T. Böhland, F. Temps and H.Gg. Wagner, Ber. Bunsenges. Phys. Chem., 89 (1985) 1013.
35. W. Hack, M. Koch, R. Wagener and H.Gg. Wagner, Ber. Bunsenges. Phys. Chem., 93 (1989) 165.
36. JANAF Thermochemical Tables, 2nd edn., D.R. Stull and H. Prophet (eds.), NSRDS-NBS 37, Washington D.C., 1971.
37. N. Al-Rubaiey and R.Walsh, J. Phys. Chem., 98 (1994) 5303.
38. B.S. Rabinovitch and M.C. Flowers, Quart. Rev., 18 (1964) 122.
39. G. Levin, P.K. Das, C. Bilgrien and C.L. Lee, Organometallics, 8 (1989) 1206.
40. Reactive Intermediates in the Gas Phase. Generation and monitoring. D.W. Setser (ed.), Academic Press, New York, 1979.
41. J.E. Baggott, H.M. Frey, P.D. Lightfoot and R. Walsh, J. Phys. Chem., 91 (1987) 3386.
42. J.O. Chu, D.B. Beach and J.M. Jasinski, J. Phys. Chem., 91 (1987) 5340.
43. B.P. Ruzsicska, A. Jodhan, I. Safarik, O.P. Strausz and T.N. Bell, Chem. Phys. Lett., 113 (1985) 67.
44. J.W. Thoman Jr. and J.I. Steinfeld, Chem. Phys. Lett., 124 (1986) 35.
45. J.E. Baggott, H.M. Frey, K.D. King, P.D. Lightfoot, R. Walsh and I.M. Watts, J. Phys. Chem., 92 (1988) 4025.
46. R. Becerra, H.M. Frey, B.P. Mason and R. Walsh, unpublished results.
47. B.P. Mason, H.M. Frey and R. Walsh, J. Chem. Soc., Faraday Trans., 89 (1993) 4405.
48. J.E. Baggott, M.A. Blitz, H.M. Frey, P.D. Lightfoot and R. Walsh, Chem. Phys. Letters, 135 (1987) 39.

49. R. Becerra, H.M. Frey, B.P. Mason and R. Walsh, J. Chem. Soc., Faraday Trans., 89 (1993) 411.

50. M.A. Blitz, H.M. Frey, F.D. Tabbutt and R. Walsh, J. Phys. Chem., 94 (1990) 3294.

51. G. Herzberg and R.D. Verma, Can. J. Phys., 42 (1964) 395.

52. P. Ho, W.G. Breiland and R.W. Carr, Chem. Phys. Lett., 132 (1986) 422.

53. B.P. Ruzsicska, A. Jodhan, I. Safarik, O.P. Strausz and T.N. Bell, Chem. Phys. Lett., 139 (1987) 72.

54. J.E. Baggott, H.M. Frey, P.D. Lightfoot and R. Walsh, Chem. Phys. Lett., 125 (1986) 22.

55. G. Herzberg, Molecular Spectra and Molecular Structure, Van Nostrand Reinhold, New York, 1966.

56. I. Dubois, G. Herzberg and R.D. Verma, J. Chem. Phys., 47 (1967) 4262.

57. J.M. Jasinski and J.O. Chu, J. Chem. Phys., 88 (1988) 1678.

58. M.J. Michalczyk, M.J. Fink, D.J. De Young, C.W. Carlson, K.L. Welsh and R. West, Silicon, Germanium, Tin and Lead Compounds, 9 (1986) 75.

59. J.E. Baggott, M.A. Blitz and P.D. Lightfoot, Chem. Phys. Lett., 154 (1989) 330.

60. J.M. Jasinski, J. Phys. Chem., 90 (1986) 555.

61. P. John and J.H. Purnell, J. Chem. Soc., Faraday Trans., 69 (1973) 1455.

62. D.S. Rogers, K.L. Walker, M.A. Ring and H.E. O'Neal, Organometallics, 6 (1987) 2313.

63. M.S. Gordon, J. Chem. Soc., Chem. Comm., (1981) 890.

64. R.S. Grev and H.F. Schaefer III, J. Chem. Soc., Chem. Comm., (1983) 785.

65. A. Sax and G. Olbrich, J. Am. Chem. Soc., 107 (1985) 4868.

66. M.S. Gordon, D.R. Gano, J.S. Binkley and M.J. Frisch, J. Am. Chem. Soc, 108 (1986) 2191.

67. J.E. Baggott, H.M. Frey, K.D. King, P.D. Lightfoot, R. Walsh and I.M. Watts, J. Phys. Chem., 92 (1988) 4025.

68. N. Al-Rubaiey and R. Walsh, unpublished results.

69. B.P. Mason, H.M. Frey and R. Walsh, unpublished results.

70. K.F. Roenigk, K.F. Jensen and R.W. Carr, J. Phys. Chem., 91 (1987) 5726.

71. J.E. Baggott, H.M. Frey, P.D. Lightfoot, R. Walsh and I.M. Watts, J. Chem. Soc., Faraday Trans., 86 (1990) 27.

72. T.R. Dietrich, S. Chiussi, M. Marek, A. Roth and F.J. Comes, J. Phys. Chem., 95 (1991) 9302.

73. R. Becerra, H.M. Frey, B.P. Mason, R. Walsh and M.S. Gordon, J. Am. Chem. Soc., 114 (1992) 2751.

74. J.G. Martin, M.A. Ring and H.E. O'Neal, Int. J. Chem. Kinet., 19 (1987) 715.

75. R. Becerra, H.M. Frey, B.P. Mason, R. Walsh, unpublished results.

76. A.J. Vanderwielen, M.A. Ring and H.E. O'Neal, J. Am. Chem. Soc., 97 (1975) 993.

77. I.C. Carpenter and R. Walsh, unpublished results.
78. R. Becerra and R. Walsh, unpublished results.
79. C.D. Eley, M.C.A. Rowe and R. Walsh, Chem. Phys. Lett., 126 (1986) 153.
80. B.A. Sawrey, H.E. O'Neal, M.A. Ring and D. Coffey Jr., Int. J. Chem. Kinet., 16 (1984) 31.
81. G. Raabe, H. Vancik, R. West and J. Michl, J. Am. Chem. Soc., 108 (1986) 671.
82. H. Shizuka, H. Tanaka, K. Tonokura, K. Murata, H. Hiratsuka, H. Ohshita and M. Ishikawa, Chem. Phys. Lett., 143 (1988) 225.
83. I.M.T. Davidson and N.A. Ostah, J. Organometal. Chem., 206 (1981) 149.
84. J.E. Baggott, M.A. Blitz, H.M. Frey and R. Walsh, J. Am. Chem. Soc., 112 (1990) 8337.
85. H.E. O'Neal, M.A. Ring, W.H. Richardson and G.F. Licciardi, Organometallics, 8 (1989) 1968.
86. I. Safarik, B.P. Ruzsicska, O.P. Strausz and T.N. Bell, Chem. Phys. Lett., 113 (1985) 71.
87. A. Freedman, K.E. McCurdy, J. Wormhoudt and P.P. Gaspar, Chem. Phys. Lett., 142 (1987) 255.
88. C. Sosa and H.B. Schlegel. J. Am. Chem. Soc., 106 (1984) 5847.
89. R.L. Jenkins, A.J. Vanderwielen, S.P. Ruis, S.R. Gird and M.A. Ring, Inorg. Chem., 12 (1973) 2968.
90. M. Ishikawa, T. Takaoka, and M. Kumada, J. Organometal. Chem., 42 (1972) 333.
91. M.S. Gordon, private communication.
92. S. Sakai and M. Nakamura, J. Phys. Chem., 97 (1993) 4960.
93. R.C. Dobson, D.M. Hayes and R. Hoffmann, J. Am. Chem. Soc., 93 (1971) 6188.
94. S. W. Benson, Adv. Photochem., 2 (1964) 1.
95. G. Trinquier, J. Chem. Soc., Faraday Trans., 89 (1993) 775.
96. R. Walsh, Acc. Chem. Res., 14 (1981) 24.
97. R. Walsh, Chap. 5 in The Chemistry of Organic Silicon Compounds (Part 1), S. Patai and Z. Rappoport (eds.), Wiley, New York, 1989, p. 371.
98. M. Barber, A.M. Doncaster and R. Walsh, Int. J. Chem. Kinet., 14 (1982) 669.
99. K.N. Houk, N.G. Rondan and J. Mareda, Tetrahedron, 14 (1985) 1555.
100. V.J. Tortorelli, M. Jones Jr., S. Wu and Z. Li, Organometallics, 2 (1983) 759.
101. R.T. Conlin and P.P. Gaspar, J. Am. Chem. Soc., 98 (1976) 3715.
102. N. Al-Rubaiey, H.M. Frey, B.P. Mason, C. McMahon and R. Walsh, Chem. Phys. Lett., 204 (1993) 301.
103. J.A. Boatz and M.S. Gordon, J. Phys. Chem., 93 (1989) 3025.
104. D.A. Horner, R.S. Grev and H.F. Schaefer III, J. Am. Chem. Soc., 114 (1992) 2093.
105. D.H. Berry, private communication.

106. J.R. Fisher and F.W. Lampe. J. Photochem. Photobiol. A, Chem., 58 (1991) 173.
107. F. Anwari and M.S. Gordon, Israeli J. Chem., 23 (1983) 129.
108. N. Al-Rubaiey and R. Walsh, unpublished results.
109. R. Becerra, H.M. Frey, B.P. Mason and R. Walsh, J. Chem. Soc., Chem. Comm., (1993) 1050.
110. R. Becerra and R. Walsh, Int. J. Chem. Kinet., 26 (1994) 45.
111. J.W. Erwin, M.A. Ring and H.E. O'Neal, Int. J. Chem. Kint., 17 (1985) 1067
112. I.M.T. Davidson, J. Organometal. Chem., 437 (1992) 1.
113. N. Al-Rubaiey and R. Walsh, unpublished results.
114. J.E. Baggott, M.A. Blitz, H.M. Frey, P.D. Lightfoot and R. Walsh, J. Chem. Soc., Faraday Trans., 84 (1988) 515.
115. D.S. Rogers, H.E. O'Neal and M.A. Ring, Organomet., 5 (1986) 1467.
116. P.P. Gaspar, S. Konieczny and S.H. Mo, J. Am. Chem. Soc, 106 (1984) 424.
117. R-J. Hwang and P.P. Gaspar, J. Am. Chem. Soc, 100 (1978) 6626.
118. J.A. Boatz, M.S. Gordon and L.R. Sita, J. Phys. Chem., 94 (1990) 5488.
119. R. Becerra and R. Walsh, unpublished results.
120. D. Lei, R-J. Hwang and P.P. Gaspar, J. Organometal. Chem., 271 (1984) 1.
121. M.P. Clarke and I.M.T. Davidson, J. Chem. Soc., Chem. Comm., 3 (1988) 241.
122. M.A. Blitz, Ph.D. Thesis, University of Reading, 1990.
123. W. Hack, M. Koch, H.Gg. Wagner and A. Wilms, Ber. Bunsenges. Phys. Chem., 92 (1988) 674.
124. A.P. Dickenson, H.E. O'Neal and M.A. Ring, Organometallics, 10 (1991) 3513.
125. H.M. Frey and R. Walsh, Chem. Rev., 69 (1969) 103.
126. G.R. Gillette, G.H. Noren and R. West, Organometallics, 6 (1987) 2617.
127. G.R. Gillette, G.H. Noren and R. West, Organometallics, 8 (1989) 487.
128. W. Ando, K. Hagiwari and A. Sekiguchi, Organometallics, 6 (1987) 2270.

129. W. Ando, A. Sekiguchi, K. Hagiwari, A. Sakakibari and H. Yoshida, Organometallics, 7 (1988) 558.
130. K. Raghavachari, J. Chandrasekhar, M.S. Gordon and K.J. Dykema, J. Am. Chem. Soc., 106 (1984) 5853.
131. D. Tzeng and W.P. Weber, J. Am. Chem. Soc., 102 (1980) 1451.
132. T.Y. Gu and W.P. Weber, J. Am. Chem. Soc., 102 (1980) 1641.
133. K.P. Steele and W.P. Weber, Inorg. Chem., 20 (1981) 1302.
134. J.E. Baggott, M.A. Blitz, H.M. Frey, P.D. Lightfoot and R. Walsh, Int. J. Chem. Kinet., 24 (1992) 127.
135. S. Su and M.S. Gordon, Chem. Phys. Lett., 204 (1993) 306.
136. K.D. King, W.D. Lawrance and W.S. Staker, private communication.
137. R. Becerra, I.M. Carpenter and R. Walsh, unpublished results.
138. N. Al-Rubaiey and R. Walsh, unpublished results.

139. J.O. Chu, D.B. Beach, R.D. Estes and J.M. Jasinski, Chem. Phys. Lett., 143 (1988) 135.
140. V. Sandhu, A. Jodhan, I. Safarik, O.P. Strausz and T.N. Bell, Chem. Phys. Lett., 135 (1987) 260.
141. V. Sandhu, O.P. Strausz and T.N. Bell, Res. Chem. Intermediates, 13 (1990) 43.
142. V. Sandhu, I. Safarik, O.P. Strausz and T.N. Bell, Res. Chem. Intermediates, 11 (1989) 19.
143. V. Sandhu, O.P. Strausz and T.N. Bell, Res. Chem. Intermediates, 11 (1989) 235.
144. R. Becerra, H.M. Frey, B.P. Mason and R. Walsh, Chem. Phys. Lett., 185 (1991) 415.
145. A.C. Stanton, A. Freedman, J. Wormhoudt and P.P. Gaspar, Chem. Phys. Lett., 122 (1985) 190.
146. T.P. Hamilton and H.F. Schaefer III, J. Chem. Phys., 90 (1989) 1031.
147. C.A. Arrington, J.T. Petty, S.E. Payne and W.C.K. Haskins, J. Am. Chem. Soc., 110 (1988) 6240.
148. M-A. Pearsall and R. West, J. Am. Chem. Soc., 110 (1988) 7229.
149. C.B. Moore, J. Biedrzycki and F.W. Lampe, J. Amer. Chem. Soc., 106 (1984) 7761.
150. T. Dohmaru and F.W. Lampe, J. of Photochem., 41 (1988) 275.
151. I.M.T. Davidson and R.J. Scampton, J. Organometal. Chem., 271 (1984) 249.